# 7th International Conference on Advanced Composite Materials and Manufacturing Engineering (ACMME 2020)



Held online due to COVID-19

IOP Conference Series: Materials Science and Engineering
Volume 914

Yunnan, China
20 - 21 June 2020

ISBN: 978-1-7138-3188-4
ISSN: 1757-8981

**Printed from e-media with permission by:**

Curran Associates, Inc.
57 Morehouse Lane
Red Hook, NY 12571

**Some format issues inherent in the e-media version may also appear in this print version.**



**Additional copies of this publication are available from:**

Curran Associates, Inc.
57 Morehouse Lane
Red Hook, NY 12571 USA
Phone: 845-758-0400
Fax: 845-758-2633
Email: curran@proceedings.com
Web: www.proceedings.com

# TABLE OF CONTENTS

## 1. ENVIRONMENTALLY FRIENDLY MATERIALS


THE INFLUENCING FACTORS OF BACILLUS MEGATERIUM AFFECT THE GROWTH OF THE ANTAGONISTIC STRAIN AND THE ACTIVITIES OF ANTIBACTERIAL SUBSTANCES ........ 1
*Ran Sun, Yang Xiao, Bokai Guo, Zhu Li, Ke Wan, Long Yang, Jinyi Zhao*

SAND AND DUST HAZARD RISK ZONING FOR NINGXIA EXPRESSWAY ........ 6
*Wanli Wu, Xiaohui Guo, Yao Liu, Lei Ma*

AN OVERVIEW OF SUSTAINABLE HIGHWAY INFRASTRUCTURE DEVELOPMENT IN YUNNAN, CHINA ........ 17
*Junjie Zhang*

STUDY ON MAKING FORMULATION OF JUJUBE DIETARY FIBER TABLEWARE ........ 22
*Shaocheng Wang, Shijun Liu, Huanhuan Sun, Yanping Lv*

RESEARCH ON STRUCTURAL DESIGN OF FRICTION TEST BASED ON FINITE ELEMENT ANALYSIS HEORY ........ 28
*Xiao Liu*

RESEARCH ON ELECTROCHEMICAL MOLECULAR IMPRINTING SENSOR OF P-FECO NANOPARTICLES ........ 35
*Lei Li, Bixuan Zheng, Mingguo Tang, Junchuang Feng, Kun Huang, Yile Liu*

RESEARCH PROGRESS ON THE PHOTOELECTRIC PROPERTIES OF INDIUM-DOPED CADMIUM OXIDE TRANSPARENT CONDUCTIVE FILMS ........ 41
*Xiongfei Shi*

ISOLATION AND IDENTIFICATION OF BIRD VIRUSES IN WETLAND ECOLOGY ........ 50
*Yong Li, Hanping Zhang, Guang Chen*

ANTI-TUMOR EFFECT AND MECHANISM OF BLETILLA RHIZOMA BASED ON NETWORK PHARMACOLOGY ........ 55
*Xiusheng Tang, Rongze Fang, Ling Li, Jun Luo*

STUDY ON THE MECHANISM OF ANTI-CANCER ACTION OF ROSA ROXBUGHII TRATT BASED ON NETWORK PHARMACOLOGY ........ 62
*Ying Mei, Xiang Pu, Xiangling Qu*

APPLICATION AND RESEARCH BASED ON GREEN LANDSCAPE MATERIALS IN INTERIOR DESIGN ........ 72
*HuiJuan Li*

HETEROLOGOUS EXPRESSION AND ENZYMATIC PROPERTIES OF EXOCELLULASE CELA ........ 78
*Ming Sui, Jiachao Rong, Yang Zhang, Peng Qu, Yusong Zhu*

STUDY ON THE VARIATION OF MICROBIAL DIVERSITY DURING THE FERMENTATION OF POTATO LIQUOR BASED ON BIOLOG ECO TECHNOLOGY ........ 83
*Liuwei Yang, Qiao Lin, Yuxin Yang, Shudong Zhao, Yanling Ren, Yao Chen, Bangzhou Luo*

STUDY ON THE PREPARATION TECHNOLOGY OF SANQI HEMOSTATIC SPONGIA FOR PERCUTANEOUS ABSORPTION ..... 92
*Shijun Liu, Lin Li, Zhishu Tang, Yu Zhang, Jing Sun, Zhongxing Song, Hongbo Xu, Chunli Cui, Hongbo Liu*

POTATO WINE BREWING PROCESS OPTIMIZATION BY RESPONSE SURFACE ANALYSIS ..... 96
*Qiao Lin, Ting Qiu, Xiaoyan Liu, Junhua Liu, Cheng Li*

FATIGUE AND RECOVERY OF WUSHU ATHLETES BASED ON FATIGUE DAMAGE MODEL ..... 106
*Xue Yu*

ADSORPTION CHARACTERISTICS OF ETHYL PARABEN FROM AQUEOUS SOLUTION USING RICE HUSK BIOCHAR ..... 111
*Meng Guo, Wenjing Chen, Han Ru, Lili Yang, Yu Zhang*

RESEARCH ON THE EVALUATION MODEL OF MULTI-OBJECTIVE OPTIMAL ALLOCATION OF WATER RESOURCES IN GUIZHOU PROVINCE ..... 115
*Huan Wang, Zaimei Mei*

## 2. GREEN MANUFACTURING

DESIGN AND IMPLEMENTATION OF CODE SYSTEM FOR MEASURING INSTRUMENTS ..... 121
*Chang Hu, Yanan Wu, Ying Zheng, Xiao Zhang*

A NEW INSULATE METHOD OF SHIP-SWAYING INTERFERENCE ..... 128
*Min Tao, Hao Zhang, Yu Wang*

DESIGN AND IMPLEMENTATION OF VIBRATION DETECTING SYSTEM OF SHIP-BORNE RADAR ..... 134
*Min Tao, Hao Zhang, Yu Wang*

REAL-TIME OBJECT DETECTION BASED ON DIVERGENT ACTIVATION ..... 140
*Hui Liang, Yueyuan Wang, Yuan Ma, Kewen Liu*

ECONOMIC ADAPTATION GRADE EVALUATION OF INTEGRATED TRANSPORT SYSTEM BASED ON MATTER-ELEMENT EXTENSION WITH ANP ..... 147
*Jinhui Zhang*

8-BIT NCL ASYNCHRONOUS MULTIPLIER BASED ON RADIX-4 BOOTH ALGORITHM ..... 151
*Weixiang Deng, Wenxiang Cheng, Jie Cheng, Leibin Ni, Anping He*

DESIGN OF EMERGENCY EVACUATION SCHEME FOR LOUVRE ..... 159
*Zhuoqun Zhang, Chen Wang, Yinzhe Wang*

STUDY ON THE ESTABLISHMENT AND OPTIMIZATION OF SANDCASTLE INFRASTRUCTURE MODEL ..... 169
*Zhixuan Han, Yuqiu Gao, Shuyu Ren, Depin Chen*

APPLICATION OF ARTIFICIAL INTELLIGENCE IN DATA PROCESSING ..... 176
*Xingnan Jiang, Yueying Wang, Bowen Deng*

APPLICATION OF NEURAL NETWORK BASED ON SELF-ORGANIZING INCREMENTAL LEARNING IN ANTI-INTERFERENCE ..... 180
*Shaodong Hu, Xuechen Wu, Bo Xue*

CASE-BASED ANALYSIS OF DISCRIMINATION IN POLICE SURVEILLANCE SCENE REGARDING FACIAL RECOG .......... 186
*Tingyao Chen*

DATA-PARALLEL CLUSTERING ALGORITHM BASED ON MUTUAL INFORMATION MINING OF JOINT CONDITION .......... 191
*Changjiang Huang*

THE OPTIMAL ENVIRONMENT SELECTION STRATEGY OF RISK MODEL WITH PERTURBED DIFFUSION .......... 199
*Shumei Guo, Lili Zhang, Jie Guo, Xinna Li, Jiandong Yang*

STUDY ON THE PREPARATION TECHNOLOGY OF SANQI SHANGYAO SMEARED FILMS FOR PERCUTANEOUS ABSORPTION .......... 205
*Shi jun Liu, Pan Zhang, Zhi shu Tang, Yu Zhang, Jing Sun, Zhong xing Song, Chun li Cui, Hong bo Liu, Hong bo Xu*

ACOUSTIC TARGET RECOGNITION ALGORITHM BASED ON PARTICLE SWARM NEURAL NETWORK .......... 209
*Yalei Liu, Xiaohui Gu*

DESIGN AND RESEARCH ON TEMPERATURE CONTROL SYSTEM OF ELECTRIC FURNACE BASED ON PLC TOUCH SCREEN INTEGRATED MACHINE .......... 215
*Haiyu Zhao, Jiashuo Zheng*

ELECTRONIC CIRCUIT DESIGN OF STEPPING MOTOR BASED ON 51 SINGLE CHIP MICROCOMPUTER .......... 221
*Jianwei Jia, Bo Xue, Yimeng Wang*

INFLUENCE OF IRRIGATION CANAL CONSTRUCTION ON THE STABILITY OF A NATURAL HIGH AND STEEP SLOPE .......... 227
*Yanhui Guo, Jin He*

A COMPARISON OF RIDE PERFORMANCE OF HYDRO-PNEUMATIC SUSPENSION SYSTEM WITH THOSE OF RUBBER AND LEAF SUSPENSION SYSTEMS .......... 233
*Le Xuan Longa, Le Van Quynh, Nguyen Van Liemb, Bui Van Cuong, Hoang Anh Tand*

STATUS AND DEVELOPMENT TREND OF ACTIVE SUB-ARRAYS STRUCTURE DESIGN IN ACTIVE PHASED ARRAY ANTENNA .......... 240
*Rui Xiao, Shouli Jianga, Jianfeng Zhongb*

HEALTH MANAGEMENT INFORMATION SYSTEM BASED ON REGRESSION ANALYSIS .......... 246
*Dong Wang, Xiujuan Chen, Honghe Wei*

RESEARCH ON DESIGN AND APPLICATION BASED ON BRIDGE INSPECTION AND MAINTENANCE SCHEME .......... 252
*Lingye Tan, Ziyang Zhang, Xin Wang*

INFLUENCE RESEARCH OF WEAK LAYER ON SLOPE SLIDING MODE AND STABILITY .......... 256
*Ang Li, Wenli Li*

RESEARCH ON RAINFALL INFILTRATION LAW AND SLOPE STABILITY ANALYSIS .......... 260
*Ang Li, Lingbo Meng*

PAPER COUNTING DISPLAY DEVICE BASED ON STM32F103 SINGLE CHIP COMPUTER .......... 264
*Lizhen Gao, Fanmao Jiang, Shihao Wang*

**Author Index**

IOP Conf. Series: Materials Science and Engineering **914** (2020) 011001 doi:10.1088/1757-899X/914/1/011001

# Preface

The organizing committee of ACMME 2020 is proud to present the proceedings of the 2020 7th International Conference on Advanced Composite Materials and Manufacturing Engineering (ACMME 2020), held on June 20-21, 2020 in Xishuangbanna, Yunnan, China.

ACMME 2020 intends to be a global forum for researchers and engineers to present and discuss recent innovations and new techniques related to composite materials, materials processing technologies, functional materials and manufacturing engineering, as well as their analytical techniques and related environmental issues.

In light of the prevention and control of COVID-19 and travel restrictions worldwide, ACMME 2020 was held online. So, we adjust the format to online meeting to avoid personnel gatherings, and the conference could go ahead as scheduled. The conference model was divided into two parts, including keynote speeches and online discussion. In the first part, keynote presentations were delivered in 40 minutes for each talk, all of which were conducted via Zoom. The second part was held through a communication group via WeChat for experts and scholars to discuss more about the talks and presentations. The WeChat discussion lasted for 60 minutes.

By submitting a paper to ACMME 2020, the authors agreed to the review process and understood that papers would undergo a peer-review process. Manuscripts were reviewed by qualified experts in the field selected by the conference committee. The committees invested great efforts in reviewing the papers submitted to the conference and organizing the sessions to enable the participants to gain maximum benefit.

We do believe your great participation will make our conference success and we would appreciate your participation. We express our heartfelt thanks to IOP publishing house. Hopefully, all participants and other interested readers will benefit or find inspirations from this proceeding.

With our warmest regards,

Conference Organizing Chair
Xishuangbanna, Yunnan, China

IOP Conf. Series: Materials Science and Engineering **914** (2020) 011002 doi:10.1088/1757-899X/914/1/011002

# Committees

**Conference General Chair**
Prof. D. J. Fang, Wuhan University, China

**Steering Committee Chairs**
Prof. Kwon Myung Jang, Pukyong National University, South Korea
Prof. Q. S. Han, South-Central University for Nationalities, China

**Program Committee Chairs**
Prof. P. Zhao, Beijing Material Research Center, China
Prof. H. V. Lee, Universiti Sultan Zainal Abidin, Malaysia
Prof. F. L. Zhang, Shanxi University, China

**Technical Committees**
Prof. Y. D. Zhang, Wuhan Institute of Technology, China
Prof. G. Q. Xia, Wuhan Institute of Technology, China
Prof. X. Han, Wuhan Institute of Technology, China
Prof. Z. G. Fan, Wuhan University, China
Prof. H. Y. Qiu, Wuhan University, China
Prof. S. W. Wei, Shanxi University, China
Prof. K. Wei, Shanxi University, China
Prof. Z. X. Hou, Hainan University, China
Prof. H. L. Yan, Yunnan University, China
Prof. M. Xu, Yanshan University, China
Prof. S. D. Wei, Northeastern University, China
Prof. A. M. Qian, Beijing Material Research Center, China
Prof. L. G. Sun, Beijing Material Research Center, China
Prof. D. L. Mo, South-Central University for Nationalities, China
Prof. W. J. Han, South-Central University for Nationalities, China
Prof. C. K. Feng, South-Central University for Nationalities, China
Prof. S. P. Narasimha, Universiti Sultan Zainal Abidin, Malaysia
Prof. G. L. Miao, University of Science and Technology of China, China
Senior Engineer, H. Yang, China Petroleum Pipeline Engineering Co., Ltd, China
Dr. L. Y. Cai, Wuhan Institute of Technology, China
Dr. Xu Han, Hunan University, China
Dr. Danghuang Xu, Nanjing University of Aeronautics and Astronautics, China
Dr. G. Tao, Industry and Information Technology Department of Hebei Province, China

IOP Conf. Series: Materials Science and Engineering **914** (2020) 011003 doi:10.1088/1757-899X/914/1/011003

| No. | First name | Surname |
|---|---|---|
| 1 | Liuwei | Yang |
| 2 | Yu | Zhang |
| 3 | Hui | Liang |
| 4 | Jinhui | Zhang |
| 5 | Weixiang | Deng |
| 6 | Tingyao | Chen |
| 7 | Changjiang | Huang |
| 8 | Yalei | Liu |
| 9 | Rui | Xiao |
| 10 | Dong | Wang |
| 11 | Lingye | Tan |
| 12 | Ying | Mei |
| 13 | HuiJuan | Li |
| 14 | Shijun | Liu |
| 15 | Xue | Yu |
| 16 | Junjie | Zhang |
| 17 | Xiao | Liu |
| 18 | Xiongfei | Shi |
| 19 | Xiusheng | Tang |
| 20 | Yanhui | Guo |
| 21 | Ang | Li |

| Institution |
| --- |
| School of Xichang University, Sichuan, China |
| Shaanxi University of Chinese Medicine, Xianyang 712083, China |
| Xi'an Satellite Control Center, Xian, China |
| School of Traffic and Transportation, Beijing Jiaotong University, Beijing 100044, China |
| School of information Science and Engineering, Lanzhou University, Lanzhou, China |
| Law School, the South China University of Technology, Guangzhou 510006, China |
| Sontan College Guangzhou University, Guangdong Guangzhou, 511370, China |
| Department of mechanical and electrical, China maritime police academy of CAPF, Zhejiang, China |
| Nanjing Institute of Electronic Technology, Nanjing, China |
| ical and Information Engineering, Shanghai Urban Construction Vocational College, Shanghai, 201415, ( |
| Faculty of Engineering, University of new south wales, New South Wales, 2032, Australia |
| School of Guizhou University of Traditional Chinese Medicine, Guizhou, China |
| Wuhan software engineering Career Academy, School of Arts and media, HuBei 430000 |
| Medicine/Shaanxi Collaborative Innovation Center of Chinese Medicinal Resource Industrialization, Xiar |
| Department of Sports Education and Research, Northwestern University, ShaanXi, Xian 710069, china |
| International Business School, Yunnan University of Finance and Economics, China |
| Shengli College China University of Petroleum, 257061, Shandong, China |
| School of northeast Normal University, Changchun 130000, China |
| The First Affiliated Hospital of Guizhou University of Traditional Chinese Medicine, Guizhou, China |
| lege of Civil and Architectural Engineering, Yunnan Agricultural University, Kunming 650201, Yunnan, Ch |
| School of Engineering, Huanghe Science and Technology College, Zhengzhou, Henan 450000, China |

| Email address |
| --- |
| 1606285355@qq.com |
| zhangyu19432@163.com |
| lianghwui@163.com |
| zhang_sdxs@126.com |
| 594766943@qq.com |
| 13824450795@163.com |
| 33946676@qq.com |
| liuyalei820@163.com |
| 250242323@qq.com |
| wd_lulu@163.com |
| tanlingyet@126.com |
| 37979960@qq.com |
| disjill02@163.com |
| l181618@126.com |
| xueyu8528@163.com |
| Zhangjj@ynufe.edu.cn |
| slytliu@163.com |
| shixf846@nenu.edu.cn |
| 850447260@qq.com |
| guoyanhui0818@163.com |
| 293513627@qq.com |

IOP Conf. Series: Materials Science and Engineering **914** (2020) 011004 doi:10.1088/1757-899X/914/1/011004

# Peer review declaration

All papers published in this volume of *IOP Conference Series: Materials Science and Engineering* have been peer reviewed through processes administered by the Editors. Reviews were conducted by expert referees to the professional and scientific standards expected of a proceedings journal published by IOP Publishing.

- **Type of peer review:** Double-blind
- **Conference submission management system:**
  Firstly, authors send their papers to the conference mailbox, and all submissions will go through a preliminary review. Secondly, papers that out of scope will be rejected outright, papers that in the scope will enter the peer review process.
- **Number of submissions received:** 59
- **Number of submissions sent for review:** 51
- **Number of submissions accepted:** 42
- **Acceptance Rate (Number of Submissions Accepted / Number of Submissions Received X 100):** 71%
- **Average number of reviews per paper:** 2.3
- **Total number of reviewers involved:** 11
- **Any additional info on review process:** None
- **Contact person for queries:** Dajing Fang (djfang@whu.edu.cn, Wuhan University)

# The Influencing Factors of Bacillus megaterium Affect the Growth of the Antagonistic Strain and the Activities of Antibacterial Substances

**Ran Sun[1], Yang Xiao[2], Bokai Guo[3], Zhu Li[1,*], Ke Wan[1], Long Yang[1], Jinyi Zhao[1]**

[1]College of Life Science/Key laboratory of Plant Resource Conservation and Germplasm Innovation in Mountainous Region (Ministry of Education), Guizhou University, Guiyang, 550025, China
[2]Institute of Products Detection & Determination of Guizhou Province, Guiyang, 550025, China
[3]Guizhou Provincial People's Hospital, Guiyang, 550025, China

**Abstract.** 164 strains of Antagonistic bacteria were isolated from 131 mature-period soil samples from eight counties in Bijie City, Guizhou Province, China. In this study, the antagonistic bacteria that inhibited *Alternaria alternata* were screened and isolated. In addition, the antagonistic strain L2 was identified as *Bacillus megaterium.* Influencing factors also demonstrated that the suitable conditions for the production of antibacterial substances by strain L2 are as follows: liquid volume 50 mL/500 mL at pH 7.0 and 30°C under oscillation culture conditions, with beef extract as the carbon, nitrogen source.

## 1. Introduction

Alternaria brown spot is caused by the genera *Alternaria* [1-2]. Tobacco brown spot, is a major disease that attacks tobacco leaves at the late maturation stage. This disease is characterized by both a short incubation period as well as rapid spreading, and it will directly impact tobacco quality and yield [3-5].

In previous decades, chemical-control methods have been the primary means of combating the brown spot. However, many of the chemical pesticides are detrimental to animal and human health alike. Accordingly, solutions have been sought to reduce the use of these chemicals while still effectively combatting the spread of pathogens. In 1978, Kloepper and Schroth first reported that plant growth-promoting rhizobacteria (PGPR) can stimulate plant growth, increase yield, and reduce pathogen infections [6]. Since then, considerable research has been conducted concerning the capability of PGPR to provide resistance against plant pathogens as well as its role in promoting crop growth [7-12]. In this study, the antagonistic bacteria that inhibited *Alternaria alternata* were screened and the Influencing factors of antibacterial action was inspected, and in order to provide microbiological resources for the researsh of biocontrol and evaluate the values of L2 applications.

IOP Conf. Series: Materials Science and Engineering **914** (2020) 012001 doi:10.1088/1757-899X/914/1/012001

## 2. Materials and Methods

### *2.1. Materials*

The antagonistic bacteria were isolated from 131 mature-period soil samples from eight counties in Bijie City, Guizhou Province, China. The antagonistic strain L2 was identified as *Bacillus megaterium*. The indicator fungi, *Alternaria alternata*, was isolated and preserved by The Institute of Fungi Resource, College of Life Science, Guizhou University.

### *2.2. Methods*

*2.2.1. Preparation of spore suspensions and seed solutions.* A. alternata was cultured on potato dextrose agar (PDA)[13] plates at 29 °C for 15 d and then placed under an ultraviolet lamp for 3 d with continuous irradiation to induce spore production. Then, 10 mL of a 1% dextrose solution was added for 15 min of immersion. Next, spores and mycelia were scraped off with sterile cotton swabs, and the mycelia were removed through filtration with absorbent cotton. The spore concentration was adjusted to $2.0 \times 10^5$ CFU/mL using 1% dextrose.

The *B. megaterium* strain L2 was activated twice on nutrient agar (NA)[13] plates and then streak-inoculated for 24 h. Next, the strain was inoculated into the nutrient broth (NB)[13] culture solution and cultivated at 29 °C at 150 rpm/min for 12 h. Then, the seed bacterial liquid was prepared.

*2.2.2. Bacteria-free filtrate preparation.* 1 mL of the fermentation solution of L2 was collected and centrifuged at 5,500 rpm/min for 25 min. Then, the supernatant was collected and vacuum-filtered through 0.22-μm filter membranes to produce bacteria-free filtrates.

*2.2.3. Influencing factors that affect the growth of the antagonistic strain and the activities of antibacterial substances.* In preliminary experimentations, 30 °C, culture time 48 hours, liquid volume 50 mL, rotation speed 150 rpm/min and pH 7.0 has been found optimal for the growth of the antagonistic strain. Based on these results, the influencing factors of the growth of the antagonistic strain and the activities of antibacterial substances were investigated. The investigated influencing factors included temperature (20, 25, 30, 35, 40, 45, and 50°C), pH (4.0, 4.5, 5.0, 5.5, 6.0, 6.5, 7.0, 7.5, 8.0, 8.5, and 9.0), liquid volume (five levels in 500-mL culture flasks: 50, 100, 150, 200, and 250 mL), and carbon, nitrogen (C, N) source (11 types, including yeast extract, beef extract, dextrose, sucrose, arginine, and cysteine). The bacterial growth was detected using the UV spectrophotometry [14], and the antibacterial activity was detected using the Oxford cup method [15].

*2.2.4. Statistical analysis.* Statistical analyses of quantification data were carried out using SPSS22 software, and were subject to one-way analysis of variance (ANOVA) analysis; means were analyzed using Dunnett's multiple range tests. In the tables, the different superscript letters within the same column indicate statistically significant differences. For each experient, three replicates were included.

## 3. Results and Analysis

### *3.1. Influencing factors on bacterial growth and antibacterial substances*

*3.1.1. Effects of pH on bacterial growth and antibacterial substance activity.* As is shown in Fig. 1, at pH 5.0-7.0, the increases were significant ($P = 5.12 \times 10^{-12} << 0.01$). The increment at pH 7.0 maximized at 1.2777, and then, as pH increased, it gradually declined. Furthermore, the activity levels of the antibacterial substances did not change greatly from pH 4.0 to 6.0, but the antibacterial substances produced at pH 6.0-7.0 had significantly greater activity levels ($P = 3.22 \times 10^{-5} << 0.01$). The activity was maximized at pH 7.0, and the diameter of the inhibition zone reached 1.47 cm. This result was consistent with the incremental changes, indicating that pH 7.0 was favorable for the

IOP Conf. Series: Materials Science and Engineering **914** (2020) 012001 doi:10.1088/1757-899X/914/1/012001

growth of strain L2 and the production of antibacterial substances. Moreover, the activity levels of the antibacterial substances gradually decreased as the pH increased further.

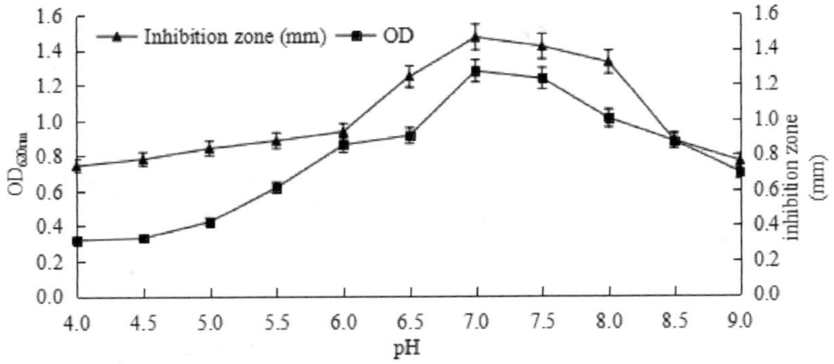


**Fig. 1** Strain L2 growth and antibacterial substances' activity levels at varying pH values.

*3.1.2. Effects of liquid volumes on L2 bacterial growth and antibacterial substances' activity levels.* As is shown in Fig. 2, when 50 mL of liquid was added, the increment was maximized, and the $OD_{620nm}$ was 1.4037. Furthermore, the activity levels of the antibacterial substances were also maximized, and the diameter of the inhibition zones was 1.46 cm. As the liquid volume increased, the increment dropped significantly ($P = 1.39 \times 10^{-5} << 0.01$). Although the activity levels of the antibacterial substances were significantly decreased as the liquid volume increased (analysis of variance, $P = 0.0025 < 0.01$), the declining amplitude was very small compared with the change in the growth mass. This result indicated that, compared with the incremental change in the strain growth, the activity levels of the antibacterial substances were less affected by the liquid volume.

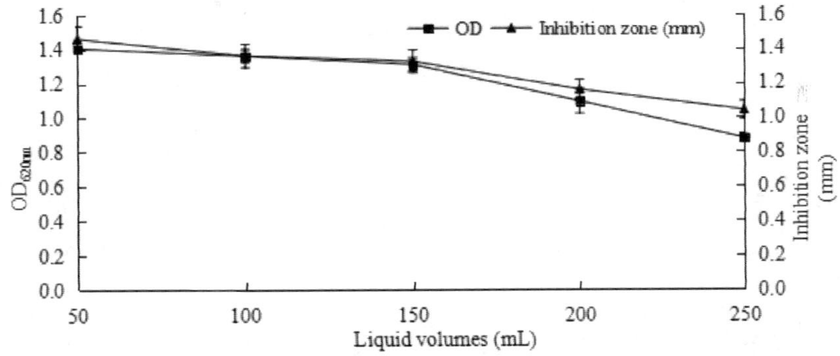


**Fig. 2** Effects of liquid volumes on strain L2's growth and the antibacterial substances' activity levels.

*3.1.3. Effects of temperature on bacterial growth and antibacterial substances' activity levels.* As is shown in Fig. 3, at both 25 and 30°C, the growth change increments were both significantly positive ($P = 4.75 \times 10^{-6} << 0.01$). The incremental change reached a maximum of 1.304 at 30°C. However, the incremental growth changes decreased as the temperature continued to rise, reaching a minimum (0.4677) at 50°C. The activity levels of the antibacterial substances produced at both 20 and 25°C were not significantly different, but the activity level at 30°C was significantly greater ($P = 3.47 \times 10^{-6} << 0.01$), with an inhibition zone diameter of 1.30 cm, which was the maximum. The activity levels of the antibacterial substances at 35 and 30°C were not significantly different, but as the temperature rose, the activity levels of the antibacterial substances gradually decreased.

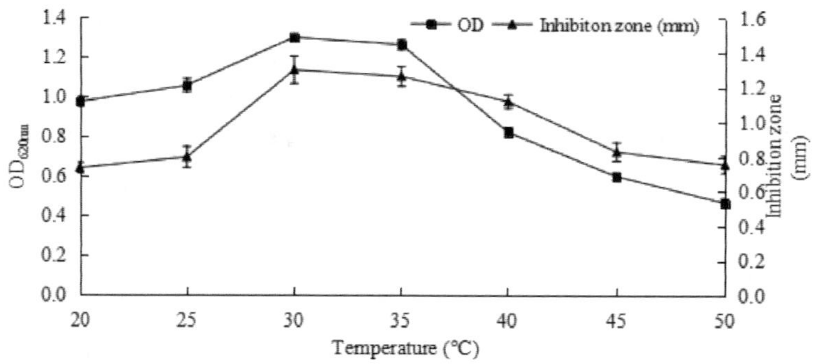


**Fig. 3** Strain L2's growth and the antibacterial substances' activity levels at varying temperatures.

*3.1.4. Effects of C, N sources on strain L2's growth and the antibacterial substances' activity levels.* As detailed in Table 1, the growth levels of strains that were inoculated into media that contained beef extract, yeast extract, sucrose, or dextrose increased greatly, with an $OD_{620nm}$ of 1.2837, and the activity levels of the antibacterial substances produced by the strain feeding on the beef extract were the greatest, producing an inhibition zone diameter of 1.51 cm. Although the strains could grow on sucrose or dextrose, no active antibacterial substances were detected, which indicated that these two C, N sources were suitable for strain growth but not for producing antibacterial substances.

**Table 1.** Effects of C, N sources on strain L2's growth and antibacterial substances' activity levels

| C, N source | $OD_{620nm}$ | Inhibitory zone diameter(cm) |
|---|---|---|
| peptone (0.5%) | 0.6937±0.0110c | 0.8367±0.0404a |
| beef extract(0.3%) | 1.2837±0.0045d | 1.5100±0.0351b |
| yeast extract(0.1%) | 1.0013±0.4225bcd | 1.0766±0.0839ab |
| sucrose(1%) | 1.0970±0.0108b | 0.7233±0.0115a |
| glucose(1%) | 1.1573±0.0021b | 0.7067±0.0058a |
| $KNO_3$(0.1%) | 0.4110±0.0080a | 0.7100±0.0100a |
| inositol(0.2%) | 0.1227±0.0025e | 0.7067±0.0115a |
| glycine(0.5%) | 0.7443±0.0032c | 0.7367±0.0351a |
| glutamic(0.5%) | 0.4177±0.0049a | 0.8200±0.0436a |
| aspartic(0.5%) | 0.3240±0.0017f | 0.7700±0.0173a |
| tyrosine(0.5%) | 0.2673±0.0032g | 0.7300±0.0264a |

## 4. Conclusion

Based on the above experimental results, we concluded that the most suitable conditions for the production of antibacterial substances by strain L2 were as follows: liquid volume of 50 mL, pH 7.0, 30 °C, oscillation culture conditions, and beef extract as the C, N source. In comparison to previous reports, this study highlighted the correlation between bacterial growth and the activity levels of the antibacterial substances. Generally, the incremental increases in the growth of the antagonistic strain L2 were positively correlated [$R^2$(pH) = 0.78, $R^2$(Liquid volumes) = 0.96, $R^2$(Temperature) = 0.45] with the activity levels of the antibacterial substances. However, while the strain grew in dextrose- and sucrose-based C, N culture solutions, the activity levels of the antibacterial substances were low. We, therefore, hypothesize that the growth of the antagonistic strain and the production of the antibacterial substances required different C, N sources. Moreover, the incremental growth changes in the antagonistic strain and the activity levels of the antibacterial substances could be detected throughout the entire tested pH range, the entire temperature gradient, and with all of the different C, N sources, indicating that the antagonistic strain possessed a strong environmental adaptability.

We then investigated its antibacterial mechanism. It was strongly inhibitory to *A. alternata* because the *A. alternata* mycelia were damaged and decomposed. Additionally, we found that the gross metabolites from L2 inhibited *A. alternata* spore germination. In the future, we will systematically and specifically probe into the prevention and treatment of tobacco brown spot and evaluate the values of L2 applications.

**Acknowledgements**
The Institute of Fungi Resource's work that is referred to here was supported by grants from The National Natural Science Foundation of China (31660533 and 31460486); the Guizhou province outstanding youth project ([2017]5613); the Open Fund Project of Guizhou Province Domestic First-Class Discipline Construction in Biology (GNYL[2017]009FX3KT20); and the Guizhou Province Science and Technology Project (NY[2014]3033, SY[2014]3053). We thank International Science Editing for editing this manuscript.

**Refrences**

[1] Perina J. Fabiano, Amaral C. D and Fernandes S. R, *et al.* "Thymus vulgaris essential oil and thymol against *Alternaria alternata* (Fr.) Keissler: effects on growth, viability, early infection and cellular mode of action." Pest Management Science 71.10 (2014): 1371-1378.
[2] Spurr W. Harvey, Main C. E. "Brown pigment formation in tobacco leaves infected with *Alternaria*." Phytopathology 64.5 (1974): 738-745.
[3] Zhongyi Zhang, Huaiqiong L, Zhiming Z, *et al.* Plant pathogenic fungi(In Chinese). Chengdu: Sichuan science and technology publishing company, (1988):376-377.
[4] Stavely J. R, Main C. E. "Influence of temperature and other factors on initiation of tobacco brown spot." Phytopathology 60.11 (1970): 1591-1596.
[5] George Blanchard Lucas. "Diseases of Tobacco (3rd edn.)" Raleigh, NC, USA: Biological Consulting Associates, (1975): 407-410.
[6] Kloepper JW, Schroth M, Kloepper J, *et al.* "Plant growth-promoting rhizobacteria on radishes." Proceedings of the 4th International Conference on Plant Pathogenic Bacteria, 2 (1978): 879-882.
[7] Van Loon L. C, Bakker P. A. H. M. "Induced systemic resistance as a mechanism of disease suppression by rhizobacteria." PGPR: Biocontrol and Biofertilization. Springer Netherlands, (2006): 39-66.
[8] D J O'Sullivan, Fergal O'Gara. "Traits of fluorescent Pseudomonas spp. involved in suppression of plant root pathogens." Microbiological Reviews 56.4 (1992): 662-676.
[9] Sivan Alex, Chet I. "Microbial control of plant diseases." Microbiological Reviews. New York (NY): Wiley-Liss, (1992): 335-354.
[10] Sutton C. John, Gang P. "Manipulation and vectoring of biocontrol organisms to manage foliage and fruit diseases in cropping systems." Annual Review of Phytopathology 31.1 (1993): 473-493.
[11] Cook R. James. "Making greater use of introduced microorganisms for biological control of plant pathogens." Annual Review of Phytopathology 31.1 (1993): 53-80.
[12] Chet Ilan, Jacob I. "Biological control of fungal pathogens." Applied Biochemistry and Biotechnology 48.1 (1994): 37-43.
[13] Zhu Li, Bokai G, and Ke W, *et al.* "Effects of bacteria-free filtrate from *Bacillus megaterium* strain L2 on the mycelium growth and spore germination of *Alternaria alternata*" Biotechnology & Biotechnological Equipment 29.4 (2015): 1062-1068.
[14] Ting Liu, Yu L and Jingjing X, *et al.* "Bacterial traits and quality contribute to the diet choice and survival of bacterial-feeding nematodes." Soil Biology and Biochemistry 115 (2017): 467-474.
[15] Wang Yu, Zhaoxin L and Hao W, *et al.* "Study on the antibiotic activity of microcapsule curcumin against foodborne pathogens International." Journal of Food Microbiology 136.1 (2009): 71-74.

IOP Conf. Series: Materials Science and Engineering **914** (2020) 012002 doi:10.1088/1757-899X/914/1/012002

# Sand and Dust Hazard Risk Zoning for Ningxia Expressway

**Wanli Wu[1, 2], Xiaohui Guo[1, 2], Yao Liu[1, 2], Lei Ma[2]**

[1]Key Laboratory for Monitoring Early Warning and Risk Management of Characteristic Agricultural Meteorological Disasters in Dry Areas, China Meteorological Administration, Yinchuan, 750002, China
[2]Ningxia Meteorological Service Center, Yinchuan, 750002, China

**Abstract.** Based on GIS and AHP methods, the risk assessment and research of expressway sand and dust hazard are carried out from three aspects: 1) dangerous hazard formative factors, 2) environmental sensitivity, and 3) hazard-affected bodies vulnerability. According to 24 stations from 1981 to 2018 in observation data, the risk zoning is given combined with geographic information data. The results show that the high-risk areas are located in Yanchi section of Qingyin expressway (G20), Yanchi section of Dingwu expressway (G2012) and Ningxia section of Yane expressway. The higher risk areas are located in Xingren-downstream section of Beijing-Tibet expressway (G6), Zhongwei section of Dingwu expressway (G2012), Gunquan section of Beijing-Tibet expressway (G6) and Qingtongxia section of Wuma expressway (G1816). The medium risk areas are located in Huinong section of Beijing-Tibet expressway (G6), southern section of Yinchuan ring expressway, Ningdong section of Qingyin expressway (G20) and Taiyangshan-Maerzhuang section of Dingwu expressway (G2012). The dust risk is generally low in Wuzhong section of Beijing-Tibet expressway (G6), Wuzhong section of Yinkun expressway (G85), Tongxin south section of Fuyin expressway (G70) and Ningxia section of Qinglan expressway (G22), etc.

## 1. Introduction

Expressway transportation is a highly meteorological sensitive industry, which is easily affected by various natural environmental factors. With the rapid development of national economy and urbanization, expressway traffic safety guarantee caused by meteorological conditions has become a key factor affecting economic development [1]. Based on the definition of meteorological observation specifications, floating dust, blowing sand and sandstorm are all atmospheric dust particles phenomenon [2]. They belong to the category of sand and dust weather, and thus they are collectively referred to as sand and dust weather. Floating dust refers to dust and fine sand floating in the air evenly, making the meteorological visibility less than 10km. Sand blowing refers to that the ground dust is blown up by strong wind, which makes the air quite turbid, and the meteorological visibility is greater than or equal to 1.0km to less than 10km. Sandstorm refers to the fact that a large amount of dust on the ground is drawn into the air by strong wind, which makes the air very turbid, and the meteorological visibility is less than 1km. Among them, the meteorological visibility less than 500m is called severe sandstorm, and the meteorological visibility less than 50m is called extremely severe sandstorm. Sand and dust weather can cause visual obstacles, accompanied by strong wind, which is easy to cause expressway vehicle collision, and has a serious impact on expressway traffic. Ningxia is located in the eastern part of northwest China and the intersection of Loess Plateau, Mongolian Plateau and Qinghai Tibet Plateau. It is surrounded by Tengger Desert, Ulan Buh Desert and Mu Us sandy land in the west, north and east, which is the only place for sand and dust supplement and transportation to the east in northwest China. From the spatial and temporal distribution of sandstorms, Ningxia is one of the most frequent sandstorms in China [4]. Its special geographical and geomorphic conditions determine that there are almost always sandstorm weather effects in spring [5]. It is one of the most serious disastrous weather in spring in Ningxia, which has great harm to industrial and agricultural production, transportation, people's life, life and property [6]-[7]. The research on sand and dust weather in Ningxia mainly focuses on the formation mechanism and climate background of sand and dust weather, the basic characteristics and change trend of severe sand and dust weather, as well as the ecological effects and control countermeasures of sand and dust hazard. With regard to disaster risk analysis, Cheng Wenhao, et al., analyzed the risk of severe sandstorm hazard in Ningxia [8]. Huang Jianwen, et al., carried out the risk zoning analysis of gale and dust meteorological hazard in Jinchang, Gansu Province [9]. Lu Tianping, et al., carried out the risk assessment model construction and zoning of gale and dust hazard in Xinjiang forestry and fruit industry [10]. Liang Fengjuan, et al., carried out the characteristics and risk zoning of hazard in Bayannaoer region [11]. Li Jinrong carried out a research on the risk assessment of sandstorm hazard in Xilin Gol League, Inner Mongolia [12]. However, there are few literatures on the risk assessment of sand and dust weather hazard on traffic. Based on the research methods at home and abroad, this work took the main expressways in Ningxia as the research object, and analyzed the temporal and spatial characteristics of sandstorm events in Ningxia expressway with the support of GIS spatial information technology. The hazard risk of expressway sand and dust weather is comprehensively analyzed from three aspects: dangerous hazard formative factors, subsquently environmental sensitivity and hazard-affected bodies vulnerability, which offers scientific basis for the early warning and forecasting service of expressway sand and dust weather.

## 2. Materials and Methods

### *2.1. Basic situation of Ningxia expressway*

By the end of 2018, the highway mileage in Ningxia was 35405 km, and the highway density was 53.3 km / 100 km$^2$. Among them, 1678 kilometers of expressways are open to traffic. All prefecture level cities have access to expressways, and all counties (cities) have one hour access to expressways. The proportion of class II and above highways, the proportion of grade roads, and the cement asphalt pavement rate of national and provincial trunk lines are in the forefront of western China. The national expressways in Ningxia include: Beijing-Tibet expressway (G6), Qingyin expressway (G20), Qinglan expressway (G22), Fuyin expressway (G70), Yinkun expressway (G85), Wuma expressway (G1816), Wuyin expressway (G1817), Yinchuan ring expressway (G2004), Dingwu expressway (G2012), etc. The provincial expressways in Ningxia include: Shizuishan ring expressway (S10), Yane expressway (S15), Guqing expressway (S30), Tonghai expressway (S40), Heihai expressway (S50), Guhui expressway (S60), Pengqing expressway (S70), etc. Among them, the Beijing-Tibet expressway (G6) is mainly in the north-south direction, and the main section is located in the north-central part of Ningxia. It is located in the Yellow River irrigation area with flat terrain. Qingyin expressway (G20) is mainly located in the northeast of Ningxia, passing through Yinchuan and Yanchi. Fuzhou-Yinchuan expressway (G70) is mainly located in the southern part of Ningxia, passing through Tongxin county, Guyuan city and Liupan mountain area. The Wuma expressway (G1816) along the Helan mountains has been completed, starting from Shizuishan city in the north to the west of Qingtongxia in the south, which is greatly affected by the environment. Dingwu expressway (G2012) is the main road in the east-west direction of the region, which is located in the central part of Ningxia. It mainly passes through Zhongwei, Zhongning and Yanchi in the central arid zone (see Fig. 1).

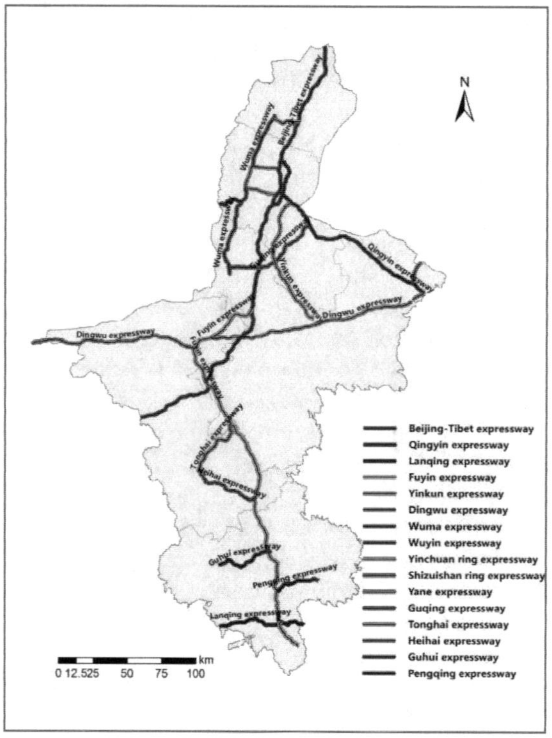


**Fig. 1** Distribution map of main expressways in Ningxia

### *2.2. Data sources*

According to the "interim provisions on early warning service of sand and dust weather" issued by China Meteorological Administration, the sand and dust weather is divided into floating dust, blowing sand and sandstorm. In this work, the sum of sandstorm days, sand blowing days and floating dust days is taken as the number of sand and dust days. The data were collected from 24 conventional weather stations in Ningxia from 1981 to 2018. The traffic network distribution and other data are from Ningxia Highway Bureau, and the traffic flow information is from the traffic reorganization data of Ningxia traffic information center in 2016. The data of GDP spatial distribution, vegetation index (NDVI) and population spatial distribution are from the Data Center for Resources and Environment Science, Chinese Academy of Sciences. SRTM 90m elevation data is from the free website of SRTM international scientific data service platform. Other basic geographic information data are from the national basic geographic information center.

### *2.3. Risk assessment model*

It is generally believed that the hazard situation of sand and dust hazard is caused by the interaction among the hazard formative factors, subsquently environment and hazard-affected bodies. Its influence degree is determined by dangerous hazard formative factors, subsquently environmental sensitivity and hazard-affected bodies vulnerability. Based on the principle of meteorological hazard, meteorological hazard risk is the comprehensive action result of dangerous hazard formative factors, subsquently environmental sensitivity, hazard-affected bodies vulnerability, etc. The calculation formula of the evaluation model is as follows:

$$V_i = \sum_{i=1}^{n} W_i \bullet X'_{ij}$$

$V_j$ is the evaluation factor value, and $W_{\_i}$ is the weight of the i-th index. $X'_{ij}$ is the normalized value of the i-th index for factor j, and n is the number of evaluation indicators.

The factors that may lead to hazard are called hazard formative factors, and the risk degree of hazard formative factors is determined by the intensity and frequency of hazard formative factors. Generally, the greater the possibility and variation intensity of hazard formative factors, the higher the risk of hazard formative factors. In this work, the days of floating dust, blowing sand and sandstorm are selected as the risk assessment index of hazard formative factors. The specific subsquently environment in a region basically determines the type of disaster in the region, which is mainly composed of geological conditions, geographical environment, water system conditions, ecological environment, etc. The subsquently environment can enlarge or reduce the risk of hazard formative factors. In this work, terrain elevation (topography), slope, river network density and vegetation coverage are selected as the subsquently environmental sensitivity evaluation indexes of expressway sand and dust hazard. Hazard-affected body is the object of hazard formative factors and the entity suffering from hazard. The nature and structure of hazard-affected body basically determines its vulnerability. For expressway sand and dust hazard, hazard-affected body vulnerability refers to the loss of expressway affected by sand and dust weather hazard when hazard comes. In this work, population density, vehicle flow, road density and GDP are selected as hazard-affected body vulnerability evaluation indexes.

In order to eliminate the dimensional difference between the indexes, the following formula is used to normalize the values of each index.

$$Xi'=(Xi- Xi, min)/( Xi, max - Xi, min )$$

$X_i$ is the i-th index value, and $X_{i'}$ is the normalized value of the ith index. $X_{i, max}$ and $X_{i, min}$ are the maximum and minimum values of the i-th index value respectively.

### *2.4. Weight coefficient and zoning method*

Analytic hierarchy process (AHP) is a decision-making method that decomposes the elements related to decision-making into objectives, criteria, schemes, etc., and carries out qualitative and quantitative analysis on this basis [13]. The analytic hierarchy process has a good accuracy in calculating the weight coefficient, and is a simple and feasible decision-making method that is easy to understand and grasp, and can solve complex multi-objective problems. This work selects 11 evaluation indexes that affect the sand and dust hazard of Ningxia expressway, including floating dust weather days, blowing sand weather days, sandstorm days, GDP spatial distribution, vehicle flow, road density, population density, terrain height, river network density, slope and vegetation coverage. According to the different indexes of hazard formative factors, subsquently environment and hazard-affected bodies for Ningxia expressway sand and dust weather hazard, AHP method is used to construct judgment matrix, and different weight coefficients are given. When all the index hierarchy judgment matrices pass the consistency test (i.e., $CR < 0.1$), it indicates that the allocation of indicators is reasonable. Finally, the weight of each index is sorted and calculated comprehensively to determine the weight value of each index, and the weights of 11 raster data layers are obtained (see Table 1). Combined with GIS spatial analysis method, it is divided into five levels based on the natural breakpoint method. The dangerous hazard formative factors, subsquently environmental sensitivity and hazard-affected bodies vulnerability are analyzed according to five levels: low, secondary low, medium, secondary high and high.

IOP Conf. Series: Materials Science and Engineering **914** (2020) 012002 doi:10.1088/1757-899X/914/1/012002

**Table 1.** Evaluation index system and weight of expressway sand and dust weather risk

| Project | Criterion layer | Evaluation layer 1 | Evaluation layer | Weight |
|---|---|---|---|---|
| Sand and dust weather risk evaluation | Dangerous hazard formative factors 0.5378 | Floating dust | Floating dust days C1 | 0.1272 |
| | | Blowing sand | Blowing sand days C2 | 0.1766 |
| | | Sandstorm | Sandstorm days C3 | 0.234 |
| | Subsquently environmental sensitivity 0.2509 | Terrain | Elevation C4 | 0.0373 |
| | | | Gradient C5 | 0.0446 |
| | | Rivers | River network density C6 | 0.0734 |
| | | Vegetation | Vegetation coverage C7 | 0.0956 |
| | Hazard-affected bodies vulnerability 0.2116 | Population | Population density C8 | 0.0542 |
| | | Road | Road density C9 | 0.0496 |
| | | Vehicles | Vehicle flowrate C10 | 0.0561 |
| | | Economic | GDP C11 | 0.0514 |

Each evaluation index layer is weighted. Finally, the risk of expressway sand and dust hazard in Ningxia is divided according to the low risk, lower risk, medium risk, higher risk and high risk, thus forming the traffic risk zoning of sand and dust hazard in Ningxia.

## 3. Results Analysis

### *3.1. Basic characteristics of sand and dust weather*

*3.1.1. Distribution characteristics of sand and dust weather days.* The sum of floating dust days, blowing sand days and sandstorm days was taken as the statistical days of sand and dust weather, and the observation data of 24 conventional weather stations in the whole region from 1981 to 2018 were counted. The annual average days of sand and dust weather in Ningxia is 29.4 days, and Yanchi is the most, reaching 79.9 days. The second was Zhongwei Xingren, which reached 62.8 days. The sand and dust weather in Pingluo, Taole, Yinchuan, Weizhou, etc., lasts 31.1~48.5 days. The least is south of Xiji, about 8 days, and the weather in other areas is 14.9-26.9 days. According to the occurrence days statistics of floating dust, sand blowing and sandstorm (see Fig. 2), the annual average days of floating dust weather are 2~27.4 days. The annual average days of sand blowing weather are 1~50.8 days, and the annual average days of sandstorm weather are 0~14.4 days. Due to the difference of topography and landform, the frequency of sandstorm has different distribution characteristics, showing a trend of "high in the middle and east, low in the south". The sandstorm prone areas in Ningxia are mainly located in Yanchi, Tongxin, Haiyuan and Taole in the north, and the mountainous area in southern Guyuan is less frequent.

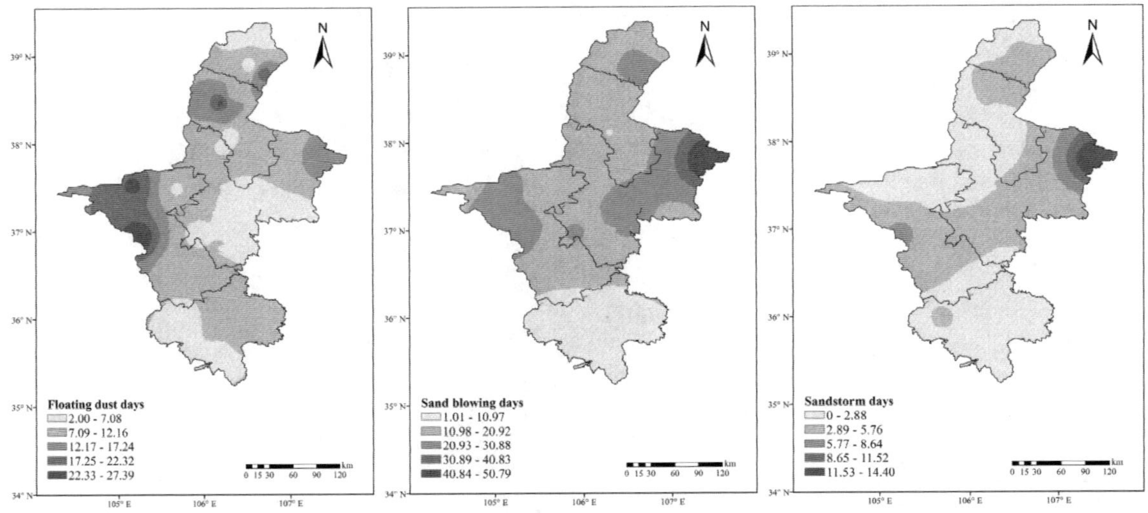


**Fig. 2** Distribution map of mean annual dust days in Ningxia autonomous region

*3.1.2. Distribution characteristics of sand and dust weather months.* Based on the statistics of monthly sand and dust weather days at various meteorological stations in Ningxia, there will be sand and dust weather in all seasons. Spring (March to May) was the most frequent occurrence, accounting for 56% of the total frequency, followed by winter (December to February), accounting for 25%. The sand and dust weather in summer (June to August) is less, accounting for 14%, and autumn is the least, accounting for only 5%. Among them, April is the most in spring, accounting for 24% of the total frequency, and autumn (September, October) is the least, accounting for 0.8% and 1% respectively.

*3.1.3. Effects of meteorological factors on sand and dust weather.* The correlation between the frequency of sand and dust weather in spring and the precipitation of last year was analyzed. It was found that the frequency of sand and dust in spring (March to May) was inversely correlated with the precipitation of the same year and last year. This is due to the fact that the vegetation growth is better in the years with more precipitation, and the bare degree of the surface is relatively low in spring, which increases the critical wind speed of sand and dust. The frequency of sand and dust weather in Ningxia is obviously opposite to that of winter temperature. When the temperature is low in winter, the frozen soil layer is thick, and the loose soil layer formed after thawing in early spring is also thicker. Therefore, it is easy to form sandstorm weather.Evaluation of dangerous hazard formative factors

The number of floating dust days, sand blowing days and sandstorm days are taken as the evaluation index of dangerous hazard formative factors, and the weight coefficients are 0.1272, 0.1766 and 0.2340 respectively. The index of dangerous hazard formative factors is calculated, and the distribution map of sand and dust hazard dangerous hazard formative factors is obtained (see Fig. 3). It can be seen from Fig. 3 that Yanchi in the eastern part of Ningxia has the highest of dangerous hazard formative factors in space, and the secondary high risk is distributed in Zhongwei city, represented by Xingren, and the low risk is mainly concentrated in the south of Guyuan.

IOP Conf. Series: Materials Science and Engineering **914** (2020) 012002 doi:10.1088/1757-899X/914/1/012002

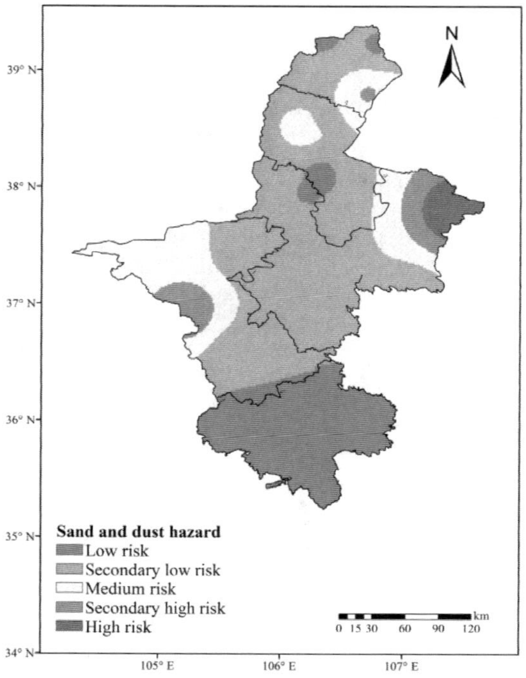


**Fig. 3** Risk zoning of sand and dust weather hazard formative factors in Ningxia expressway

### *3.2. Evaluation of subsquently environmental sensitivity*

Terrain elevation (topography), slope, river network density and vegetation coverage are selected as the subsquently environmental sensitivity evaluation indexes of expressway sand and dust weather hazard. The weights are 0.0373, 0.0446, 0.0734 and 0.0956 respectively. The subsquently environmental sensitivity index is calculated and the distribution map of subsquently environmental sensitivity is obtained (see Fig. 4). It can be seen from Fig. 4 that the overall subsquently environmental sensitivity of Ningxia expressway is relatively high. The high sensitive area is located in Huinong section, Gunquan-Zhongning section and Taoshan-Xingren section of Beijing-Tibet expressway (G6), with poor vegetation coverage and obvious topographic slope fluctuation. The secondary sensitive area is located in Zhongwei section of Dingwu expressway (G2012) and Qingtongxia section of Wuma expressway (G1816). The low sensitive areas are located in Wuzhong section of Beijing-Tibet expressway (G6) and GuQing expressway (S30), Fuyin expressway (G70), Qinglan expressway (G22), Pengqing expressway (S30), etc. These sections have high river network density, high vegetation coverage, small slope or in the mountains, which are not easy to be affected by sand and dust hazard.

IOP Conf. Series: Materials Science and Engineering **914** (2020) 012002 doi:10.1088/1757-899X/914/1/012002

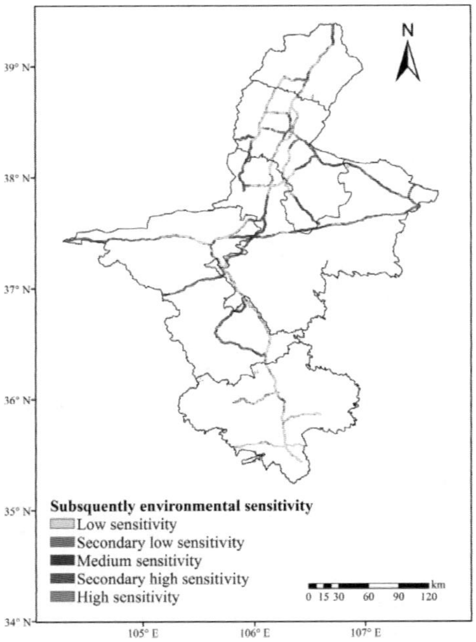


**Fig. 4** Environmental sensitivity zoning of sand and dust weather in Ningxia expressway

### *3.3. Evaluation of hazard-affected bodies vulnerability*

Population density, vehicle flow, road density and GDP are used as hazard-affected bodies vulnerability evaluation indexes of expressway sand and dust hazard. The weights are 0.0542, 0.0496, 0.0561 and 0.0514, respectively. The hazard-affected bodies vulnerability index is calculated, and the distribution map of hazard-affected bodies vulnerability in Ningxia expressway is obtained (see Fig. 5). It can be seen from Fig. 5 that the high vulnerability of hazard-affected bodies is located in Yinchuan-Yongning section of Beijing-Tibet expressway (G6) and Yinchuan-Ningdong section of Qingyin expressway (G20). Population density, vehicle flow, road density and GDP all belong to high value area. The secondary vulnerable area is located in Yongning-Wuzhong section of Beijing-Tibet expressway (G6). The medium vulnerable areas are located in Shizuishan section of Beijing-Tibet expressway (G6), Yinchuan ring city, Wuzhong-Zhongning section of Beijing-Tibet expressway (G6), Ningdong-Yanchi section of Qingyin expressway (G20), Zhongwei city of Dingwu expressway (G201 2) and Gunquan-Taoshan section of Fuyin expressway (G70). The Liupan mountain area section of Fu-Yin expressway (G70), Tonghai expressway and Guxi expressway belong to low vulnerability area due to low traffic volume and GDP.

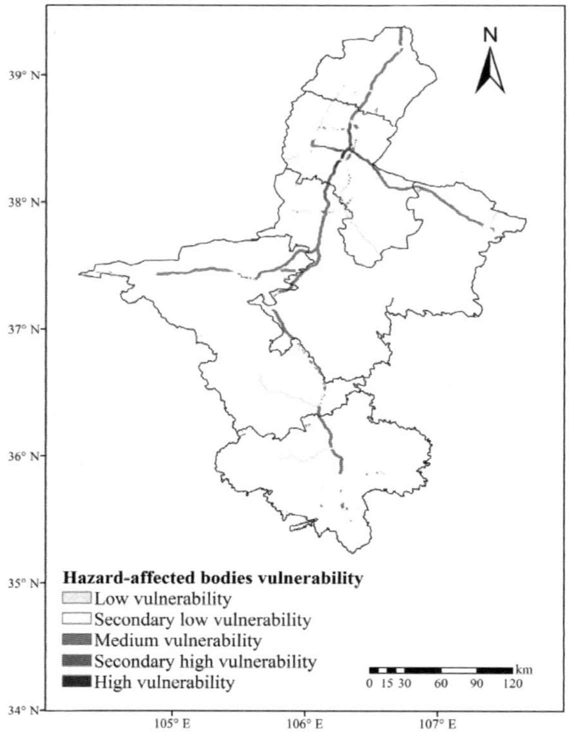


**Fig. 5** Hazard-affected bodies vulnerability zoning of sand and dust weather in Ningxia expressway

### *3.4. Risk zoning for sand and dust weather hazard in Ningxia expressway*

The work selects 11 evaluation indexes that affect the sand and dust weather hazard in Ningxia expressway, and gives different weight coefficients (see Table 1). The risk assessment and zoning analysis of sand and dust weather hazard in Ningxia expressway are carried out by using GIS fuzzy comprehensive evaluation model, and the comprehensive zoning of sand and dust weather hazard in Ningxia expressway is formed (see Fig. 6). It can be seen from Fig. 6 that the high risk areas of sand and dust hazard in Ningxia expressway are located in Yanchi section of Qingyin expressway (G20), Yanchi section of Dingwu expressway (G2012), Ningxia section of Yane expressway. The higher risk areas are located in Xingren-downstream section of Beijing-Tibet expressway (G6), Zhongwei section of Dingwu expressway (G2012), Gunquan section of Beijing-Tibet expressway (G6) and Qingtongxia section of Wuma expressway (G1816). The medium risk areas are located in Huinong section of Beijing-Tibet expressway (G6), southern section of Yinchuan ring expressway, Ningdong section of Qingyin expressway (G20) and Taiyangshan-Maerzhuang section of Dingwu expressway (G2012). The dust risk is generally low in Wuzhong section of Beijing-Tibet expressway (G6), Wuzhong section of Yinkun expressway (G85), Tongxin south section of Fuyin expressway (G70) and Ningxia section of Qinglan expressway (G22), etc.

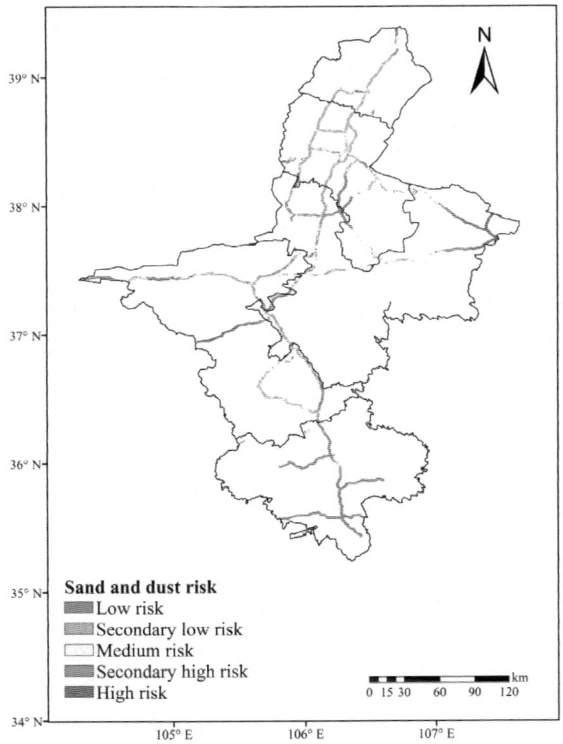


**Fig. 6** Risk zoning of sand and dust weather hazard in Ningxia expressway

## 4. Conclusion

First, the annual average days of sand and dust weather in Ningxia is 29.4 days, and Yanchi is the most, reaching 79.9 days. The second was Zhongwei Xingren, which reached 62.8 days. The sand and dust weather in Pingluo, Taole, Yinchuan, Weizhou, etc., lasts 31.1~48.5 days, and the least is south of Xiji. Based on the statistics of monthly sand and dust weather days at various meteorological stations in Ningxia, there will be sand and dust weather in all seasons. Spring (March to May) was the most frequent occurrence, accounting for 56% of the total frequency, followed by winter (December to February), accounting for 25%. The sand and dust weather in summer (June to August) is less, accounting for 14%, and autumn is the least, accounting for only 5%.

Second, there are mobile sand areas with serious soil desertification in the middle part of Ningxia, which is the area with the highest frequency of sandstorm. Zhongwei city, due to its back to the famous Tengger Desert in China, is prone to sand in case of strong wind. Most of Guyuan city is relatively humid area in Ningxia, with less gale weather, large vegetation coverage and less sandstorm.

Third, according to Ningxia expressway, this work studied the dangerous hazard formative factors, subsquently environmental sensitivity and hazard-affected bodies vulnerability based on GIS, fuzzy comprehensive evaluation and AHP method, and gave the risk zoning of sand and dust weather hazard. The results show that the high risk areas of sand and dust hazard in Ningxia expressway are located in Yanchi section of Qingyin expressway (G20), Yanchi section of Dingwu expressway (G2012), Ningxia section of Yane expressway. The higher risk areas are located in Xingren-downstream section of Beijing-Tibet expressway (G6), Zhongwei section of Dingwu expressway (G2012), Gunquan section of Beijing-Tibet expressway (G6) and Qingtongxia section of Wuma expressway (G1816). The medium risk areas are located in Huinong section of Beijing-Tibet expressway (G6), southern section of Yinchuan ring expressway, Ningdong section of Qingyin expressway (G20) and Taiyangshan-Maerzhuang section of Dingwu expressway (G2012). The dust risk is generally low in

IOP Conf. Series: Materials Science and Engineering **914** (2020) 012002 doi:10.1088/1757-899X/914/1/012002

Wuzhong section of Beijing-Tibet expressway (G6), Wuzhong section of Yinkun expressway (G85), Tongxin south section of Fuyin expressway (G70) and Ningxia section of Qinglan expressway (G22), etc.

**Acknowledgements**
Fund Project: Road Traffic and Wind and Solar Energy Meteorological Support Service Project in the Five Northwest Provinces of Silk Road Economic Belt.

**References**
[1] Mo Zhenlong. Analysis and Countermeasures on the Influence of Unfavorable Climate on Expressway Traffic Safety [J]. China Water Transport, 2013, 13 (1): 59-61.
[2] GB/T 35224-2017. Specifications for Surface Meteorological Observations [S]. Beijing: China Standard Press, 2017.
[3] Zhao Guangping, Yang Youlin, Chen Nan, et al. Satellite Remote Sensing Monitoring System for Ningxia Regional Strong Sandstorm [J]. Chinese Desert, 2004, 24 (6): 711-714.
[4] Jiang Dahai. Preliminary Study on Quantitative Assessment and Risk Management of Sandstorm Hazard [D]. Lanzhou University, 2013.
[5] Li Yanchun, Zhao Guangping, Chen Nan, et al. Research Progress of Sandstorms in Ningxia [J]. Chinese Desert, 2006, 20 (1): 137-141.
[6] Zhang Yulin, Zhao Guangping, Yang Shuping. Exploration and Research on the Formation Mechanism of Sandstorms in Ningxia [J]. Chinese Desert, 1996, 16(4): 351-355.
[7] Zhao Guangping, Zheng Guangfen, Wang Weidong. Research on Climate Background and Hazard Law of Super Strong Sandstorm in Ningxia [J]. Desert of China, 2003, 28 (3): 420-427.
[8] Cheng Wenhao, Liu Yaolong. Risk Analysis on Strong Sandstorm Hazard in Ningxia [J]. Guizhou Agricultural Sciences, 2011, 39 (7): 95-98.
[9] Huang Jianwen, Jin Mingyu, Xin Yanan, Liu Xiaoying. Analysis on Meteorological Hazard Risk Zoning of Wind and Sand Dust in Jinchang, Gansu [J]. Gansu Science and Technology, 2019, 35 (22): 37-39.
[10] Lu Tianping, Guo Jing, et al. Construction and Zoning of Risk Assessment Models for Wind and Sand Hazard in the Forest and Fruit Industry in Xinjiang [J]. Journal of Agricultural Engineering, 2016, 32 (S2): 169-176.
[11] Liang Fengjuan. Characteristics and Risk Zoning of Sandstorms in Bayannaoer Area Based on GIS [C]. Chinese Meteorological Society. S10 Meteorology and Modern Agriculture Development, 2012: 839-845.
[12] Li Jinrong. Research on Sandstorm Risk Assessment Based on RS and GIS [D]. Beijing Forestry University, 2011.
[13] Jiang Tong, Wang Yanjun, Zhai Jianqing. Technical Guide for Meteorological Hazard Risk Assessment [M]. Beijing: Meteorological Press, 2018, 16-18.

IOP Conf. Series: Materials Science and Engineering **914** (2020) 012003 doi:10.1088/1757-899X/914/1/012003

# An Overview of Sustainable Highway Infrastructure Development in Yunnan, China

**Junjie Zhang**

International Business School, Yunnan University of Finance and Economics, China

Zhangjj@ynufe.edu.cn

**Abstract**. Highway infrastructure development in Yunnan Province remains a top priority for the Chinese Government, the aim is to improve access to neighbouring countries and support economic growth and social development in the province. While the social and economic benefits of highway infrastructure projects are indisputable, sustainable performance still leaves room for improvement in Yunnan. In an effort to promote a more sustainable approach to decisions relating to highway infrastructure projects, this paper explore the critical issues associated with sustainable highway infrastructure. Firstly, reviewed the economic, social and environmental background of Yunnan, and explain the functions of highways in its socio-economic development. After a brief overview of the conditions and future demands for highway infrastructure in Yunnan, analysed the challenges of building highways in a sustainable way.

## 1. Introduction

Yunnan is a border province in southwest China, in the north it connects with the Silk Road Economic Belt, and it links the Maritime Silk Road in the south. Since China proposed Belt and Road Initiative (BRI), the government clearly defines the important position of Yunnan province in the BRI. It is regarded as the gateway of China connecting with South Asia and Southeast Asia [1]. It is known that Yunnan's bio-diversity, multiculturalism, frontier status and is less developed than other parts of China [2]. In order to improve access to neighboring countries and support economic growth and social development, the government remains the highway infrastructure construction on a top priority [3]. Meanwhile, the cost of a large number of highway construction on the environmental and social development has attracted the people's attention. The terms of sustainability and green-road have been widely promoted in the highway infrastructure development process. It is generally considered that build the sustainable highway infrastructure can not only supports to established the ecological civilization and promote local development, but also contributes to the implementation of BRI [4].

## 2. Overview of sustainable highway infrastructure development in Yunnan

### *2.1. The concept of sustainable highway infrastructure*

The 1970s saw the expansion of the concepts of sustainable development and sustainability, with a series of documents, such as 'Rio Declaration' and 'Agenda 21' formulating sustainable development strategies. Since then, sustainability has become a widely recognised and accepted development consideration. Under the guidance of the sustainable development principles, the architecture profession developed the 3Rs of building designing principles: reduce, reuse, and recycle, i.e. reduce the energy and material used, reuse or recycle waste materials [5]. The idea of the 3Rs provided a less environmentally damaging alternative to handle the growing wastes and the effects on the economy, ecological environment and society [6]. It aligned the ideas of green construction and sustainability, and demonstrated an increasing environmental awareness of the underlying dangers caused by construction.

Sustainable highway infrastructure projects share a common approach to sustainable construction used more generally for infrastructure projects [7]. According to Kibwami and Tutesigensi (2016), the sustainable construction can be found to manifest in economic, social and environmental sustainability, because it is a subset of sustainable development, its practices should address the threes pillars of sustainability [8]. The idea is same as ISO Standard 15392 (2008) as sustainable construction considering development in terms of its three-primary aspect (economic, environmental and social), while meeting the requirements for technical and functional performance [9]. Hussin et al. (2013) proposed that it is important to balance the three elements of sustainable development - environment, economic and social aspects to achieve sustainability in construction. Economic sustainability involves increasing economic profits through using resources and energy efficiently, controlling lifecycle costs, considering alternative financing mechanisms and economic impacts on local community. Social sustainability meets the needs of stakeholders during the building lifecycle, which includes enhancing stakeholders' participation, promoting the development of appropriate institutional frameworks, and considering the impacts on health, quality of life and social framework. Environmental sustainability minimises construction waste, reducing the consumption of natural resource by using new technologies, and reducing the material demand and the energy required for transforming goods and supply services [10].

Based on the above ideas, Lu (2012) identified that the satisfaction of the public for highway infrastructure is based upon its characteristics and impacts and as a result, the development of infrastructure is measured to meet human social and economic needs [11]. A similar view is expressed by Martland (2012) who suggested that whilst highway infrastructure projects should be feasible from the engineering, financial and social perspectives, they should also be built to achieve economic benefits greater than the costs without negative externalities [12]. A successful infrastructure project must deliver economic benefits to the owners and investors, but also consider the concerns of the public and users in terms of environmental and social aspects.

### *2.2. Highway infrastructure development in Yunnan*

Yunnan Province is on the Yunnan-Guizhou Plateau in China's southwestern region and has complex geographical conditions. The area of Yunnan Province covers 394 thousand square kilometres, 84% of the total area is mountain and hills, plateaus make up 10%, basins called 'Bazi' are only 6% of the province's total area. Yunnan is also home to 25 of the 56 recognised ethnic groups in China, 33.57% of its population are members of ethnic minorities [13]. The number of minority ethnic groups exceed all other provinces in China, and the multi-ethnic population makes for a distinct cultural diversity in this region [14]. From the 2019 statistics, GDP per capita is $ 5.6 thousand in Yunnan which is lower than the national average of $ 9.8 thousand. Despite annual GDP growth over 8.9% Yunnan Province started from such a low base that it is still ranked 30 out of China's 30 provinces for GDP per capita. In 2018, the poor rural population in Yunnan was 1.79 million, accounting for 10% of the country's total [1]. Although Yunnan has dramatically changed in the past 30 years, its overall development is still lower than the national average. The remoteness and poor road infrastructure are the main causes hampering the increase in the population's living standards and limiting economic growth in the Province.

In 2015, the Chinese Government launched the 'Belt and Road' initiative to develop infrastructure and inter-connections amongst neighbouring countries, with the important role of Yunnan being emphasised by the Government [15]. The cross-border cooperation strategy has resulted in pressure for highway infrastructure but has also intensified the effective demand for highway transportation

IOP Conf. Series: Materials Science and Engineering **914** (2020) 012003 doi:10.1088/1757-899X/914/1/012003

[16]. The Province shares a border with the provinces of Guizhou, Guangxi, Sichuan, Chongqing, and Tibet Autonomous Prefecture of China. It is also borders on Burma (Myanmar), Laos and Vietnam. Since ancient times, Yunnan has been an important route to Southeast Asia. The Central People's Government (CPG) is trying to use Yunnan's location to maximum advantage by pushing a regional integration scheme with neighbouring countries [17].

### *2.3. Challenges of building sustainable highway infrastructure*

As a Chinese proverb says, 'want to be rich, first build road', and this proverb is particularly important in Yunnan. Because the province is located on the plateau, with an average elevation is 2000m, the high elevation plus various forms of geological conditions have limited the development of railways and waterways, leaving road transportation as almost the only approach for transportation development in Yunnan. Table 1 compares five years' freight movement by the three main transportation modes in Yunnan.

**Table 1.** Freight by railway, highway and waterway.

| | 2018 | 2017 | 2016 | 2015 | 2014 |
|---|---|---|---|---|---|
| Railway freight volume (Ten thousand tons) | 4661 | 4568 | 5372 | 5108 | 4823 |
| Highway freight volume (Ten thousand tons) | 135321 | 124064 | 109487 | 101993 | 103161 |
| Waterway freight volume (Ten thousand tons) | 688 | 667 | 646 | 507 | 560 |
| Total (Ten thousand tons) | 140670 | 129298 | 115505 | 107608 | 108544 |

Source from: National Bureau of Statistics of China (2019)

The data shows that more than 90% freight in Yunnan is dependent on road transportation which illustrates the dominant position of highways in the province's transportation system [16]. By the end of the nation's 12$^{th}$ Five – Year Plan (2011-2015), the total highway mileage reached 236 thousand kilometres, an increase of 26 thousand kilometres from the 11$^{th}$ Five – Year Plan (2006-2010). In the 13$^{th}$ Five-Year Plan (2016-2020), the CPG intends to build and re-build 74 highway infrastructure projects in Yunnan amounting to a total mileage of about 6,640 kilometres, when the plan is completed, the coverage rate of the national highway will increase to 100% from 54%. The highway network will cover all 129 cities and districts throughout the province [14]. As Table 2 shows, the highway mileage has had solid growth in the past five years.

**Table 2.** Highway mileage in Yunnan (Year 2011-2015)

| | 2018 | 2017 | 2016 | 2015 | 2014 |
|---|---|---|---|---|---|
| Highway Mileage (Ten Thousand Kilometre) | 25.29 | 24.25 | 23.81 | 23.60 | 23.04 |

Source: National Bureau of statistics of China (2019)

In 2013, the Ministry of Transport (MOT) issued the 'Guidance of Implementing Green Highway Construction' to apply the requirements of good construction quality, environmental protection, saving energy and resources, high efficiency and improvement of service to highway infrastructure project plans, construction, operation, maintenance and management processes, and ultimately achieve sustainable development of highway infrastructure projects [18] (MOT, 2013). Since then, the idea of the 'Green Highway' has become the imperative for building highway infrastructure projects.

In the past ten years, there has been a notable advance in the development of highway infrastructure in Yunnan, which in turn, has supported significant gains in economic and social development [14]. In 2018, Yunnan invested $ 27.1 billion to build new highway with total 253,000 kilometres length. But the highway network is far from meeting the demands of social development in Yunnan. The rural population accounted for 52.19% of the population in Yunnan [1], and communities have remained relatively isolated due to poor road conditions in mountainous areas with people unable to access health facilities because of the poor infrastructure in these areas [19] (Li et al.,

IOP Conf. Series: Materials Science and Engineering **914** (2020) 012003 doi:10.1088/1757-899X/914/1/012003

2015). Currently in Yunnan, National-Grade and Provincial - Grade highway accounts for 27.05% of the road network, rural roads account for more than 49.05%, showing that the proportion of high-grade highway is low [14]. Meanwhile, the construction of highway infrastructure projects in Yunnan tends to emphasise short-term profits rather than long-term benefits, and this tendency is not consistent with the idea of sustainable development.

The main problems hampering highway infrastructure projects' sustainability in Yunnan including imbalanced development between different regions, unsuitable road network, lack of coordination among transportation modes, and poor serviceability rating of highways due to construction quality. Highway infrastructure is entering a fast-developing period and faces new challenges in Yunnan, but there is still a large gap between total highway infrastructure network size, population increase and economic growth. The diverse climatic and special geographic conditions and social features challenge the development of sustainable highway infrastructure projects in Yunnan. Meanwhile, the Government is also faced with tackling rising pollution problems through a series of strategies and initiatives such as the carbon tax initiative.

## 3. Conclusion

Highway infrastructure is the critical component for development in Yunnan, and it has gradually started to incorporate sustainability considerations into its provision. This paper firstly provided a brief overview of the concept of sustainable construction, and identified the ultimate goals of sustainable highway infrastructure. Then reviewed the local factors considered within the context of current highway construction in Yunnan.

**Acknowledgments**
I gratefully thank the Dr. Stuart Allan from the University of Greenwich for his useful comments and kind helps.

## References

[1] National Development and Reform Commission, Ministry of Foreign Affairs, and Ministry of Commerce of the People's Republic of China, Vision and Actions on Jointly Building Silik Road Economic Belt and 21st-Century Maritime Silk Road. 2015. Information on http://www.mofcom.gov.cn/article/resume/n/201504/20150400929655.shtml.
[2] Foreign & Commonwealth Office, UK. Overseas Business Risk - China. 2016. Information on: https://www.gov.uk/government/publications/overseas-business-risk-china.
[3] People's Government of Yunnan Province, 2019. The Annual Government Work Report of Yunnan Province 2019. Information on http://www.yn.gov.cn/ztgg/2019gzbg/dzd/2019gzzd/201902/t20190225_180039.html.
[4] Central Government of the People's Republic of China, Deepen the construction of four major mechanisms to promote the high-quality development of the "Belt and Road". 2019. Information on http://www.gov.cn/zhengce/2019-04/25/content_5386058.html.
[5] United States Environmental Protection Agency. Reduce, Reuse, and Recycle Construction and Demolition Materials at Land Revitalization Projects. EPA-560-F-08-242. 2008.
[6] C. Mohanty, Reduce, Reuse and Recycle (the 3rs) and Resource Efficiency as the Basis for Sustainable Waste Management. In: Synergizing Resource Efficiency with Informal Sector Towards Sustainable Waste Management. CSD-19 Learning Centre: UNCRD. 2011.
[7] G. Fernández-Sánchez, and F.Rodríguez-López, A Methodology to Identify Sustainability Indicators in Construction Project Management—Application to Infrastructure Projects in Spain. Ecological Indicators, 10(2010), pp. 1193-1201.
[8] N.Kibwami, and A. Tutesigensi, Integrating Clean Development Mechanism into the Development Approval Process of Buildings: A Case of Urban Housing in Uganda. Habitat International, 53 (2016), pp. 331-341.
[9] ISO Standard 15392. Sustainability in Building Construction - General Principles. (2008).

[10] Hussin, J. M., Rahman, I. A. and Memon, A. H.. The Way Forward in Sustainable Construction: Issues and Challenges. International Journal of Advances in Applied Sciences, 2 (2013)), pp.15-24.
[11] M. L. Lu. Construction Environment and Sustainable Development. Heihe Journal, 183 (2012), pp.16-174.
[12] C. Martland, Toward More Sustainable Infrastructure: Project Evaluation for Planners and Engineers. New York: John Wiley and Sons. 2012.
[13] Yunnan Provincial Bureau of Statistics 2016. Information on http://www.stats.yn.gov.cn/TJJMH_Model/newslist.aspx?classid=133608.
[14] J.He, A Temporal and Spatial Analysis of Regional Differences of Economic Development Quality in Yunnan Province. Journal of Yunnan Agricultural University, 8 (2014), pp.21-27.
[15] G. P. Zhu, Main Reasons of People Weak Awareness on Environmental Protection. Theoretic Observation, 7(2015), pp. 129-131.
[16] Transport Department of Yunnan. (2016a). The Long-Term Plan of Highway Network. (2016-2030). Information on: http://www.ynjtt.com/item/43879.aspx.
[17] National Development and Reform Commission. 2019. Policy measures to support Yunnan Province to accelerate the construction of a radiation centre for South Asia and Southeast Asia. Information on http://yndrc.yn.gov.cn/content.aspx?id=885832138171.
[18] Ministry of Transport of China. Guiding Opinions of Accelerating the Advancement of the Development of Green, Circular and Low-Carbon Transportation. 2013. Information on http://www.moc.gov.cn/2013wangshangzhibo/2013third/xiangguanlianjie/201512/t20151.
[19] Y. Q. Li, T. A. Yuan and Y. B. Zhang. The Study of Xi-Shi Highway on Social-Economic Impacts. Times Finance, 7 (2015), pp. 131.

IOP Conf. Series: Materials Science and Engineering **914** (2020) 012004 doi:10.1088/1757-899X/914/1/012004

# Study on Making Formulation of Jujube Dietary Fiber Tableware

**Shaocheng Wang[1, 2, 3], Shijun Liu[1, 2, 3, *], Huanhuan Sun[1, 2, 3] and Yanping Lv[1, 2, 3]**

[1]Shaanxi University of Chinese Medicine / Shaanxi Collaborative Innovation Center of Chinese Medicinal Resource Industrialization, Xianyang, Shaanxi, 712083, China
[2]State Key Laboratory of Research & Development of Characteristic Qin Medicine Resources (Cultivation), Xianyang, Shaanxi, 712083, China
[3]Shaanxi Innovative Drug Research Center, Xianyang, Shaanxi, 712083, China

*Corresponding author e-mail: 1181618@126.com

**Abstract**. Objective: To develop jujube dietary fiber tableware and optimize the best formula for jujube dietary fiber tableware. Methods: The tableware samples were made from the raw materials of jujube dietary fiber tableware according to different formulations, and the properties of anti-impact, leakage time, falling resistance, color and smell were evaluated, and the response surface method was used to establish a model for analysis and optimization, and the best formula was finally determined. Results: The results showed that the best formula of jujube dietary fiber tableware was jujube dietary fiber 4.1g, starch 5.0g, white granulated sugar 5.0g, gelatin 0.1g, carrageenan 0.1g, glycerol 1.0mL, water 12.0mL, fermentation time 23.0min. Conclusion: Through the best formula, the color and hardness of jujube dietary fiber tableware are similar to those of normal wooden tableware, and the jujube dietary fiber tableware is beautiful and practical.

## 1. Introduction

Jujube is the dried and mature fruit of *Ziziphus jujuba* Mill. (Rhamnaceae) [1]. There are a lot of jujube lost and wasted due to natural disasters and unreasonable utilization every year, so it is of great significance to carry out research on deep processing and comprehensive utilization of jujube [2]. In this study, jujube residue was used as raw materials to extract jujube dietary fiber, thus a kind of tableware rich in jujube dietary fiber was developed, which provided a reference for the development of degradable and pollution-free green tableware.

## 2. Methods and results

### *2.1. Preparation of jujube dietary fiber*

Self made in laboratory: Drying of jujube residue→Crushing→Passing sieve No. seven→Weighing of pre prepared jujube residue→Adding hydrochloric acid at 45°C water bath temperature and mixing with water→filtering dregs→washing→drying and weighing (insoluble dietary fiber) [3].

### *2.2. Preparation of tableware of jujube dietary fiber*
Prepare different jujube dietary fiber tableware according to the designed prescription [4].

### *2.3. Performance test [5]*
Through the performance test, the performance of jujube dietary fiber tableware samples made under different raw material ratio was evaluated. The performance evaluation standard are shown in Table 1.

**Table 1.** Performance evaluation standard.

| Standard | Evaluation standard |
|---|---|
| Anti-impact (20 scores) | No change, 15~20; There are cracks, 8~14; Splintering, 0~7 |
| leakage time (20 scores) | ≥60min, 15~20; 30~45min, 10~14; 0~30min, 0~9 |
| Falling resistance (20 scores) | No change, 15~20; There is a crack but not a fracture, 10~14; Fracture, 0~9 |
| Color (20 scores) | Brown, 15~20; Black brown, 8~14; Gray brown, 0~7 |
| Smell (20 scores) | It has the smell of jujube, 15~20; Basically no smell of jujube, 8~14; There is a roasting smell, 0~7 |

### *2.4. Result analysis*

#### *2.4.1. Single factor experimental analysis.*
(1) Effect of different addition ratio of jujube dietary fiber to starch on tableware

The different addition ratio of jujube dietary fiber and starch has great influence on jujube dietary fiber tableware. Dietary fiber has better water retention. Adding dietary fiber to the dough can increase the water absorption of the dough, adjust the extensibility and flexibility of the dough, increase the plasticity of the dough and improve the product quality [6]. In the preliminary experiment, it was found that when the ratio of jujube dietary fiber and starch was between 0.6:1 and 1:1, the dough had better ductility and plasticity, which was beneficial to sample preparation.

(2) Effect of water quantity on tableware

In the early pre experiment, it was found that adding too much water would make the dough difficult to form, and the hardness of the sample made was not enough and easy to deform; Adding too little water would make the dough viscosity was not enough, the ductility and flexibility was too low, and the sample made was slightly rough. When the amount of water added was between 10-16mL, the dough had good ductility and flexibility, which was beneficial to sample preparation.

(3) Effect of different fermentation time on tableware

In the previous pre-experiment, it was found that if the fermentation time was too short, the dough would be too sticky and the hardness of the sample would be insufficient; Too long fermentation time would reduce the malleability of the dough, which was not conducive to forming. When the fermentation time was between 20~30min, the water content of the dough was more suitable, and the dough was sticky, which was beneficial to the preparation of the sample.

#### *2.4.2. Experimental analysis of response surface.*
(1) Experimental design

Through the single factor experiment, it was known that jujube dietary fiber addition, the amount of water added and fermentation time had great effect on the production technology of tableware. Design-Export 8.0.6 was used in this experiment. The Box-Behnken program in the software carries on the experimental design of three factors and three levels. The experimental factors and levels were shown in Table 2.

IOP Conf. Series: Materials Science and Engineering **914** (2020) 012004 doi:10.1088/1757-899X/914/1/012004

**Table 2.** Test factors and level.

| levels | Factors | | |
|---|---|---|---|
| | A / Jujube Dietary fiber addition /g | B / The amount of water added /mL | C / Fermentation time / min |
| -1 | 3 | 10 | 20 |
| 0 | 4 | 13 | 25 |
| 1 | 5 | 16 | 30 |

(2) Evaluation standard

The samples of jujube dietary fiber tableware were scored according to the performance evaluation standard in Table 1.

(3) Response surface test results

The formulation of tableware samples was optimized by Design Expert V 8.0.6 software. The test scheme and results are shown in Table 3.

**Table 3.** Test design and results.

| Std | Run | Factor | | | Response |
|---|---|---|---|---|---|
| | | A / g | B / mL | C / min | |
| 8 | 1 | 5.00 | 13.00 | 30.00 | 71 |
| 7 | 2 | 3.00 | 13.00 | 30.00 | 70 |
| 2 | 3 | 5.00 | 10.00 | 25.00 | 75 |
| 13 | 4 | 4.00 | 13.00 | 25.00 | 80 |
| 12 | 5 | 4.00 | 16.00 | 30.00 | 74 |
| 17 | 6 | 4.00 | 13.00 | 25.00 | 82 |
| 9 | 7 | 4.00 | 10.00 | 20.00 | 79 |
| 1 | 8 | 3.00 | 10.00 | 25.00 | 69 |
| 11 | 9 | 4.00 | 10.00 | 30.00 | 54 |
| 14 | 10 | 4.00 | 13.00 | 25.00 | 84 |
| 10 | 11 | 4.00 | 16.00 | 20.00 | 56 |
| 5 | 12 | 3.00 | 13.00 | 20.00 | 74 |
| 4 | 13 | 5.00 | 16.00 | 25.00 | 63 |
| 6 | 14 | 5.00 | 13.00 | 20.00 | 72 |
| 15 | 15 | 4.00 | 13.00 | 25.00 | 82 |
| 3 | 16 | 3.00 | 16.00 | 25.00 | 69 |
| 16 | 17 | 4.00 | 13.00 | 25.00 | 80 |

Final Equation in Terms of Coded Factors:

$Y=+81.60-0.12*A-1.87*B-1.50*C-3.00*A*B+0.75*A*C+10.75*B*C-3.30*A^2-9.30*B^2-6.55*C^2$

Through the analysis of variance of the regression equation, the results are shown in Table 4.

**Table 4.** Analysis of variance of regression equation.

| Source | Sum of squares | Degree of freedom | Mean square | F | P | Significant |
|---|---|---|---|---|---|---|
| Model | 1194.17 | 9 | 132.69 | 42.31 | < 0.0001 | Significant ** |
| A | 0.13 | 1 | 0.13 | 0.040 | 0.8474 | |
| B | 28.13 | 1 | 28.13 | 8.97 | 0.0201 | * |
| C | 18.00 | 1 | 18.00 | 5.74 | 0.0478 | * |
| AB | 36.00 | 1 | 36.00 | 11.48 | 0.0116 | * |
| AC | 2.25 | 1 | 2.25 | 0.72 | 0.4250 | |
| BC | 462.25 | 1 | 462.25 | 147.41 | < 0.0001 | ** |
| A^2 | 45.85 | 1 | 45.85 | 14.62 | 0.0065 | ** |
| B^2 | 364.17 | 1 | 364.17 | 116.14 | < 0.0001 | ** |
| C^2 | 180.64 | 1 | 180.64 | 57.61 | 0.0001 | ** |
| Residual error | 21.95 | 7 | 3.14 | | | |
| Lack of Fit | 10.75 | 3 | 3.58 | 1.28 | 0.3949 | not significant |
| Pure error | 11.20 | 4 | 2.80 | | | |
| Cor Total | 1216.12 | 16 | | | | |
| R-Squared | 0.9820 | | | | | |
| C.V.% | 2.44 | | | | | |
| Adeq Precision | 19.862 | | | | | |

Note: "*" (P<0.05), Significant; "**" (P<0.01), Extremely significant

It can be seen from table 4 that the P value of the model is (< 0.0001) < 0.05, indicating that the model is significant and the fitting accuracy is good, so the approximate model of the response surface can be used for optimal design, and the P value of lack of fit is 0.3949 > 0.05, that is, it is not significant, indicating that the approximate model fits well in the whole regression area under study; $R^2 = 0.9820$, indicating that the model fits well; C.V. = 2.44% < 10%, indicating that the reliability and accuracy of the test are both high. Adeq precision, i.e. accuracy = 19.862 > 4, indicates that the model is reasonable [7].

(4) Factor interaction analysis

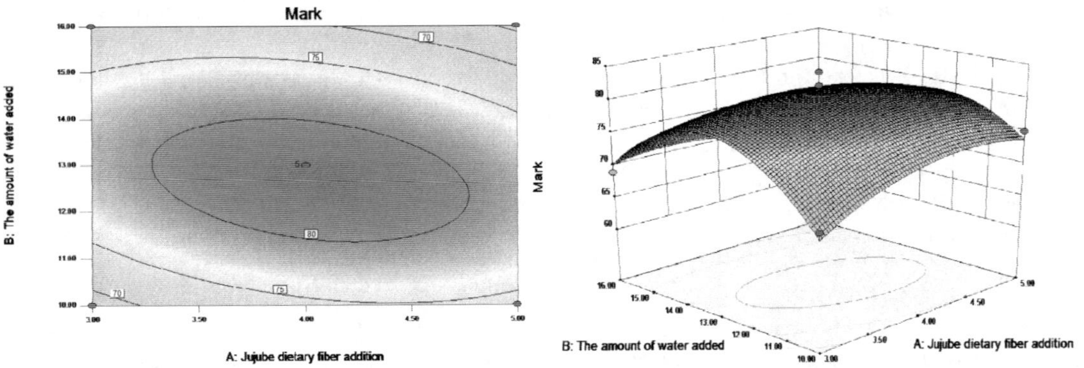


**Figure 1.** Contour map and response surface map of jujube dietary fiber addition and the amount of water added to the score of tableware samples.

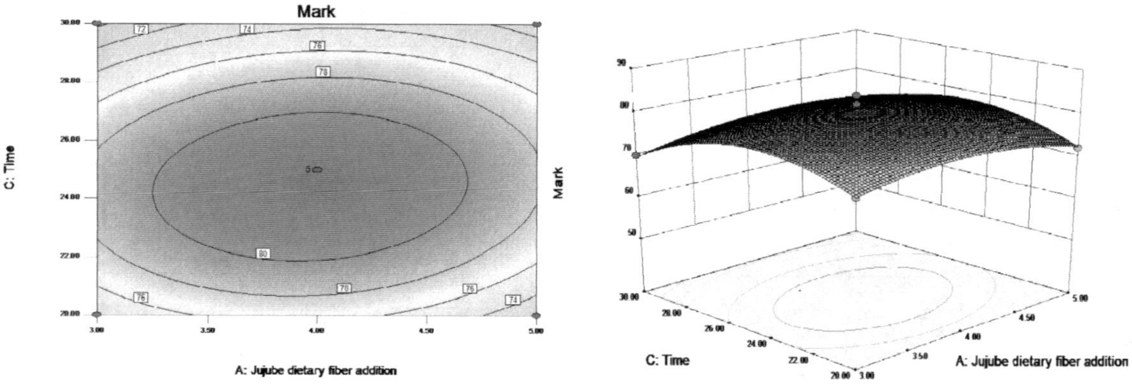


**Figure 2.** Contour map and response surface map of jujube dietary fiber addition and fermentation time to the score of tableware samples.

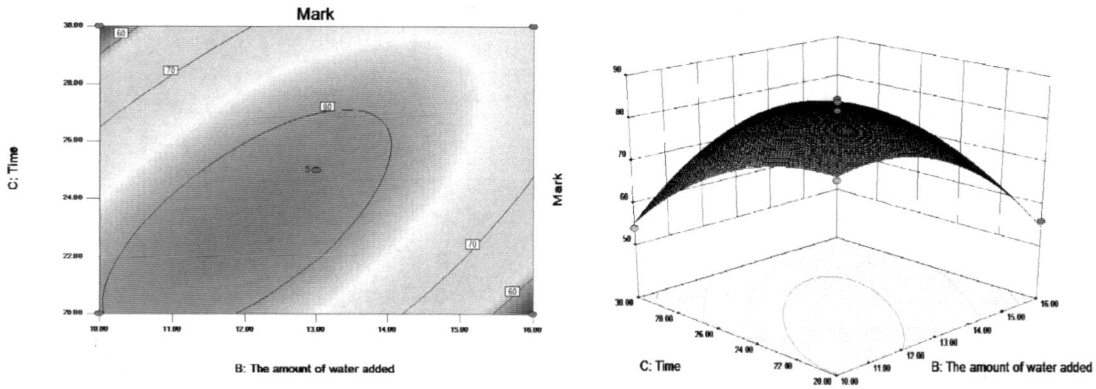


**Figure 3.** Contour map and response surface map of the amount of water added and fermentation time to the score of tableware samples.

From Fig.1, Fig.2, Fig.3, we can see that the maximum range of jujube dietary fiber addition, the amount of water added and fermentation time.

## 3. Verification test

Through the optimization analysis by using Design Expert V 8.0.6 software, the best formula of jujube dietary fiber tableware was obtained as follows: jujube dietary fiber 4.09g, starch 5.0g, white granulated sugar 5.0g, gelatin 0.1g, carrageenan 0.1g, glycerol 1.0mL, water 11.98mL, fermentation time 23.06min. Considering the maneuverability of tableware sample production, the dietary fiber of jujube was changed to 4.1g, the amount of water added was changed to 12.0mL, and the fermentation time was changed to 23.00min. According to the formula, it was verified by repeated experiments for three times under the same conditions that the tableware sample had good hardness and beautiful color. Therefore, it can be considered that the selection of the test model is reasonable and the test results are ideal.

## 4. Conclusion

Through the single factor test and response surface method design test, and scored from the aspects of impact resistance, leakage time, fall resistance, color and smell, the best formula of jujube dietary fiber tableware was as follows: jujube dietary fiber 4.1g, starch 5.0g, white granulated sugar 5.0g, gelatin 0.1g, carrageenan 0.1g, glycerol 1.0mL, water 12.0mL, fermentation time 23.0min. Under this formula, the color and hardness of tableware are similar to those of wooden tableware. At present, the state advocates the construction of ecological civilization, and the emergence of environmentally friendly tableware meets the needs of the people and is conducive to the construction of ecological civilization. Although jujube dietary fiber tableware needs to be improved in terms of hardness, preservation time and water resistance, it is believed that it will have a broad market prospect through the continuous improvement of researchers.

**Acknowledgments**
This work was supported by Innovation and entrepreneurship training plan for college students (201805099); Shaanxi Project (Grant no. 2015KTCL03-14, 15JF001, ZYMS 025, 2018TD-005, 2018SF-285) and Shaanxi Education Finance ((2013)171); Key disciplines of traditional Chinese Medicine Pharmaceutical Engineering of Shaanxi Provincial Administration of traditional Chinese Medicine (2018).

**References**
[1] Liu SJ, Tang ZS, Cui CL, et al. Advances in Studies on Chemical Constituents of Ziziphus jujuba [J]. Journal of Yunnan University of Traditional Chinese Medicine, 2015, 38 (3): 96-100.
[2] Wang X, Wu JF, Li L, et al. Study on the Development Countermeasures of Jujube Industry in Shaanxi Province [J]. Shaanxi Journal of agricultural sciences, 2010, 56 (5): 104-107.
[3] Liu SJ, Gao Sen, Tang ZS, et al. Study on the preparation and physicochemical properties of jujube dietary fiber [J]. Shaanxi Journal of Agricultural Sciences, 2016, 62 (10): 72-73, 89.
[4] Li HY, Huang JH, Ji XG, et al. Method for preparing environment-friendly edible chopsticks from wheat gluten [P]. CN107595064A, 2018-01-19.
[5] Zhao YF, Xie JZ, Gao JT, et al. Study of the New Environmental Protection of Corncob's Property and Manufacture [J]. Farm Products Processing, 2017 (14): 9-11, 18.
[6] Song H, Ming J, Zhao GH. Overview on Effects of Addition of Dietary Fiber on Qualities of Dough and Products of Wheat Flour [J]. Food Science, 2008, 29 (2): 493-496.
[7] Li L, Zhang S, He Q, et al. Application of Response Surface Methodology in Experiment Design and Optimization [J]. Research and exploration in Laboratory, 2015 (8): 49-53.

IOP Conf. Series: Materials Science and Engineering **914** (2020) 012005 doi:10.1088/1757-899X/914/1/012005

# Research on Structural Design of Friction Test Based on Finite Element Analysis Heory

**Xiao Liu**

Shengli College China University of Petroleum, 257061, Shandong, China

slytliu@163.com

**Abstract**. The thesis uses the ring upsetting method to study the determination of the friction coefficient in volume forming. Use finite element analysis to determine a set of friction coefficient calibration curves, and use the ring upset test to determine the percentage change rate of the ring inner diameter. The relationship curve between the ratio and the compression percentage of the ring height. By comparing the curves, the friction coefficient between the mold material and the forming material is obtained. The friction coefficient between the 6A02CZ material and the steel was determined when the 6A02CZ material was volume shaped in the steel mold. The friction coefficients of the 6A02CS material in the absence of lubricants and lubricants were 0.325 and 0.3, respectively. Studies have shown that the method of combining the finite element method and the ring upsetting test is more suitable for the finite element analysis in plastic volume forming.

## 1. Introduction

In the actual production of plastic forming, the friction coefficient between parts and equipment is often needed to guide the actual production. Because there are many factors that affect the coefficient of friction, and these factors are difficult to quantitatively determine the impact on the coefficient of friction. At the same time, in the plastic forming process, various influencing factors are constantly changing. Therefore, the friction coefficient also changes throughout the plastic forming process. Therefore, the friction coefficient measured by any method is only an approximate average value. Before the ring upsetting experiment, according to the original size of the ring and the possible size after deformation, the theoretical formula of the ring upsetting deformation (to find the position of the neutral layer) must be used to draw the ring height h and the inner diameter d The relationship curve between the friction factor m of the contact surface is the theoretical calibration curve.

At present, the finite element theory is basically mature, but there are still some problems in the application process, which affects the speed of popularization of the technology. For example, the selection of friction model and the determination of friction coefficient in the finite element analysis of plastic forming, the accurate description of the material model, etc. In this paper, the friction coefficient between 6A02CS and steel was determined by the ring upsetting method combined with finite element analysis [1].

IOP Conf. Series: Materials Science and Engineering **914** (2020) 012005 doi:10.1088/1757-899X/914/1/012005

## 2. Transient temperature field analysis of friction contacts of friction plate

### 2.1. *Brake upper belt-upper belt friction plate finite element model*

*2.1.1. Material thermophysical parameters.* The thermal conductivity of the brake upper belt is 48W/(m·K), the specific heat is 460J/(kg·K), and the thermal expansion coefficient is $11.5\times10^{-6}$m/K; the thermal conductivity of the upper friction plate is 100W/(m·K), specific heat is 9.8J/(kg·K), thermal expansion coefficient is $2\times10^{-6}$m/K [2].

*2.1.2. Braking heat flux density.* The kinetic energy of the band brake during operation is converted into a large amount of heat energy. This heat is distributed among the friction pair parts in the form of heat flow. Ignoring the effect of wear, the band brake heat flux density can be expressed as:

$$q(t) = \mu \cdot p(t) \cdot v(t) \tag{1}$$

In the formula, $\mu$ is the friction coefficient; $p(t)$ is the specific pressure on the friction surface ($N/m^2$); v(t) is the relative moving speed of the brake belt (m/s). Among them, the ratio on the friction surface the pressure p(t) can be obtained from reference. This article assumes that its value has nothing to do with time, but only with spatial position.

In actual work, the relative moving speed of the brake belt is the instantaneous rotation speed of the anchor sprocket. During braking, it changes with time. When the ship drops anchor, in the process of the anchor chain gradually falling, part of the anchor the gravitational potential energy of the chain and anchor is converted into the kinetic energy of the anchor chain participating in the movement, the kinetic energy of the anchor sprocket, and the kinetic energy of the main axis of the anchor chain. According to the law of conservation of energy and the 72% of the machinery for efficiency loss, the instantaneous speed of the anchor sprocket before braking is 18.58 m/s.

*2.1.3. Heat flow distribution coefficient.* The heat flow distribution coefficient is:

$$\gamma = \sqrt{k_b c_b \rho_b / k_p c_p \rho_p} \tag{2}$$

G$\rho$=11.01. In the formula, $\rho$ is the density, c is the specific heat, k is the thermal conductivity coefficient, the subscripts b and p denote the brake upper belt and the upper belt friction plate, that is $\rho_b$ is the density of the brake band, and $\rho_p$ is the density of the friction plate. Bring the thermal properties of the material into equation (2), we get: G=11.01.

*2.1.4. Determination of convection heat transfer coefficient and friction surface heat conduction coefficient on the brake.* The average heat exchange coefficient of the band brake and the sea breeze is obtained from the empirical formula:

$$\alpha = 0.664 \times (\lambda / l) \times (\varpi_\infty l / v)^{1/2} \times (P_r)^{1/3} \approx 4.5 \tag{3}$$

Among them, it is assumed that the sea breeze wind speed $\varpi_\infty$=10m/s; the length of the belt brake $l$=7.5m. The thermophysical parameters of the sea breeze are as follows: kinematic viscosity $v=15.78\times10^6 m^2/s$; Prandtl constant $P_r$=0.71; thermal conductivity $26.1\times10^3$W/(m·K). Reference, the contact heat transfer coefficient between the brake upper belt and the upper belt friction plate is $2\times10^4$W/(m·K).

*2.1.5. Brake upper belt-upper friction plate finite element model.* Apply the heat flux density that changes with space and time to the inner surface of the upper friction plate, define the two contact bodies of the upper friction plate and the brake upper belt, and specify the contact heat transfer properties: ambient temperature, convection heat release coefficient with the ambient medium The convective heat release coefficient between the contact body and the contact body during contact heat transfer. At the same time, the initial temperature of the upper friction plate and the upper brake belt is defined as the ambient temperature of 20°C.

### *2.2. Transient thermal analysis results*

In order to observe the transient temperature distribution, some nodes were selected on the inner and outer surfaces of the brake upper belt and the upper belt friction lining. The selected nodes on the inner surface of the upper belt friction lining are shown in Figure 1, and the time-varying curve is shown in Figure 2 [3].

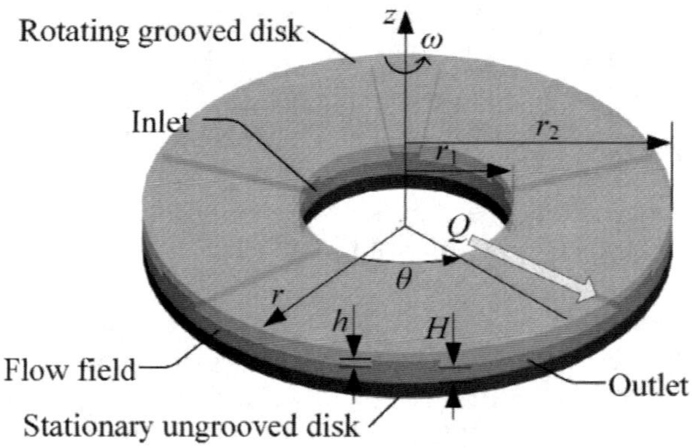


**Figure 1.** Selected nodes on the inner surface of the friction plate.

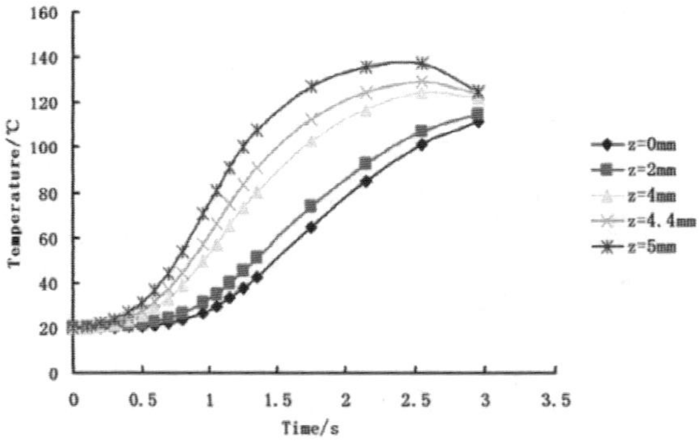


**Figure 2.** The temperature variation curve of selected nodes on the inner surface of the upper friction plate with time.

It can be seen from Figure 2: In the initial stage of braking (about 0s ~ 0.1s), due to the high speed of the anchor sprocket, the heat flux density input as a thermal boundary condition is also very large. And under emergency braking conditions, due to The braking time is very short, and the frictional heat is too late to contact the heat conduction to the brake, so the temperature of each selected point rises quickly; in the middle of braking (about 0.1s ~ 0.4s), as the speed of the anchor sprocket decreases, The input heat flux density also continues to decrease. At the same time, due to the contact heat conduction between the upper friction plate and the brake upper belt and the convection heat exchange between the upper friction plate and the air, the temperature at each selected point begins to slowly

IOP Conf. Series: Materials Science and Engineering **914** (2020) 012005 doi:10.1088/1757-899X/914/1/012005

decrease. The main reason is that the upper friction plate is organic asbestos-free, is a good conductor of heat, and has a high thermal conductivity. The effect of contact heat conduction and convection heat transfer on the moving belt is greater than the effect of the heat flux density input as a thermal boundary condition, and the temperature drop on the inner surface of the upper friction plate accelerates [4].

## 3. Establishment and solution of the finite element model of the ring upset friction plate

Design Modeler is abbreviated as DM, which is the modeling platform of ANSYS Workbench. DM can be fully parameterized for solid modeling, and has the functions of model creation, CAD model repair, CAD model simplification, and conceptual model creation. In order to facilitate the follow-up thermal analysis, this article is modeled in the DM that comes with ANSYS. The model is shown in Figure 3. Import the 3D model of the ring upset friction plate into the ANSYS transient thermal analysis module for solution. The specific solution steps are as follows:

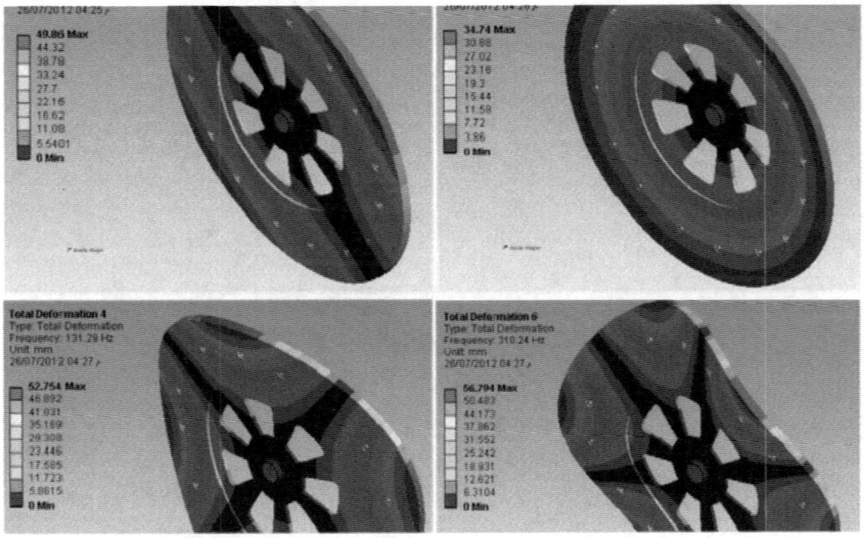


**Figure 3.** Finite element model of circular upset friction plate.

1. Set the friction plate density, heat capacity and other material properties; 2. Mesh the friction plate; 3. Add boundary conditions and initial conditions, including temperature, heat flow, thermal conductivity, etc.; 4. Insert the solution term, and set the solution time step to solve.

### *3.1. Friction torque and pressure at the joint surface*

During the upsetting process of the ring, due to the existence of friction on the contact surface of the friction plate, it is subjected to a rotational torque perpendicular to the axial direction. In the finite element analysis, according to the calculation result of the friction force, each joint surface the frictional force at each node is loaded at the corresponding position to simulate the rotational torque experienced by the friction plate. According to the dynamic analysis result of the ring upset joint characteristic, the corresponding pressure load is applied on the joint surface of the friction plate finite element model. Simulate the pressure of the friction plate when the ring is upset.

### *3.2. Solid modeling*

Take the outer friction plate as an example. The material of the friction plate is No. 15 steel. Create a geometric model, select materials, set the mesh element size, and generate mesh elements for finite element analysis. The Pro / MECHANICA finite element module uses adaptive P-method technology to automatically mesh. In the P-method technique, the displacement equations of each finite element are high-order polynomials, and the displacement equations of each finite element are linear, quadratic,

IOP Conf. Series: Materials Science and Engineering **914** (2020) 012005 doi:10.1088/1757-899X/914/1/012005

or cubic. Therefore, the P-method technique can be used to accurately fit the geometric shape and large stress gradient, and in the actual calculation process, the element first performs a preliminary calculation at a lower order, and then where the stress gradient is larger and the calculation accuracy is required Higher places automatically increase the order of the stress equation of the element, thereby ensuring the accuracy and efficiency of the calculation [5].

The P-method is used to divide the finite element mesh, and some optional features in the model must be simplified. In the outer friction plate model, the features such as rounded corners and chamfers between the plane and the surface, spline and surface are removed, and these features are simplified into one point. In this way, on the one hand, it facilitates meshing, on the other hand, it reduces the total number of generated cells, which reduces the workload of finite element calculation. After the meshing is completed, a total of 1767 units are generated. The finite element mesh is shown in Figure 4.

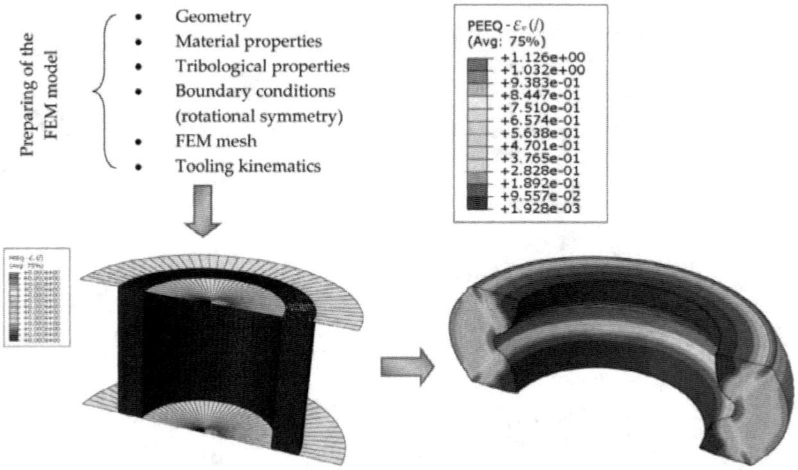


**Figure 4.** Finite element analysis model.

## 4. Experimental research

### *4.1. The effect of μ at the time of occlusion on the final forming*

First, when μ is small, the inner diameter change trend is increasing. The amount of increase in inner diameter when considering occlusion is smaller than the amount of increase in inner diameter when not considering occlusion, which is equivalent to hindering the increase in inner diameter. Second, when μ is larger, the change trend of the inner diameter is reduced. The reduction in inner diameter when considering bite is larger than the reduction in inner diameter when not considering bite, which is equivalent to promoting the reduction in inner diameter [6].

In general, the larger μ' in the bite stage (because the actual μ in the bite is larger than the average μ in the forging process. Therefore, the effect of $\mu' = \mu * 1.5$ in this experiment) is to reduce the inner diameter the trend of. In this experiment, the process of occlusion was 1 unit time and the depression process was 59 unit times; a total of 60 unit times and the depression amount was 6 units. Although the μ action time in the occlusal phase is very short, it has a great influence on the entire plastic deformation results. After the initial occlusal phase, this effect will continue and expand. Therefore, using the finite element model to draw the friction calibration curve needs to refer to the lubrication mechanism in actual production to modify the model, as shown in Figure 5.

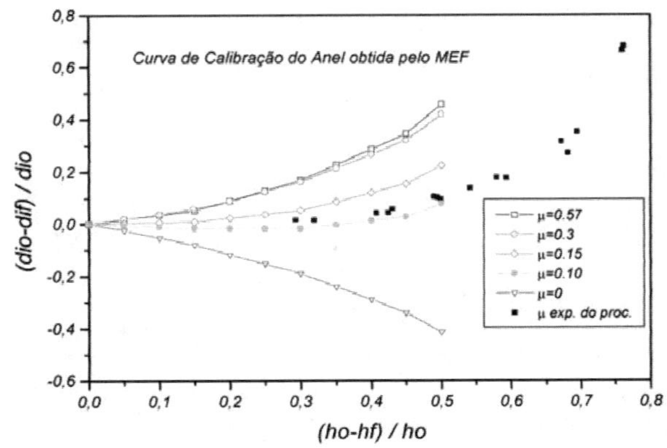


**Figure 5.** The effect of μ at the time of occlusion on the final forming.

From the study in Figure 5, it is found that when μ=0.05, the inner diameter increases first and then decreases, indicating that the neutral layer is near the initial inner diameter in this case, and the inner diameter basically does not change after large deformation. When μ>0.05, the inner diameter of the ring shrinks obviously. After μ reaches 0.3, it can be observed that the gap between the curves is not large, indicating that when μ=0.3, 0.4, the forming is close to the limit, and the deformation is difficult. When μ<0.05, the distance between the curves is large, and the specimen has a large amount of deformation, showing good plasticity.

### *4.2. Thermal finite element analysis*

The thermal deformation of the friction plate after heating is mainly along the axial and radial directions, and its value gradually increases from the middle position to both sides, and its maximum value is on the outer edge of the friction plate. After being heated, the two ends are deformed in the axial direction away from the friction surface, but because the friction plate is fixed on the driven disk by the diaphragm spring, a wave-shaped deformation is generated. This deformation exacerbates wear. In order to reduce the amount of deformation, the shape of the friction plate is designed to be slightly concave when designing, so that when the thermal load acts, it can offset the axial deformation at both ends, so that the friction plate is better combined with the pressure plate and the flywheel [7].

## 5. Conclusion

The finite element method was used to analyze the transient temperature field of friction plates with different oil groove widths, and the curve of maximum temperature with time and radial temperature distribution were obtained. Due to the influence of the oil groove, the radial temperature of the friction plate shows a stepwise distribution. The temperature difference between the bottom of the oil groove and the friction surface is large during the bonding period. The friction coefficient of 6A02CS material was measured without lubricant and lubricating oil, and the calibration curve was checked through the ring upsetting test to determine that the friction coefficient of 6A02CS material in both cases was 0.325 and 0.3.

## References

[1] Richard A. Albert, & John W. Rudnicki. Finite element simulations of tennessee marble under plane strain laboratory testing: effects of sample–platen friction on shear band onset. Mechanics of materials, Vol. 33 (2016) No. 1, p. 47-60.
[2] Wang, Y., Chen, X., Young, R., & Kinloch, I. Finite element analysis of effect of inter-yarn friction on ballistic impact response of woven fabrics. Composite Structures, Vol. 135 (2016) No. 1, p.8-16.
[3] Jain, R., Pal, S. K., & Singh, S. B. Finite element simulation of temperature and strain

distribution during friction stir welding of aa2024 aluminum alloy. Journal of the Institution of Engineers, Vol. 98 (2017) No. 1, p. 37-43.

[4] Nam, J., Choi, H., & Kang, J. Finite element analysis for friction noise of simplified hip joint and its experimental validation. Journal of Mechanical Science and Technology, Vol. 30 (2016) No. 8, p.3453-3460.

[5] Zhao, X., & Li, Z. A solution of transient rolling contact with velocity dependent friction by the explicit finite element method. Engineering Computations, Vol. 33 (2016) No. 4, p.1033-1050.

[6] Djoko, J. K., & Mbehou, M. Finite element analysis of the stationary power-law stokes equations driven by friction boundary conditions. Journal of numerical mathematics, Vol. 23 (2015) No. 1, p. 21-40.

[7] Lai, V. V. Chielo, O., Brunel, Jean-François, & Dufrénoy, Philippe. Full finite element models and reduction strategies for the simulation of friction-induced vibrations of rolling contact systems. Journal of Sound & Vibration, Vol. 444 (2019) No. 5, p. 197-215.

IOP Conf. Series: Materials Science and Engineering **914** (2020) 012006 doi:10.1088/1757-899X/914/1/012006

# Research on Electrochemical Molecular Imprinting Sensor of P-Feco Nanoparticles

**Lei Li, Bixuan Zheng, Mingguo Tang, Junchuang Feng, Kun Huang and Yile Liu**
College of science, Sichuan Agricultural University, Ya'an, Sichuan, 625014

**Abstract**. The electrochemical sensor is based on the principle of electrochemistry, which connects the concentration change of the tested component with the electrochemical signal. So as to provide real-time information of chemical components in the detected system. When electrochemical method is used to detect the tested object, it has the advantages of relatively simple detection process, fast detection speed, high sensitivity and low cost, so it has become a research hotspot in the field of analysis. In this paper, five preparation methods of MIPs are introduced, which have the characteristics of high affinity and stability. Then, the electrochemical sensor is constructed by molecularly imprinted technology combined with nanomaterials. It is a simple and convenient electrochemical sensor to detect the metal synthetic materials closely related to people's health in life. The surface current density of Nb2O5 / GO composite was increased by using FeCo nanoparticle material, and molecular imprinting technology was used. On the one hand, the carrier properties of composite materials are used. On the other hand, FeCo nanoparticles, which enhance the junction of molecularly imprinted and composite materials, are a kind of nanocomposites with many excellent properties. It has high activity of electrochemistry catalysis, high current density on the surface of electrode and long-term stability.

## 1. Introduction

Electrochemical sensor is a device that interacts with analyte and produces a specific electrical signal, and the change of electrical signal changes with the concentration of analyte [1]. Electrochemical sensors use different electrical analysis techniques, such as open circuit potential method, ampere method, cyclic voltammetry, differential pulse voltammetry, differential pulse voltammetry. Impedance sensing. Various electrochemical technologies have been widely used in different research fields because of their relative maturity, simplicity and convenience. Molecular imprinting technology (MIT) combined with electrochemical sensor has the advantages of specific recognition ability, high sensitivity and fast reaction of electrochemical analysis method, so the electrochemical sensor based on molecularly imprinted polymer has a very large development space [2].

Bisphenol A, scientific name 2,2-bis (4-hydroxyphenyl) propane, referred to as bisphenol propane (BPA), is an organic compound, chemical formula is (CH3) 2C (c6h4oh) 2. It is a white crystal, melting point 156-158°C, insoluble in water, soluble in organic solvent. Bisphenol A is an important organic chemical raw material, an important derivative of phenol and acetone, with two phenolic hydroxyl groups. Bisphenol A is used to synthesize polycarbonate (PC) and epoxy resin in industry. Since the 1960s, it has been widely used in the manufacture of plastic (milk) bottles, suction cups for children, and inner coating of food and beverage (milk powder) cans. Every year, about 27 million tons of BPA containing plastics are produced worldwide. But BPA can lead to endocrine disorders, which seriously threaten the health of fetus and children. Obesity caused by cancer and metabolic disorders is also thought to be related. The data show that bisphenol A is a low toxic chemical [3]. In animal experiments, bisphenol A was found to have the effect of simulating estrogen. Even at a very low dose, bisphenol A could make the animal produce premature puberty, decrease the number of sperm and increase the prostate gland. In addition, some data show that bisphenol A has certain embryotoxicity and teratogenicity, which can significantly increase the incidence of animal ovarian cancer, prostate cancer, leukemia and other cancers [4]. At the same time, studies have shown that BPA is associated with asthma in mice, and preliminary human experiments have shown that pregnant women affected by BPA in early pregnancy may lead to asthma infection in infants [5].

At present, the detection methods of bisphenol A are: high performance liquid chromatography, gas chromatography-mass spectrometry, liquid chromatography-mass spectrometry, etc., but the general analysis speed is slow, the sensitivity is low, and the operation is tedious [6]. Electrochemical analysis has the advantages of low cost, high sensitivity and easy operation, which can detect BPA well. However, the electrochemical response of BPA on bare electrode is poor, and it is easy to cause passivation on the electrode surface.

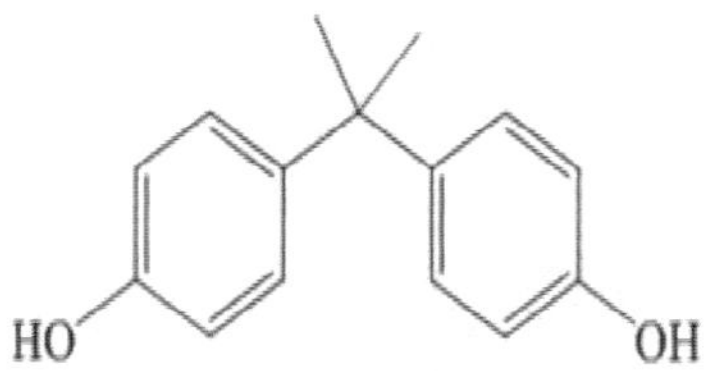


**Figure 1.** Structure of bisphenol A.

Molecularly imprinted electrochemical sensor is one of the electrochemical sensors. Molecularly imprinted is a new method developed in the later stage, which can provide people with molecular combination with desired structure and properties. Molecularly imprinting is the process of synthesizing polymers in the presence of template molecules and forming molecular recognition sites in polymers. In other words, when there are template molecules in the system, the functional monomers can be polymerized to immobilize the template molecules in a complementary way. After polymerization, the template molecules can be removed, so in this process, the changes of the system can be recorded by electrochemical signals. Thus, the obtained molecular assembly can specifically bond the template molecules and their analogues.

## 2. Preparation of E-MIPS

The complementary interaction between functional monomers and template molecules is maintained in spatial arrangement through polymerization process, and further stabilized through polymer crosslinking. The advantages of many molecularly imprinted polymers (MIPs) make more people interested in MIPS More and more interested, their high affinity and selectivity are similar to those of natural receptors. If they are to be considered as real alternative receptors, this feature is essential; their stability is enhanced, superior to the stability of natural biomolecules; the preparation is simple, so that the receptors of interested analytes are much faster than those produced by antibody; the versatility of templates, MIPS It can recognize analyte, make antibody become a challenging target antigen, and it is easy to adapt to practical application of special analysis and sensor.

The fixing of MIPs film on the surface of signal converter is one of the important factors that affect the output electric signal and sensitivity of sensor. In order to ensure the good stability of the sensor during the test, MIPs film must be evenly and tightly attached to the transducer surface. Therefore, according to the different combination mode between MIPS and transducers, the preparation methods of E-MIPS can be classified into the following five cateGOries:

(1) Surface coating method.

Surface coating is an indirect film forming method. MIPs are dispersed in low boiling solvent, and MIPS suspension is fixed on the surface of transducer by drop coating, dip coating or spin coating. MIPS polymer film is formed after solvent volatilization. Although the E-MIPS prepared by this method have good specific recognition ability for the target, due to the tedious preparation process, time-consuming.

(2) In situ initiated polymerization.

Different from surface coating, in-situ polymerization is a direct film-forming method. It refers to coating the mixed droplets containing template molecules, functional monomers, initiators and porogens on the surface of the transducer, and directly forming MIPS film on the surface of the transducer through initiation polymerization. The initiation process can be divided into light or heat initiation. The preparation method is simple, and the imprinted film can be fixed on the surface of the transducer well and is not easy to fall off. As an initial method for the preparation of E-MIPS, the polymerization method has been greatly developed. However, due to the excessive reactants remaining in the polymer film during the polymerization process, the elution of template molecules is troubled and the quantitative test of template molecules is disturbed. Therefore, the development of this method is limited.

(3) Electropolymerization

Electropolymerization refers to the method that in the presence of target molecules, functional monomers, crosslinking agents and porogens, the surface of transducer is polymerized by cyclic voltammetry, potentiostatic method or galvanostatic method to form imprinted membrane, and then the target is eluted in a certain way to obtain E-MIPS. Electropolymerization is a relatively simple direct film-forming method, which can be realized only by using electrode potential at room temperature. The polymer film obtained by this method is uniformly and firmly attached to the surface of transducer, and the thickness of E-MIPS film can be accurately controlled by adjusting the polymerization parameters. The obtained membrane has stable structure and can be used in any medium, and the imprinted holes are not suitable for deformation and good reproducibility. By controlling the kinds of monomers in the polymerization process, multi-component copolymerization films can be formed. Therefore, electropolymerization is considered to be the most practical E-MIPS polymerization method.

(4) Self assembly method

The self-assembly technology refers to the process that the target molecule and the monomer spontaneously form the target monomer complex by intermolecular force (electrostatic force, van der Waals force, hydrogen bond, hydrophobic lipophilic force, etc.) and then get the polymer membrane by cross-linking polymerization. Different from the above methods, the self-assembly method uses compounds containing sulfur functional groups to carry out chemical adsorption on the surface of metals and metal oxides, which can form stable two-dimensional single-layer films.

## 3. Preparation of molecularly imprinted sensor

### *3.1. Preparation principle*

Graphene oxide is thinner and harder. Secondly, it can transfer electrons faster at room temperature, so its conductivity is better. In addition, niobium pentoxide also has a good conductivity. Therefore, the composite of niobium pentoxide and graphene oxide can not only conduct electricity well, but also make full use of the carrier performance of the composite, enhance the combination degree of molecular imprinting and the composite, so as to improve the detection performance and detection limit of bisphenol a.

FeCo nanoparticle materials have a series of unique physical and chemical properties, which are widely used in sensitive materials, battery materials and other fields. The conductivity of FeCo

IOP Conf. Series: Materials Science and Engineering **914** (2020) 012006 doi:10.1088/1757-899X/914/1/012006

nanoparticles is significantly improved due to the synergistic effect of CO and Fe elements. Phosphorus doped FeCo nanoparticles have the advantages of high sensitivity, fast response and low detection limit. For this reason, FeCo nanoparticles have a good prospect in electrochemical research.

The difficulty of protein elution and diffusion can be effectively solved by Western blotting. It refers to the imprinting method that modifies the imprinting site on the surface of the substrate with certain bearing capacity, so as to overcome the steric hindrance caused by protein entering and leaving the polymer layer. Nano materials have the advantages of simple preparation, large specific surface area, good thermal stability and strong mechanical properties, which are applied in various fields. Therefore, MIPs based on simple and easy to get nanomaterials have the characteristics of large surface area, fast binding speed and high binding capacity.

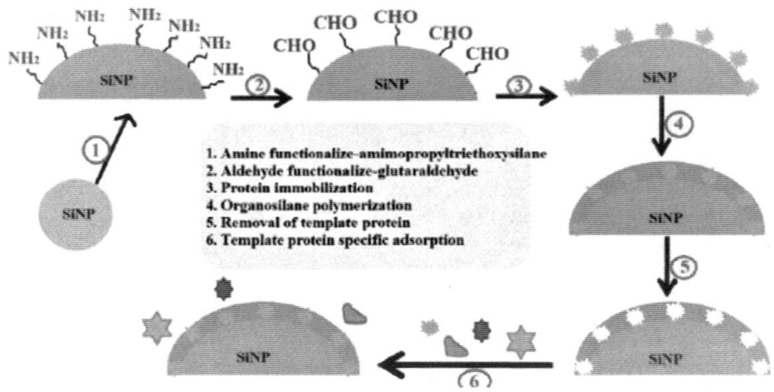


**Figure 2.** Western blotting principle of mimic antibody recognition site.

*3.2. Experimental reagents*

(1) Natural graphite powder (AR), bisphenol A (AR), 4-vinyl pyridine (AR), concentrated sulfuric acid (AR), concentrated nitric acid (AR), concentrated hydrochloric acid (AR), potassium persulfate (AR), niobium pentachloride (AR), potassium permanganate (AR), hydrogen peroxide (30%).

*3.3. Experimental equipment*

Ultrasonic cleaner, magnetic stirrer, thermostatic water bath, analytical balance, suction filter, high-speed centrifuge, vacuum drying oven, infrared spectrometer, electrochemical workstation, X-ray diffractometer, scanning electron microscope

*3.4. Preparation of P-FeCo nanomaterials*

① Synthesis of FeCo nanoparticles: weigh a certain amount of cobalt chloride hexahydrate and ferric chloride hexahydrate and add them into 30ml ultra pure water. After ultrasonic treatment for 2h, the pH value of the solution was adjusted to 10 by NaOH, and then transferred to the autoclave. The product was collected by freeze-drying at 100°C for 6h, and then transferred to the precursor FeCo nano material.

② Synthesis of phosphorus doped FeCo nanoparticles: take a certain amount of sodium dihydrogen hypophosphite, mix it evenly with the precursor FeCo nanomaterials, put it into a tubular furnace with nitrogen, keep it at 800°C for 3 hours, and then obtain the phosphorus doped FeCo nanoparticle composite.

③ Carry out structural features and element analysis.

*3.5. Preparation of phosphorus doped FeCo nanoparticles supported on Nb2O5 / GO*

① Synthesis of Nb2O5 / GO: niobium chloride was dispersed in ethanol with graphene oxide dissolved, refluxed overnight in a 45°C water bath, then centrifuged ($t \geq 40°C$) and washed with ethanol three times ($t \geq 40°C$, preheated with water bath first). The obtained solid powder was dried overnight, burned in a tubular furnace for one hour, and the temperature was controlled at 350°C to obtain Nb2O5 / GO material.

② Preparation of phosphorus doped FeCo nanoparticles loaded with Nb2O5 / GO: take a proper amount of Nb2O5 / GO and phosphorus doped FeCo nanoparticles, mix them in ultra pure water, stir for 24 hours, centrifugate them, wash them with ultra pure water several times, dry them in an oven, and obtain the composite of phosphorus doped FeCo nanoparticles loaded with Nb2O5 / GO.

### *3.6. Results and discussion*

The samples were characterized by TEM, SEM and XRD. The specific surface area and pore structure of the composite were measured by the specific surface area analyzer, and the thermal stability of the composite was studied by thermogravimetry. Bisphenol A and 4-vinyl pyridine were added to 10ml acetonitrile for dissolution, the modified electrode was placed in it, the electrode system was connected with the instrument in the range of - 0.4-1.2v potential, and the modified electrode was obtained by scanning 20 times at a scanning rate of 100mV/s.

The morphology of imprinted electrode and non imprinted motor was characterized by SEM. There are a lot of small particles on the surface of the non eluted imprinted electrode, but not on the eluted imprinted electrode, which may be caused by the adhesion of template molecules on the surface of the polypyrrole membrane. However, some specific shape structures are formed on the imprinted membrane, which provides a large adsorption area for the template molecules. There are a lot of small protuberances and globular particles on the surface of the non imprinted electrode, which may be that FeCo is coated on the imprinted polymer film; and some rod like structures on the surface of the non imprinted electrode are attributed to Nb2O5 / GO coated on the imprinted polymer.

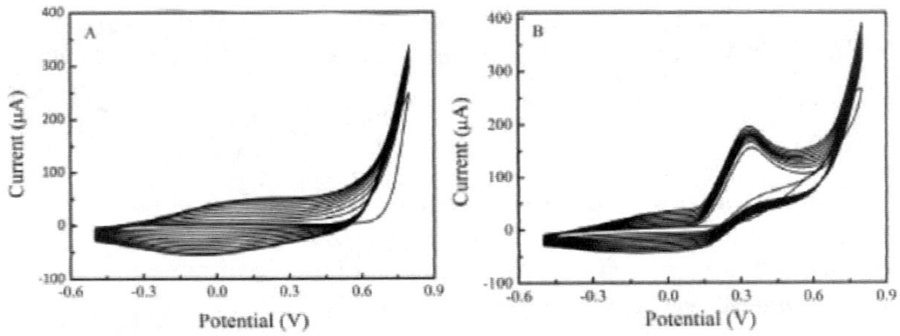


**Figure 3.** Cyclic voltammogram of electropolymerization for preparation of non imprinted electrode and imprinted electrode.

The morphology and size of FeCo, P-FeCo and Nb2O5 / GO loaded phosphorus doped FeCo were characterized by SEM. The results are shown in the figure below.

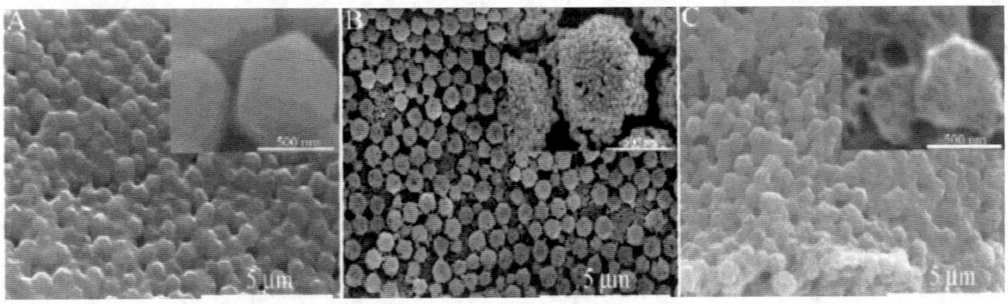


**Figure 4.** (A) FeCo nanoparticles; (B) Phosphorus doped FeCo nanoparticles supported on Nb2O5 / GO; (C) P-FeCo nanoparticles.

IOP Conf. Series: Materials Science and Engineering **914** (2020) 012006 doi:10.1088/1757-899X/914/1/012006

FeCo nanoparticles are uniform regular dodecahedrons, and the effective specific surface area of P-FeCo nanoparticles is 0.52cm$^2$. After high-temperature calcination, FeCo nanoparticles with Nb2O5 / GO loaded phosphorus doped were formed, which were porous nanospheres with a diameter of 500nm. These nanospheres are evenly and closely arranged on the surface, and the formation of porous nanospheres may be related to a large number of gases released during the calcination process. The effective specific surface area of Nb2O5 / GO integrated electrode increased to 0.87cm$^2$, which was significantly larger than that of FeCo nanoparticles. The image shows that the surface of P-FeCo nanoparticles is covered by a film, which is formed by electrodeposition. After elution, the pores of template molecules were formed, so as to detect BPA. Due to the existence of molecularly imprinted membranes, the electron transport rate may be reduced, and the specific surface area of the membranes may be reduced to about 0.66cm$^2$.

Cyclic voltammetry was used to characterize the template molecularly imprinted on the polymer membrane, as shown in the figure below. The cyclic voltammograms of electropolymerization with imprinted electrode and non imprinted electrode were prepared respectively, which was consistent with the experimental results reported previously. It can be seen from the figure that a very obvious oxidation peak is about 0.35V, which indicates that the template molecule has been successfully imprinted into the imprinted polymer.

## 4. Conclusion

In this paper, FeCo nanoparticle material was used to modify Nb2O5 / GO composite, which increased the current density of the electrode surface, and combined with molecular imprinting technology to make FeCo nanoparticle composite. It has the characteristics of large surface area, is conducive to the diffusion of objects and the combination of electrolyte, and successfully constructed a high-performance composite. A simple and convenient electrochemical sensor for the detection of metal synthetic materials which are closely related to people's health in our life has been constructed, which has great value in practice.

## References

[1] Lei Dongwei. Research on electrochemical sensing technology and application of nanomaterials [J]. Information recording materials, 2019, 20 (06): 47-48.

[2] Shen Aiping. Electrochemical sensing technology and application of nano materials [J]. Modern salt chemical industry, 2018, 45 (05): 91-92.

[3] Lei, B., et al., In vitro profiling of toxicity and endocrine disrupting effects of bisphenol analogues by employing MCF-7 cells and two-hybrid yeast bioassay. Environmental toxicology, 2017. 32 (1): p. 278-289.

[4] Vom Saal, F., Triennial reproduction symposium: environmental programming of reproduction during fetal life: effects of intrauterine position and the endocrine disrupting chemical bisphenol a. Journal of animal science, 2016. 94 (7): p. 2722-2736.

[5] Roy, A., et al., Neither direct nor developmental exposure to bisphenol A alters the severity of experimental inflammatory colitis in mice. Journal of immunotoxicology, 2013. 10 (4): p. 334-340.

[6] Lee, E.-H., et al., Simple and rapid detection of bisphenol A using a GOld nanoparticle-based colorimetric aptasensor. Food chemistry, 2019. 287: p. 205-213.

IOP Conf. Series: Materials Science and Engineering **914** (2020) 012007 doi:10.1088/1757-899X/914/1/012007

# Research Progress on the Photoelectric Properties of Indium-Doped Cadmium Oxide Transparent Conductive Films

**Xiongfei Shi**

School of northeast Normal University, Changchun 130000, China

shixf846@nenu.edu.cn

**Abstract**. Indium-doped cadmium oxide (CdO: In) thin films have excellent photoelectric properties, and have broad application prospects in the fields of photoelectric devices, enhanced spectroscopy, and the like. In recent years, the influence law and internal mechanism of various preparation technologies and process conditions on the photoelectric performance of CdO: In thin films are still one of the research hotspots. The preparation methods of CdO: In transparent conductive thin films were reviewed, the effects of various preparation process conditions on the photoelectric performance of the thin films were discussed, and the future research directions of CdO: In thin films were given.

## 1. Introduction

Transparent conductive oxides (TCO) such as cadmium oxide, indium oxide, and zinc oxide have low resistivity and high light transmittance, and thus have broad application prospects in the fields of optoelectronic materials and the like. For example, transparent conductive electrodes, surface acoustic wave devices, varistor, etc. Among these TCOs, CdO is an ideal transparent electrode material for photovoltaic solar panels due to its high transmittance and low resistivity in the visible region of the solar spectrum. Doping CdO thin films with other elements can significantly improve their optical and electrical properties. At present, the main elements of CdO thin film doping are AI, Ti, Sn, In, etc. According to Kawazoe's calculations, In-doped CdO thin films are likely to have superior photoelectric performance. Subsequently, the lowest value of the CdO: In film resistivity prepared by Gupta et al. Using the PLD method can reach $2.9 \times 10^{-5}$Ωcm, and the average transmittance in the visible light range is up to 85%, which also experimentally confirms that CdO: In is expected to become high Quality TCO film material. CdO has a simple cubic crystal structure. Doped $In^{3+}$ replaces $Cd^{2+}$ in the CdO lattice and becomes a donor, releasing more free electrons in the conduction band, thereby reducing the resistivity and widening the optical band gap. Past studies have shown that the photoelectric properties of CdO: In are closely related to the process conditions such as In content, substrate temperature, and annealing treatment. This paper refers to a large number of domestic and foreign literatures in recent years, summarizes the research progress of the optical and electrical properties of CdO: In films, and gives prospects for future research.

## 2. Preparation Technology of CdO: In Nano Thin Film Materials

At present, there are many methods for preparing CdO: In nano film, including magnetron sputtering method, pulse laser deposition method, metal organic chemical vapor deposition method, filtered cathode vacuum arc method, etc. Each method has its advantages and disadvantages, and the corresponding preparation method needs to be selected according to actual needs.

Magnetron sputtering (MS). Magnetron Sputtering (MS) is currently the most mature thin film preparation method. Because of its good repeatability and simple operation, it is widely used to deposit oxide thin films. The basic principle of the magnetron sputtering method is: under the action of the electric field and the magnetic field, the high-voltage electric field is used to accelerate the inert ions to strike the atoms on the surface of the target material, so that they escape from the lattice bond and escape, and the sputtering atoms or atomic groups are deposited on the substrate The surface reacts with oxygen atoms to form an oxide film. According to the different power sources selected, the magnetron sputtering method can be divided into three types: DC sputtering, RF sputtering and intermediate frequency sputtering. Among them, DC sputtering and RF sputtering are the most common methods. The experimental procedure of preparing CdO: In by magnetron sputtering mainly includes the following three parts: (1) A baffle is set between the indium-cadmium alloy target and the substrate, the sputtering chamber is evacuated to the required vacuum degree, and a flow meter is set Display the number and fill with argon gas to the working vacuum; (2) After setting the power and working time of the working power supply, perform pre-sputtering; (3) After pre-sputtering for a period of time, open the baffle and use the mass flow controller to Oxygen is released into the room and coating begins.

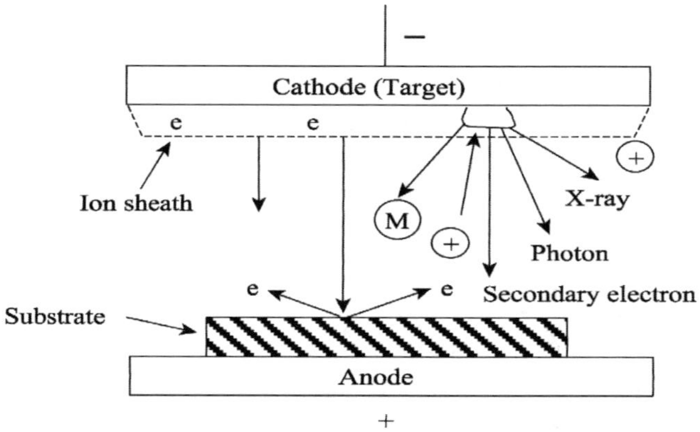


**Figure 1.** Schematic diagram of magnetron sputtering device.

Babu et al. Used the DC magnetron sputtering method, using cadmium (34%)-indium (66%) alloy as the target, setting the target-base distance to 65mm, and using a sputtering pressure of 4pa on a glass substrate at a temperature of 532K A CdO: In thin film was prepared with a low resistivity of $1.2 \times 10^{-3}$ and a visible light transmittance of 87%. Coutts et al. Obtained a CdO: In thin film with excellent photoelectric properties by radio frequency magnetron sputtering. The minimum resistivity of the CdO: In thin film is about $2.3 \times 10^{-4}$, and the average transmittance in the visible light region is as high as 90%.

Compared with other preparation methods, the magnetron sputtering method has the advantages of uniform film thickness, high repeatability of film quality, and strong adhesion between the film and the substrate. However, in the process of depositing thin films by reactive sputtering, problems such as anode disappearance and cathode poisoning often occur, which makes it difficult to control the sputtering process, which limits its application.

Pulsed laser deposition (PLD). Pulsed laser deposition (Pulsed Laser Deposition, PLD) is a new type of vacuum physical deposition film technology. The basic principle of PLD technology is shown in Figure 2. The focused laser enters the vacuum chamber through the entrance window to irradiate the target at a 45° angle, so that the target sputters plasma, and the particles in the plasma (including many particles, ions, atoms, and atomic clusters, etc.) form a Gaussian distribution along the target it expands to the substrate and deposits a film on its surface.

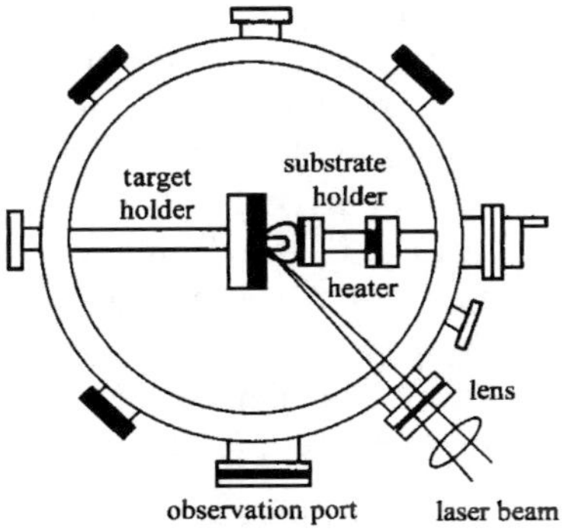


**Figure 2.** Schematic diagram of PLD device.

The experimental procedure of preparing CdO: In film by PLD technology mainly includes the following three parts: (1) pumping the air pressure in the vacuum chamber to a preset vacuum degree; (2) introducing oxygen to increase the air pressure to the working pressure and ionizing the oxygen; (3) Set the substrate temperature, deposition time, target material rotation speed and target substrate distance, and start deposition.

PLD technology has the following advantages: (1) It can ensure that the composition of the target and the film is highly consistent, and realize the transfer of the chemical stoichiometric ratio of the material; (2) The optical system and the vacuum environment are used for deposition without introducing impurities; (3) Deposit at low temperature.

Gupta et al. Used the PLD technique to deposit CdO: In films on quartz substrates, and studied the effects of growth temperature and oxygen partial pressure on the structure, optical and electrical properties of the films. Studies have shown that the optical transmittance of the film mainly depends on the growth temperature, and the oxygen partial pressure has almost no effect on the transmittance. In addition, the electrical properties of the film are closely related to the growth temperature and oxygen partial pressure. Zheng et al. Irradiated a Cd-In alloy target with pulsed laser in an oxygen atmosphere (10pa), and grown a CdO-In thin film on a quartz glass substrate at 300°C. The film shows a cubic crystal structure of CdO, but because cadmium is replaced by indium atoms, the lattice constant changes slightly. The results also show that when the mass fraction of indium in the film reaches 3.9%, the resistivity of the CdO: In film is lower ($5.95 \times 10\text{-}5$) and the direct band gap energy (2.97eV) compared to undoped CdO Higher, higher transmittance in the visible region. Subsequently, Segura prepared CdO: In thin film using PLD technology to study the relationship between the photoelectric performance of the thin film and In doping concentration and substrate material. The results show that the optical band gap of the film can reach more than 3.2eV when doped with 0.5% In. The minimum film resistivity on c-axis sapphire substrate is $4.5 \times 10\text{-}5$. On a single crystal MgO (100) substrate, the film resistivity is about $3.2 \times 10\text{-}5$. Sun Xiaonu focused on the effects of In content and substrate temperature on the structure and photoelectric properties of CdO: In films. The study found that In doping weakened the dominant growth of the thin film (200) crystal plane. When the concentration of In reaches 3.9wt%, the photoelectric performance of the thin film is the best. When the concentration of In increases to 5.6wt%, In2O3 is formed in the thin film, and the photoelectric performance of the whole thin film decreases. The surface of the film deposited at a substrate temperature of 300 ° C is the densest.

The above results show that the PLD technology can be used to prepare thin films with precise composition, simple operation and wide application range. However, PLD technology also has the disadvantage of poor uniformity of large-area deposited films, which is difficult to achieve industrial production.

Metal Organic Chemical Vapor Deposition (MOCVD). Chemical vapor deposition (CVD) is a technology in which one or several gaseous reactants undergo a chemical reaction on the surface of a substrate to deposit a film. This technology has been widely used to prepare semiconductor films (crystalline and amorphous), insulators, and metals. Metal-Organic Chemical Vapor Deposition (MOCVD) is one of chemical vapor deposition methods. The organometallic precursor used in MOCVD technology (shown in Figure 3) will decompose to form a metal oxide when heated, and then deposited as a thin film. Under appropriate growth conditions, an epitaxial layer can be formed. Oxygen plasma assisted technology (OPA) MOCVD can achieve an oxide film with extremely low defect density and high chemical purity grown under high vacuum. In many cases, MOCVD does not necessarily require a high vacuum, and indeed deposition can occur at moderate pressures (2 to 100kPa). Currently, MOCVD has become the preferred method for manufacturing compound semiconductor films.

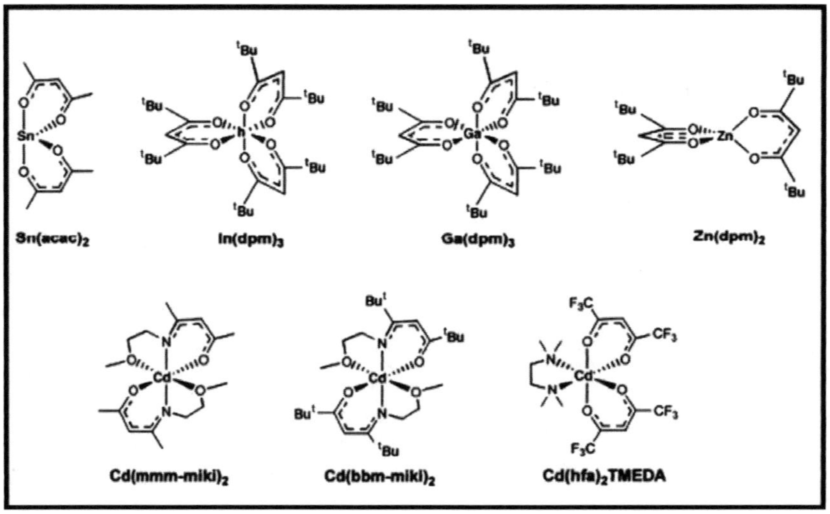


**Figure 3.** Collection of volatile organic metal precursors used for TCO thin film MOCVD processing.

Wang et al. Prepared a 150nm thick CdO: In thin film on a glass substrate and studied the effect of In doping concentration on the photoelectric properties of the thin film. The highest electron mobility of the film is 69cm2 / Vs, and the lowest resistivity is about 6 × 10-5. The CdO: In thin film deposited by Jin et al. On single crystals on MgO (100) has excellent optical transparency, with an average transmittance in the visible range > 80%. When the In doping concentration is 2.6%, the room temperature film conductivity is 5 × 10-5, and the optical band gap is 3.18eV.

The films obtained by MOCVD technology can be of high quality both chemically and structurally, but this technology also has limitations, which are related to the purity of the required metal organic precursor and the cost of preparation. In many cases, metal-organic precursors tend to be extremely reactive, and therefore require careful handling and purification through thorough recrystallization or sublimation procedures. In addition, most metal organics are volatile and often highly toxic, so special handling procedures are required.

Filter cathode vacuum arc method (FCAD). Filtered Cathodic Arc Deposition (FCAD) is a technique for preparing various high-quality metal oxide thin films. The principle of FCAD technology is: first generate and excite plasma on the surface of the cathode target, then use the magnetron elbow to filter the uncharged large particles, after filtering the plasma with a certain energy reaches the surface of the substrate and finally condenses into a film As shown in Figure 4).

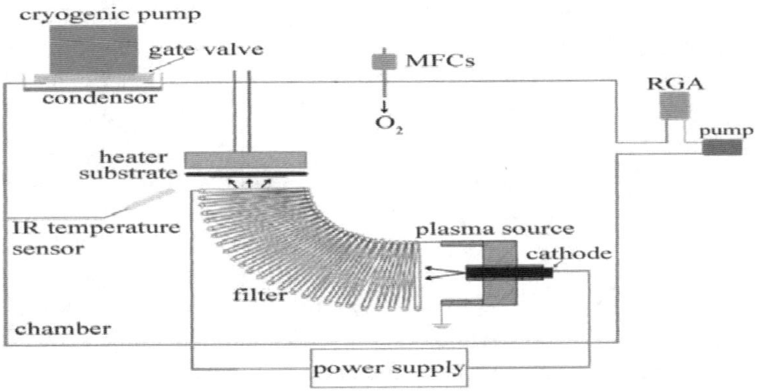


**Figure 4.** Schematic diagram of the filtered cathode vacuum arc film deposition system.

Yuankun Zhu used FCAD technology to prepare high-quality CdO: In films. The effects of substrate temperature, In doping concentration, oxygen partial pressure and other factors on the film structure and photoelectric properties were studied in detail. The films prepared under the doping conditions of 230 °C, 7mtorr and 1.2at.% In have the best photoelectric performance.

FCAD technology has the advantages of fast deposition rate, low substrate temperature, easy adjustment of composition, and large area deposition. The prepared film has good surface flatness, uniform thickness and uniform doping, and is suitable for preparing high-quality films.

The above results indicate that CdO: In films prepared by different methods have different photoelectric properties. Among these methods, the films prepared by PLD technology and FCAD technology have the best performance, but these two technologies also have their own shortcomings. Therefore, it is of great significance to further study the influence of the preparation process on the performance of the film for the application of CdO: In film.

## 3. Effects of process conditions on the photoelectric properties of CdO: In thin films

The photoelectric properties of CdO: In films are closely related to the process conditions such as In content, substrate temperature, and substrate materials in the films.

Effect of In doping amount on the photoelectric properties of CdO: In thin films. During the preparation of CdO: In thin films, changing the amount of In doping can significantly affect the photoelectric properties of the thin films. Studies have shown that with the increase of In doping content, In3 + replaces Cd2 + in the lattice to provide extra electrons, so the carrier concentration continues to increase. When the In doping concentration is low, the grain boundary scattering in the film dominates, and free electrons accumulate in the grain boundary to form a charge shield, which in turn reduces the barrier height in the grain boundary and the electron mobility gradually increases. When the content of In doping is high, ionization scattering in the film replaces grain boundary scattering, and the interaction between ionized impurities and electrons hinders the movement of free electrons, which reduces the electron mobility.

As the In doping content increases, the carrier concentration increases, and the Fermi level of the degenerate semiconductor enters the conduction band, and the Burstein-Moss effect occurs, and the forbidden band width of the film becomes larger.

Zheng et al. Used high-purity cadmium (99%) and indium (99%) powders with indium contents of 3wt%, 5wt%, and 7wt%, respectively, and fully mixed them to make targets for PLD deposition. The film was deposited at a fixed temperature of 300 ° C and an oxygen partial pressure of 10pa. Studies have shown that the resistivity of the thin film decreases to 5.95 × 10-5 as the indium content increases to 3.9wt%, and slightly increases as the indium content increases to 5.6wt%. The change in resistivity can be explained by the doping concentration of In and its presence in CdO. When the doping concentration of In3 + is less than the solid solution limit of In3 + replacing Cd2 +, the doped In3 +

replaces Cd2 + and releases more electrons to increase the carrier concentration, and the resistivity decreases accordingly. When the indium doping concentration exceeds the solid solution limit, excess In will exist in the gap In3 + or form an In-O bond (such as In2O3). The increase of the gap In3 + (acting as the donor center) will cause the receptor vacancy V2-cd to increase significantly to maintain the charge balance in the crystal, which is one of the reasons for the decrease in carrier concentration. In addition, non-conductive In2O3 clusters as carrier traps will also reduce the carrier concentration of the film. Combining the above two effects, when the concentration of doped In exceeds the solid solution limit, the carrier concentration decreases and the film resistivity increases.

Le et al. Prepared CdO: In thin films on quartz substrates by RF magnetron sputtering at room temperature. The effect of In doping concentration on the photoelectric properties of thin films was studied in detail. Studies have shown that compared with undoped CdO films, when the In doping concentration is low (0.012 at.%), The carrier concentration and electron mobility of the film increase simultaneously. The experiment confirmed that the increase in mobility is attributed to the decrease in oxygen vacancy (VO) concentration and the increase in grain size in the film. At the same time, it was found that the optical band gap of CdO: In film increases with the increase of In doping concentration, which is the increase of carrier concentration (Burstein-Moss effect) and the composition of CdO: In film (Vegard rule). The result of the role.

Zhu et al. Prepared CdO: In thin films on borosilicate glass using pulsed filter cathodic arc deposition (PFCAD). The substrate temperature was maintained at 230°C, the oxygen pressure was maintained at 0.931pa, the film was annealed in air and nitrogen atmosphere, and maintained at 450°C for 1 hour, and then naturally cooled to room temperature. Studies have shown that compared to undoped CdO thin films (refractive index of 2.52 at 500nm), the refractive index of CdO: In films decreases (refractive index of 2.28 at 500nm), and extinction coefficient also decreases Due to the increase in optical band gap.

Flores et al. Used sol-gel technology to mix CdO and In2O3 precursor solution to obtain CdO: In thin film, and explored the effect of In doping concentration on the photoelectric properties of the thin film. The experimental results show that at a sintering temperature of 450°C, the resistivity of the film decreases first and then increases with the increase of the doping amount. The resistivity is the smallest when the In doping amount is 5at%. The transmittance of the thin film in the wavelength range greater than 600nm is kept above 85%.

Effect of substrate temperature on the photoelectric properties of CdO: In thin films. In the preparation of CdO: In thin films, the temperature of the substrate is a very important process parameter. The substrate temperature will change the crystallinity, surface roughness, Cd oxidation degree and other properties of the film, and then affect the film carrier concentration and mobility changes, and ultimately determine the photoelectric properties of the film. At present, the influence of the substrate temperature on the CdO: In film is a hot issue.

Gupta et al. Studied the effect of substrate temperature on the optical and electrical properties of thin films. Studies have shown that the resistivity increases with the substrate temperature, the carrier concentration decreases with the substrate temperature, and the electron mobility increases with the substrate temperature. The decrease in carrier concentration may be caused by annealing of point defects and interstitial impurities. The annealing of point defects and interstitial impurities leads to a decrease in impurity scattering and an increase in mobility.

Using PLD technology, Zheng deposited a transparent CdO: In thin film on a quartz glass substrate and studied the effect of substrate temperature on the carrier concentration and Hall mobility of the thin film. The results show that with the increase of the substrate temperature, the resistivity tends to decrease first and then increase, showing the minimum resistivity (1.15 × 10-4) and the maximum carrier concentration (5.35 ×) at 300 °C 1020cm-3) and electron mobility (101.43cm2 / (V·s)). When the substrate temperature is greater than 300°C, the decrease in carrier mobility may be related to the increase in surface roughness and particle size of the film with increasing temperature, which leads to an increase in structural defects. In addition, as the substrate temperature increases from 100 to 300 °

C, the optical band gap increases, reaching a maximum (2.91 eV) at 300°C, and then decreases as the substrate temperature increases, which can be explained by the Burstein-Moss effect .

Zhu Yuankun used filtered cathode vacuum arc technology (FCAD) to prepare CdO: In thin films on quartz glass substrates. Studies have shown that as the substrate temperature increases, both the resistivity and surface resistance of the film decrease, the electron mobility gradually increases, and the carrier concentration decreases first and then increases. The lowest resistivity of the 125nm thick film prepared at 425°C was 5.9 × 10-5, the highest electron mobility was 112cm2 / (V·s), and the corresponding carrier concentration was 9.4 × 1020cm-3. As the substrate temperature increases, the energy of the atoms on the surface of the film increases, and these atoms are more likely to migrate to the optimal growth position, so that it is easier to crystallize to form larger grains, which reduces the grain boundaries in the film. This will facilitate the transport of free electrons, so the electron mobility of the thin film increases with increasing temperature. The carrier concentration of the film is related to the concentration of internal defects. When the substrate temperature increases from room temperature to 230°C, the defects in the film decrease, and the carrier concentration also decreases. When the temperature is further increased, the gap In3 + is more likely to replace Cd2 + to provide excess electrons, and the carrier concentration increases accordingly. In addition, with the change of the substrate temperature, the transmission curve of the film also shows changes in the absorption cut-off blue shift and near infrared plasma resonance wavelength shift. These changes are closely related to the changes in carrier concentration.

Effect of Annealing Treatment on the Photoelectric Properties of CdO: In Thin Films. Studies have shown that the photoelectric properties of CdO: In films are closely related to the crystal structure and crystal quality of the films. Appropriate annealing treatment can make the crystallinity of the film improved to a certain extent, thereby improving the photoelectric performance of the film. At present, the research on the annealing process mainly focuses on the annealing atmosphere and annealing temperature.

Flores et al. Used two solutions of CdO and In2O3 as precursors to prepare CdO: In thin films on glass substrates at room temperature using sol-gel technology. The effects of annealing temperature and annealing atmosphere on the photoelectric properties of thin films were studied. Studies have shown that after annealing, the electrical resistivity of the film is significantly reduced, which may be related to the increase in the oxygen vacancy concentration of the film. Compared with the annealing treatment in three different atmospheres of H2, N2 / H2, and vacuum, the resistivity of the film annealed in vacuum is the lowest, reaching 2 × 10-3; the direct band gap energy is the highest, reaching 3.6eV (the annealing temperature is 600°C, the annealing time is 1h, the indium concentration in the precursor solution is 55 at%).

Yang et al. Used DC magnetron sputtering technology to prepare CdO: In films on glass substrates, and studied the effect of annealing process on the structure and electrical properties of the films. The results of the study showed that the resistivity of all samples decreased after annealing. The resistivity of the film prepared at an oxygen concentration of 4.29% was 2.95 × 10-4, and after annealing treatment, the resistivity decreased to 2.00 × 10-4. On the one hand, the annealing process can provide energy for grain aggregation and reduce the grain boundaries in the film; on the other hand, the annealing process can increase the oxygen vacancy concentration in the film and release dissolved residual oxygen, thereby capturing free electrons on the film surface. Therefore, the decrease in resistivity of the film after annealing is the result of the combined action of these factors.

## 4. Summary

The transparent conductive film has both good conductivity and high transmittance in the visible light region, so it has great research significance and broad application prospects. CdO: In film is one of the candidates to occupy the transparent conductive film market due to its superior photoelectric performance. At present, the photoelectric performance of CdO: In thin films obtained by various preparation methods can reach the following levels: the transmittance in the visible light region is more than 80%, and the resistivity is on the order of 10-4.

However, there are still some deficiencies in the current research on CdO: In thin films. The main problems are as follows:

(1) There is a lack of systematic research on the relationship between the structure of CdO: In film and the photoelectric performance. The nucleation and growth process of the thin film should be studied more deeply to prepare CdO: In thin film with better performance.

(2) The optical band gap values of CdO: In films obtained by different methods are large. The extended misuse of some images in related research makes the obtained optical band gap value lack physical significance.

(3) Lack of low-cost and high-quality CdO: In thin film preparation technology. Other preparation technologies should be actively explored to meet the requirements of new optoelectronic devices.

## References

[1] J.Santos-Cruz, G.Torres-Delgado, R.Castanedo-Pe'rez, S.Jime'nez-Sandoval, J.Ma'rquez-Mari'n, O.Zelaya-Angel, Sol. Energy Mater. Sol. Cells 909 (2006) 2272.

[2] C.H.Kwon, H.K.Hong, D.H.Yun, K.Lee, S.T.Kim, Y.H.Roh, B.H.Lee, Sens. Actuators B 25 (1995) 610.

[3] R. Ferro, J.A. Rodriguez, O. Vigil, A. Morales-Acevado, G. Contreras-Puente, Phys. Stat. Solidi A 177 (2000) 477.

[4] R.Gupta, K. Ghosh, R. Patel, et al. Preparation and Characterization of Highly Conducting and Transparent Al Doped CdO Thin Films by Pulsed Laser Deposition. Current Applied Physics [J]. 2009, 9 (3): 673-677.

[5] B. Saha, R. Thapa, K. Chattopadhyay. Bandgap Widening in Highly Conducting CdO Thin Film by Ti Incorporation through Radio Frequency Magnetron Sputtering Technique. Solid State Communications [J]. 2008, 145 (1-2): 33-37.

[6] M. Yan, M. Lane, C. R. Kannewurf, et al. Highly Conductive Epitaxial CdO Thin Films Prepared by Pulsed Laser Deposition. Applied Physics Letters [J]. 2001, 78(16): 2342-2344.

[7] In doped CdO films:Electrical,optical, structural and surface properties.

[8] H. Kawazoe, K. Ueda. Transparent Conducting Oxides Based on the Spinel Structure. Journal of the American Ceramic Society [J]. 1999, 82 (12): 3330-3336.

[9] R. K. Gupta, K. Ghosh, R. Patel, et al. Structural, Optical and Electrical Properties of In Doped CdO Thin Films for Optoelectronic Applications. Materials Letters [J]. 2008, 62 (19): 3373-3375.

[10] Preparation methods and research progress of ITO transparent conductive films.

[11] Transparent conductive cadmium indate thin films prepared by dc magnetron sputtering. Mohan Babu.

[12] T. J. Coutts, X. Wu, W. P. Mulligan. High Performance Transparent Conducting Films of Cadmium Indate Prepared by RF Sputtering. Materials Research Society Symposium Proceedings [J]. 1996, 426: 479-489.

[13] Xu Wanjin. Development and application of magnetron sputtering technology.

[14] Research progress of functional films deposited by Cheng Yong and pulsed laser.

[15] Gupta,Structual and electrical properties of In doped CdO thin films for optoelectronic applications.

[16] Zheng.Optical and electrical properties of In-doped CdO thin films fabricated by pulse laser deposition.

[17] A. Segura, J.F. Sánchez-Royo, B. García-Domene, G. Almonacid, Current underestimation of the optical gap and Burstein-Moss shift in CdO thin films: A consequence of extended misuse of -versus- plots, Appl. Phys. Lett. 99 (15) (2011) 151907.

[18] Sun xiaonu. Preparation of indium doped cadmium oxide film by pulsed laser deposition and its photoelectric properties.

[19] G. Exarhos, X. Zhou. Discovery-Based Design of Transparent Conducting Oxide Films. Thin Solid Films [J]. 2007, 515 (18): 7025-7052.

[20] A. Wang. Indium-Cadmium-Oxide Films Having Exceptional Electrical Conductivity and Optical Transparency: Clues for Optimizing Transparent Conductors. Proceedings of the National Academy of Sciences [J]. 2001, 98 (13): 7113-7116.
[21] S. Jin, Y. Yang, J. E. Medvedeva, et al. Tuning the Properties of Transparent Oxide Conductors Dopant Ion Size and Electronic Structure Effects on CdO-Based Transparent Conducting Oxides Ga- and In-Doped CdO Thin Films Grown by MOCVD. Chemistry of Materials [J]. 2008, 20: 220-230.
[22] B. Tay, Z. Zhao, D. Chua. Review of Metal Oxide Films Deposited by Filtered Cathodic Vacuum Arc Technique. Materials Science and Engineering: R:Reports [J]. 2006, 52 (1-3): 1-48.
[23] Study on the properties of indium doped cadmium oxide transparent conductive films prepared by Zhu Yuankun filtered cathodic arc.
[24] Investigation of material properties and defect behavior in In-doped CdO films.
[25] L.M. Guia, V. Sallet, S. Hassani, M.C. Martínez-Tomás, V. Muñoz-Sanjosé, Effect of Growth Temperature on the Structural and Morphological Properties of MgCdO Thin Films Grown by Metal Organic Chemical Vapor Deposition, Cryst. Growth Des. 17 (12) (2017) 6303–6310.
[26] Influences of indium doping and annealing on microstructure and optical properties of cadmium oxide thin films.
[27] Flores. Optical, electrical and structural properties of indium-doped cadmium oxide films obtained by the sol–gel technique.
[28] Properties of R.F. magnetron sputtered cadmium–tin–oxide and indium–tin–oxide thin films.
[29] Zheng 2013 Influence of substrate temperature on the structural and properties of In-doped CdO films prepared by PLD.
[30] M. A. Flores Mendoza, R. Castanedo Pérez, G. Torres Delgado, et al.Cadmium Indate Thin Films, as Transparent Conducting Oxides, Obtained by the Sol–Gel Technique. Solar Energy Materials and Solar Cells [J]. 2010, 94 (1): 80-84.
[31] F.F. Yang, L. Fang, S.F. Zhang, K.J. Liao, G.B. Liu, J.X. Dong, L. Li, G.Z. Fu, Optical properties of CdIn2O4 thin fifilms prepared by DC reactive magnetron sputtering, Journal of Crystal Growth 297 (2006) 411–418.

IOP Conf. Series: Materials Science and Engineering **914** (2020) 012008 doi:10.1088/1757-899X/914/1/012008

# Isolation and Identification of Bird Viruses in Wetland Ecology

**Yong Li[1], Hanping Zhang[2] and Guang Chen[1]**

[1]The Monitoring center wildufe Disease and Resource in Hubei Province Wuhan 430075

[2]Computer Department, Wuhan Polytechnic, Wuhan 430074, China

**Abstract**. In order to study the occurrence of Newcastle disease virus (NDV) in wild birds of Honghu wetland in Hubei, we collected wild bird feces samples in the wild bird habitat of Honghu in Hubei in the winter and spring of 2017 ~ 2019. By SPF chicken embryo isolation, hemagglutination-positive samples were identified by RT-PCR. A total of 38 NDV and 3 mixed viruses including NDV and avian influenza virus (AIV) were detected. The results showed that the NDV positive rate in wild birds was about 1%. These monitoring results lay the foundation for further understanding the status of NDV carried by wild birds and formulating corresponding preventive measures.

## 1. Introduction

The vast and numerous wetlands have a variety of ecological functions and are rich in natural resources. They are known as the "kidneys of the earth", species reservoirs, climate regulators, and water purifiers. They are important immune systems on the planet. It plays an irreplaceable role in protecting the ecological environment, maintaining biodiversity, and developing the economy and society. [1-5] Wetland ecosystems are water ecosystems. Its biological community is composed of aquatic and terrestrial species, which has active material circulation, energy flow, and species migration and evolution, and has high ecological diversity, species diversity, and biological productivity. Wild birds are the most important and active species in wetlands. Their species and quantity are important indicators of the quality of wetlands. Wild birds often carry unpredictable viruses during migration, which directly threaten animal safety, ecological safety, and affect to human production and life. Based on the fecal samples collected from Honghu Wetland during the wild bird migration and wintering period in 2017 ~ 2019, this paper uses biological laboratories to isolate and identify, and uses monitoring data to study the change rule of Newcastle disease virus positive rate, and to explore the trend about infection of NDV and pathogenic host range.

With the gradual immunization of vaccines and the evolution of strains, clinical necropsy symptoms of Newcastle disease are becoming increasingly atypical, and the diagnosis needs to rely on laboratory tests. The routine laboratory testing methods include fluorescent RT-PCR, RT-PCR, HI tests, virus isolation and identification, and virus gene sequence determination, etc. [6] In this experiment, chicken embryo isolation and RT-PCR were used to study the carrying of NDV from wild bird droppings collected in Honghu wetland.

IOP Conf. Series: Materials Science and Engineering **914** (2020) 012008 doi:10.1088/1757-899X/914/1/012008

## 2. Materials and Methods

### 2.1. *Main reagents and instruments*

*2.1.1. Main reagent.* SPF Chicken Embryo (Beijing Merial Vital Laboratory Animal Technology Co., Ltd.), RNA extraction kit, proteinase K, M-MLV reverse transcriptase, RRNA inhibitor, DEPC water, dNTP Mixture, 2×Tsingke Master Mix (Xi'an Tianlong Technology Co., Ltd.)

*2.1.2. Main instrument.* Keyu brand microcomputer incubation equipment, eppendorf Centrifuge 5810R, G100 gene amplification instrument, nucleic acid extraction instrument (Xi'an Tianlong Technology Co., Ltd.), Autoclave (HIBAYAMA HVE-50), Vortex QL-901, ElectrophoresisApparatus (DYY-6B Type constant voltage and steady flow ElectrophoresisApparatus), FR-980A biological electrophoresis image analysis system.

*2.1.3. Buffer and solution formulations.*
(1) Sampling solution: 15% glycerol, 425mL of 1×PBS, 1.5g of 0.3% BSA, make up to 500mL.

(2) TAE buffer: 48.4g of Tris, 11.42mL of glacial acetic acid, 6.624g of EDTA, make up to 10L.

(3) Anticoagulant: 5g of sodium citrate, 0.9g of sodium chloride, make up to 100mL with distilled water, filter and sterilize.

(4) Physiological saline: 0.9g of sodium chloride, make up to 100mL with distilled water, and autoclave at 121°C for 15min.

(5) 1% red blood cell: the mixed blood of SPF or non-immune healthy chicken is collected, injected into a 50mL centrifuge tube with an appropriate amount of anticoagulant, equilibrated on a balance and centrifuged in a centrifuge, with 1500rpm, and centrifuged for 15min. Then we separate out the upper serum, and repeat the washing 4 ~ 5 times by adding normal saline, and finally make up to 1% red blood cells stored at 4°C for later use [7].

### 2.2. *Primer synthesis*
Designing primers with reference to the full sequence of NDV strains published in GenBank, as follows:

ClassI_NDV_F: CACCAAGCTGGAGAAGGGCATAC,
ClassI_NDV_R: CAGTATGTTTGCAGCATTCTGGTTGG,
ClassII_NDV_F: CCATTGCTAAATACAATCCTTTCA,
ClassII_NDV_R: CTGCCACTGCTAGTTGTGATAATCC.

Class I NDV product size is 727bp, and Class II NDV is 708bp.

### 2.3. *Experimental method*

*2.3.1. Sample processing.* The collected cotton swabs were thawed on ice, centrifuged at 8000rpm for 5 min at 4°C after mixing with vortex, and 200μL of the upper serum was taken. 20μL of 1000×dual antibody (penicillin: streptomycin = 1: 1) was added to the upper serum. Finally inoculating chicken embryos after shaking and mixing.

*2.3.2. Chicken embryo inoculation.* Take 9 to 11-day-old chicken embryos to streaking in a dark room, avoiding large blood vessels, and line drawing between the two blood vessels along the junction of the air chamber and the liquid surface. After streaking, we place the chicken embryos on an egg tray and take them into the biosafety cabinet, then wipe the scribe line with an alcohol cotton ball. With burning the hole punch on the alcohol lamp, the egg shell can be poked gently at the upper part of the scribe line, then burn the hole punch again, and beat the chicken embryos in turn. hole. After punching, we use a syringe to aspirate the treated sample, then inject it into the allantoic cavity of the chicken embryo, and seal the pinhole with paraffin to mark it, and place it in a 37°C incubator.

*2.3.3. Chicken embryo breeding and allantoic fluid collection.* Observed with lighted eggs every day, the dead chicken embryos were taken out and placed in refrigerator at 4°C, and all chicken embryos were placed in refrigerator at 4°C for 8h or overnight after 72h. Then collect the chicken embryo, and wipe the eggshell with an alcohol cotton ball. Then break the eggshell with one end of the tweezers, and remove the membrane, and use a pipette tip with a filter element to suck 50μL of allantoic fluid onto the hemagglutination plate, and then add 50μL of 1% chicken erythrocytes, waiting for 20 minutes, to observe positive blood clotting, then collect the allantoic fluid of hemagglutination-positive chicken embryos, and store it at -80°C.

*2.3.4. Extraction and reverse transcription of viral RNA.* Melt the hemagglutination-positive allantoic fluid on ice. Use a RNase-free pipette tip to add 200μL to the corresponding well of the kit in a biosafety cabinet. Then add 20μL proteinase K respectively. After the addition, put them into the nucleic acid extraction instrument. When the procedure is completed, we collect RNA in the corresponding well (approximately 90μL per well), and insert it into the ice box quickly after collection. Take 10μL of the extracted RNA, then add 1μL each of the corresponding 10umol/l primers Tuni-12 and Tuni-13, and mix and centrifuge in the PCR machine. After heating at 70°C for 5 minutes, then place them in ice bath for 2 minutes. Add the following reagents to each tube: 10μL of 5×RT buffer, 25μL of DEPC, 1μL of dNTP Mixture, 1μL of RRI, 1μL of M-MLV, gently mix them with 42°C 1h, 95°C 2min, 4°C hold. Then use the reverse transcription product as a template for PCR.

*2.3.5. PCR identification of NDV virus.* Take the above cDNA and add them into the reaction system on ice, including 9.5μL of ddH2O, 12.5μL of 5×Tsingke Master Mix, 0.5μL each of the forward and reverse primers, and 2μL of the template. Put them in the PCR instrument and run the program at 95°C for 3min, then at 95°C for 1 min, at 51°C for 45s, at 72°C for 2min and 30s, for 35 cycles, and finally at 72°C for 10mins. After the procedure is completed, 5μL of the reaction product is tested on 1.2% agarose gel electrophoresis.

## 3. Results and discussion

Chicken embryo samples collected from Honghu Wetland in Hubei from 2017 to 2019 were used for chicken embryo isolation, hemagglutination detection, RNA extraction, reverse transcription and PCR and agarose gel electrophoresis. (the PCR product has 727bp band, indicating NDV CLASS I positive. A band of 708bp indicates a positive NDV CLASS II.) The following results were obtained.

**Table 1.** Surveillance statistics of Honghu wild bird Newcastle disease virus from 2017 to 2019.

| Sampling year | Sampling month | Number of sampling swabs (parts) | NDV positive | NDV positive rate (%) |
|---|---|---|---|---|
| 2017 | 12 | 250 | 0 | 0 |
| 2018 | 01 | 974 | 1 | 0.10% |
| | 02 | 908 | 15 | 1.65% |
| | 11 | 390 | 0 | 0 |
| | 12 | 1128 | 25 | 2.22% |
| 2019 | 01 | 491 | 0 | 0 |
| | 02 | 650 | 0 | 0 |

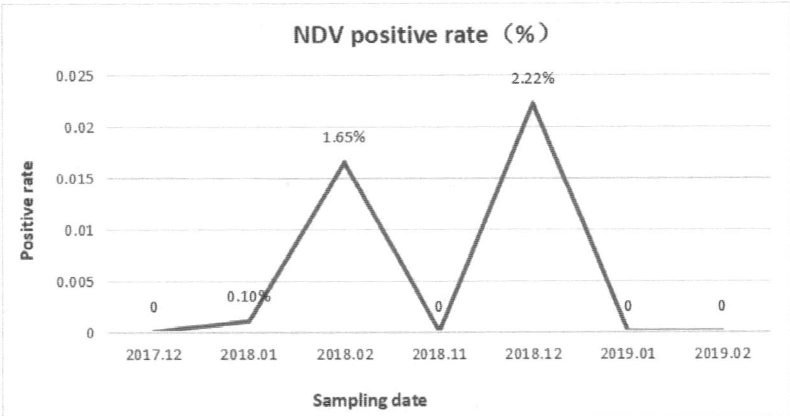


**Figure 1.** 2017-2019 Honghu wild bird Newcastle disease virus positive rate.

## 4. Summary

The specific sampling information from 2017 to 2019 is shown in Table 1. The positive rate in January 2018 was 0.10% in the months tested positive. The positive rate in February 2018 was 1.65%, and the positive rate in December 2018 was 2.22%. It can be seen that the separation rate of NDV is slightly different in different months. The above data remind us that we should strengthen the detection of NDV to detect and control its spread as soon as possible. In recent years, the range of NDV infection and pathogenic hosts has expanded. Some wild bird NDV and wild bird type A bird flu virus attenuated strains can be transformed into virulent strains after passaging in chickens [8~9], thus endangering animal safety and the development of poultry breeding. Studies have shown that the evolution rate of the NDV genome and the frequency of recombination between different genotypes and different gene fragments are gradually increasing [10]. Some NDV isolated in this study may have cross infection between wild birds and poultry. Therefore, strengthening the monitoring of wild bird-sourced NDV is essential to protecting animal safety and ecological safety, promoting the healthy development of poultry farming, and preventing and controlling the large-scale epidemic of Newcastle disease.

## References

[1] Yin Zhen, editor of Liu Jinghua. Animal Virology [M]. Beijing: Science Press, 1997.
[2] Di Chang, Xia Zhang, Qiong Liu, Ge Gao, and Yue Wu. "Location based robust audio watermarking algorithm for social TV system." In Pacific-Rim Conference on Multimedia, pp. 726-738. Springer, Berlin, Heidelberg, 2012.
[3] Di Chang, Xia Zhang, and Yue Wu. "A Multi-Source Steganography for Stereo Audio." Journal of Wuhan University (Natural Science Edition), 2013 (3): 277-284.
[4] XIa Zhang, Di Chang, et al. "An Audio Steganography Algorithm Based on Air-Channel Transmitting." Journal of Wuhan University (Natural Science Edition) 57, no. 6 (2011): 499-505.
[5] Xia Zhang, Di Chang, et al. "Tree-like Dimensionality Reduction for Cancer-informatics." In IOP Conference Series: Materials Science and Engineering, vol. 490, no. 4, pp. 042028. IOP Publishing, 2019.
[6] Lu Huijun, Jin Jianshi, Jin Ningyi. Mutation, evolution and cross-species infection of Newcastle disease virus [J]. Chinese Poultry, 2010, 32 (7): 1-5.
[7] Cui Ronghua, Zhong Jian, Miao Xiaobin. Isolation and Identification of Newcastle Disease Virus [J]. Friends of Farmhouse: Theory Edition, 2011 (1).
[8] CHONG Y L, PADHI A, HUDSON P J, et al. The Effect of Vaccination on the Evolution and Population Dynamics of Avian Paramyxovirus-1 [J]. Plos Pathogens, 2010, 6 (4): e1000872.

[9] ZEIEITLIN G A, MASLOW M J. Avian influenza [J]. Current Allergy & Asthma Reports, 2006, 6 (2): 163-170.
[10] Yu Shengqing, Ding Chan, NORIKOKISHIDA, et al. Study on the transformation of a waterfowl isolate of Newcastle disease virus from non-pathogenic type to rapid type after passaging of chickens [J]. Chinese Journal of Preventive Veterinary Medicine, 2003, 25 (1): 59-64.

IOP Conf. Series: Materials Science and Engineering **914** (2020) 012009 doi:10.1088/1757-899X/914/1/012009

# Anti-Tumor Effect and Mechanism of Bletilla Rhizoma Based on Network Pharmacology

**Xiusheng Tang[1, a], Rongze Fang[2, b], Ling Li[1, c] and Jun Luo[1, *]**

[1]The First Affiliated Hospital of Guizhou University of Traditional Chinese Medicine, Guizhou, China

[2]School of Guizhou University of Traditional Chinese Medicine, Guizhou, China

*Corresponding author e-mail: 53011402@qq.com, [a]850447260@qq.com, [b]2753779015@qq.com, [c]1654226588@qq.com

**Abstract**. *Bletilla* Rhizoma is a traditional Chinese medicine used for hemostasis. Modern research shows that *Bletilla* Rhizoma has anti-tumor effect, but its mechanism is not clear. We obtained the active chemical components and targets of *Bletilla* Rhizoma through the systematic pharmacology database of Traditional Chinese Medicine (TCMSP), and established the data set of the intersection of active components and disease targets. The protein-protein interaction network (PPI) was used to analyze the protein interaction network of *Bletilla* Rhizoma active components target and tumor target. The biological information annotation database (DAVID) was used to perform gene ontology (GO) function enrichment analysis and based on Kyoto encyclopedia of genes and genomes (KEGG) pathway enrichment analysis to predict related mechanisms of anti-tumor effects of *Bletilla* Rhizoma. 15 active components were obtained from TCMSP, and 40 potential targets were screened out through the evaluation of the network topology. The DAVID database was used to perform gene GO functional enrichment analysis and KEGG pathway enrichment analysis on 40 potential target targets to screen for participation. The first 20 biological processes and signaling pathways involved in the anti-tumor effect of *Bletilla* Rhizoma were screened out. Through network pharmacology, we have drawn the conclusion that the anti-tumor effect of *Bletilla* Rhizoma is mainly regulated by pathways such as insulin resistance, legionelosis and proteoglycans in cancer, among which the proteoglycans in cancer pathway has a strong correlation with cancer, and its anti-cancer pathway is closely related to TNF, a key target.

## 1. Introduction

*Bletilla* Rhizoma is the tuber of *Bletilla* striata (Thunb.) Reichb.f., which has the functions of astringent, hemostasis, detoxification, tonic deficiency. It is commonly used in the treatment of traumatic bleeding, ulcer bleeding, hematemesis [1]. Modern pharmacological studies have found *Bletilla* Rhizoma has anti-tumor, anti-inflammatory, anti-gastric ulcer activities [2-4]. In recent years, *Bletilla* Rhizoma has been used in the treatment of tumors, including uterine fibroids, liver cancer and esophageal cancer [5]. *Bletilla* Rhizoma can inhibit the growth and proliferation of human gastric cancer cells (SGC), human ovarian cancer cells (A2780), hepatocarcinoma cells (HepG2) [6-7], and can significantly induce A549 human lung adenocarcinoma cell cycle to block in the G0/G1 phase, which has potential activity in the treatment of lung cancer [8]. In addition, *Bletilla* Rhizoma can significantly inhibit S180 sarcoma and prolong the survival time of H22 ascites hepatoma mice [9]. At present, the anti-tumor effect of *Bletilla* Rhizoma is mainly based on animal and in vitro cell experiments, and the treatment and mechanism of human tumor still need to be further explored.

Tumors have multi-factor, multi-link pathogenic mechanisms, and the multi-target, multi-link, and various regulation methods of traditional Chinese medicine can be applied to multiple links of tumor occurrence and development. Traditional Chinese medicine has the characteristics of multi-component and multi-target, so it is difficult to define the material basis and mechanism of efficacy. Traditional research methods are difficult to carry out systematic research from the level of molecule, cell, tissue and organ to the whole [10]. Network pharmacology is different from the concept of "one drug, one target, one disease" in traditional pharmacology research methods. It constructs a multi-level network by retrieving relevant databases, combining high-throughput group data analysis and computer simulation. Based on the "disease gene target drug" interaction network, we can predict drug targets as a whole, and comprehensively and systematically study the law and mechanism of drug body interaction. Therefore, this article used the network pharmacology method to excavate the main anti-tumor active ingredients and action targets of *Bletilla* Rhizoma, so as to provide a theoretical basis for the in-depth development of experimental research and guidance of clinical rational drug use.

## 2. Materials and Methods

### *2.1. Screening of active ingredients*

By using the Traditional Chinese Medicine Taiwan Database (TDT, http://tcm.cmu.edu.tw/), the BATMAN-TCM Database, the Traditional Chinese Medicine Comprehensive Database (TCMID, http://www.megabionet.org/tcmid/search/) and the Traditional Chinese Medicine System Pharmacological analysis platform (TCMSP, http://lsp.nwu.edu.cn/tcmsp.php), retrieves all the chemical components contained in *Bletilla* Rhizoma. According to the TCMSP database [8], ADME parameters (OB≥30% and DL≥0.18) were used as the standard [9], or the *Bletilla* Rhizoma components reported in the literature were used to screen the active chemical components.

### *2.2. Establish target data set*

By the comprehensive analysis of TCMSP database, Chinese medicine target database and BATMAN-TCM database [10], the targets of *Bletilla* Rhizoma active components were screened, and the component targets data set was established. Based on the similarity comparison, the active ingredients corresponding to the screening target were passed through the Chinese herbal medicine active ingredient database (HIT, http://lifecenter.sgst.cn/hit/) and the therapeutic target database (TTD, http://bidd.nus.edu.sg/group/cjttd/) to screen potential targets for active ingredients. A potential target data set of *Bletilla* Rhizoma and active components was established, and genes and protein targets related to tumors were screened through the Genecards [11] database, and a disease-drug common target data set was established.

### *2.3. Build PPI network*

The selected targets were imported into string (https://string-db.org/) to construct PPI, the protein type was set as "homosapiens (human)", the confidence degree was 0.4, and the PPI was obtained. The PPI was saved as TSV file, imported into Cytoscape 3.6.1, analyzed with its "network analyzer" plug-in, and the core anti-tumor targets of *Bletilla* Rhizoma were determined.

### *2.4. KEGG signal pathway and GO biological process enrichment analysis*

The common targets of disease and drug were introduced into the human genome annotation database David 6.8 (https://david.ncifcrf.gov/) for GO enrichment analysis and KEGG signal pathway analysis, and $P<0.05$ was used as the critical value of significant function and pathway to screen the target genes.

## 3. Results

### 3.1. Active ingredients

15 active chemical components and 318 potential targets were selected, and the results are shown in Table 1.

**Table 1.** Bletilla Rhizoma active compounds and target number.

| No. | Component | Target |
|---|---|---|
| 1 | PHB | 15 |
| 2 | protocatechuic acid | 27 |
| 3 | physcion | 25 |
| 4 | HBA | 4 |
| 5 | caffeate | 22 |
| 6 | blespirol | 2 |
| 7 | cinnamic acid | 9 |
| 8 | sitosterol | 46 |
| 9 | 2, 4, 7 – trimethoxy - 9, 10 - dihydrophenanthrene | 58 |
| 10 | 2, 3, 4, 7 - tetramethoxyphenanthrene | 9 |
| 11 | 1 - (4 - hydroxybenzyl) - 4 - methoxy - 9, 10 - dihydrophenanthrene - 2, 7 - diol | 21 |
| 12 | 1, 3 – dimethoxy -5 - [2- (3 - methoxyphenyl) ethyl] benzene | 28 |
| 13 | 1, 3 - dimethoxy - 5 - (2 - phenylethyl) benzene | 25 |
| 14 | 4, 7 - dihydroxy - 1 - p - hydroxybenzyl - 2 - methoxy - 9, 10 - dihydrophenanthrene | 21 |
| 15 | 3, 7 - dihydroxy - 2, 4 - dimethoxyphenanthrene - 3 - O - glucoside | 6 |

### 3.2. Screening for disease and drug targets

With "tumour" as the key word, the tumor-related target genes were retrieved in Genecards. A total of 2568 related target genes were searched in OMIM, which were matched with related targets of *Bletilla* Rhizoma active components and Venn diagram was drawn, as shown in Figure 1. There are 40 common targets (TNF, EGFR, CASP3, ESR1, AR, NFKB1, PPARG, etc.) involved in the common system of white and active components—anti-tumor, which are key targets, as shown in Figure 2.

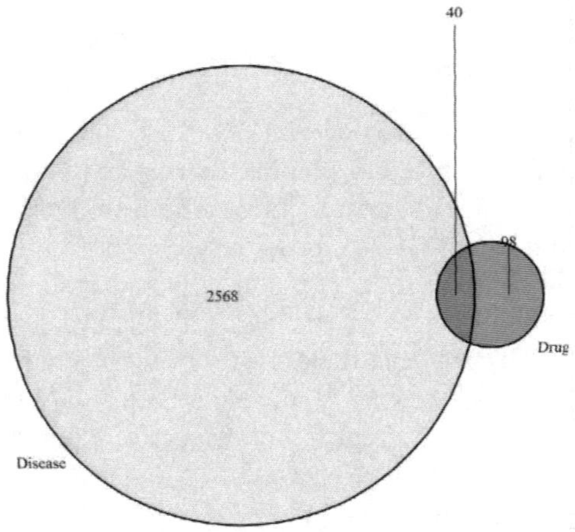


**Figure 1.** Venn diagram of matching Bletilla Rhizoma and disease target genes.

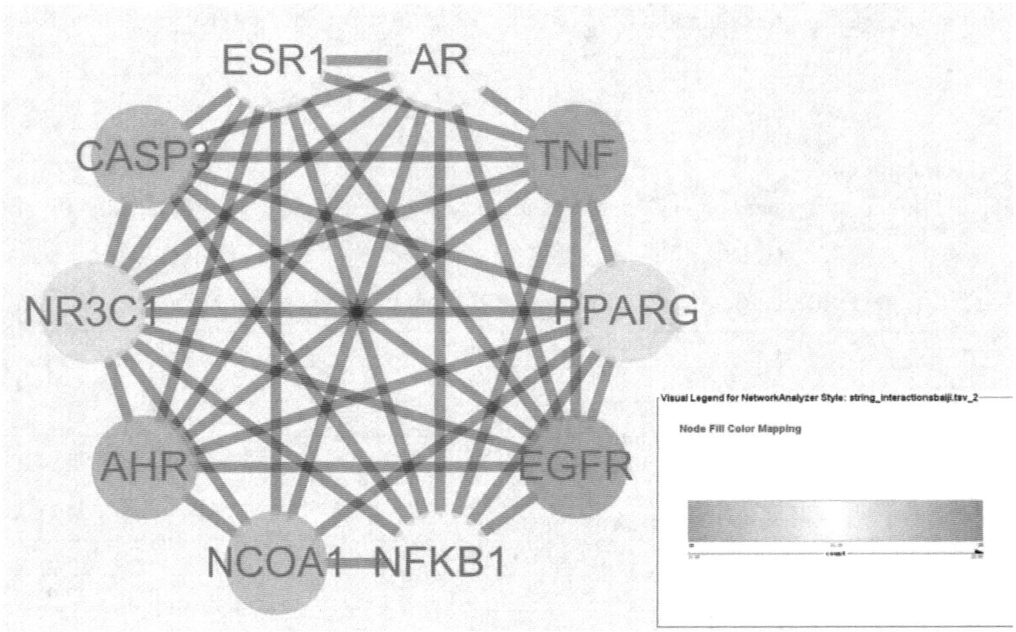


**Figure 2.** Top 10 proteins with strong associations.

**Table 2.** Topological parameters related to the direct target of Bletilla Rhizoma and its anti-tumor active components.

| UniProt ID | Protein names | name | Degree | Closeness Centrality | Betweenness Centrality |
|---|---|---|---|---|---|
| P01375 | tumor necrosis factor | TNF | 20 | 0.6923 | 0.2305 |
| P00533 | epidermal growth factor receptor | EGFR | 20 | 0.6792 | 0.1995 |
| P42574 | Caspase - 3 | CASP3 | 19 | 0.6667 | 0.1538 |
| P03372 | estrogen receptor | ESR1 | 15 | 0.6102 | 0.0445 |
| P10275 | androgen receptor | AR | 15 | 0.6000 | 0.0579 |
| P19838 | nuclear factor NF - kB p105 subunit | NFKB1 | 15 | 0.6102 | 0.0419 |
| P37231 | peroxisome proliferator-activated receptor γ | PPARG | 14 | 0.5806 | 0.0425 |
| P04150 | glucocorticoid receptor | NR3C1 | 14 | 0.5902 | 0.0221 |
| Q15788 | nuclear receptor coactivator 1 | NCOA1 | 12 | 0.5000 | 0.0198 |
| P35869 | aryl hydrocarbon receptor | AHR | 11 | 0.5455 | 0.0137 |

### *3.3. Go biological function enrichment analysis*

GO enrichment analysis was performed on 40 common targets of *Bletilla* Rhizoma and tumor-related targets. The top 3 enriched targets include transcription initiation from RNA polymerase II promoter, positive regulation of transcription from RNA polymerase II promoter and signal transduction. The yellow circle represents the biological process of the adjusted $P<0.0001$, and the size of the circle indicates the enrichment of related targets in the pathway. The darker the color of the circles, the stronger the enrichment effect, reflecting that the anti-tumor mechanism of white and *Bletilla* Rhizoma may involve multiple biological processes in vivo, and suggesting that *Bletilla* Rhizoma and active ingredients may exert anti-tumor effects by regulating these biological processes.

### *3.4. KEGG pathway enrichment analysis*

By using the software of Cytoscape 3.6.1, 40 common genes, diseases and components were visualized and analyzed, and *Bletilla* Rhizoma and anti-tumor interactive network was constructed, as shown in Figure 3. After selecting the corresponding interactive proteins and visualizing them with different colors and shapes, we can intuitively see the network relationship between the active chemical components and the target. There are 15 in red, representing the active components of the drug, and 40 in green, representing the active targets. The hexagon color represents the disease, and the purple color represents the drug. The larger the connectivity, the larger the shape, as shown in Figure 4.

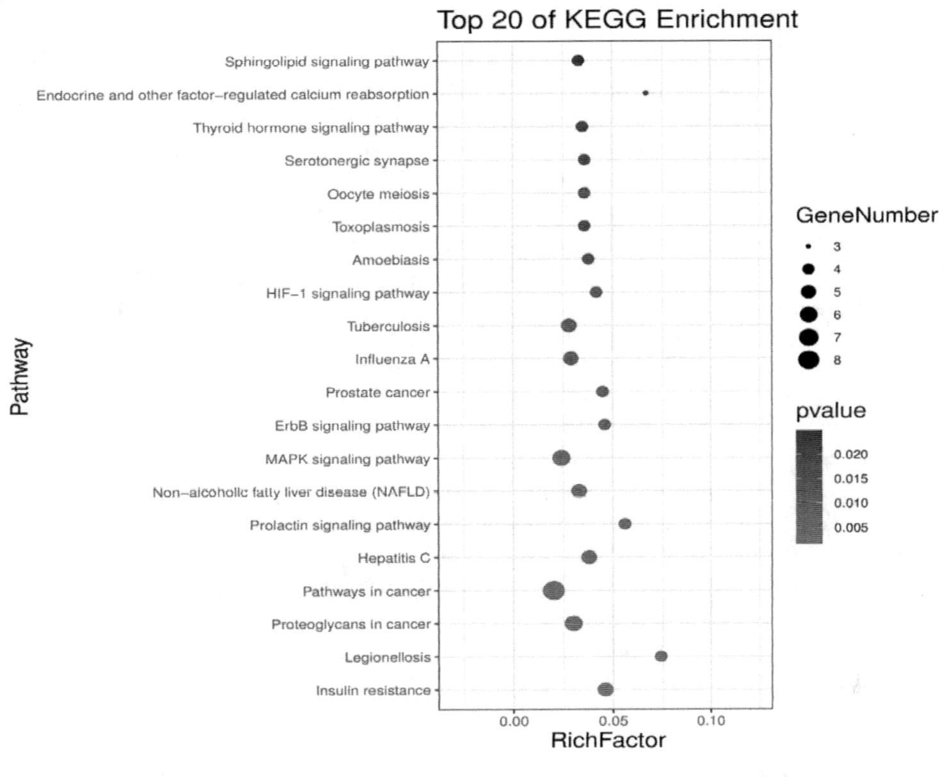


**Figure 3.** Bubble chart of enrichment analysis of KEGG pathway.

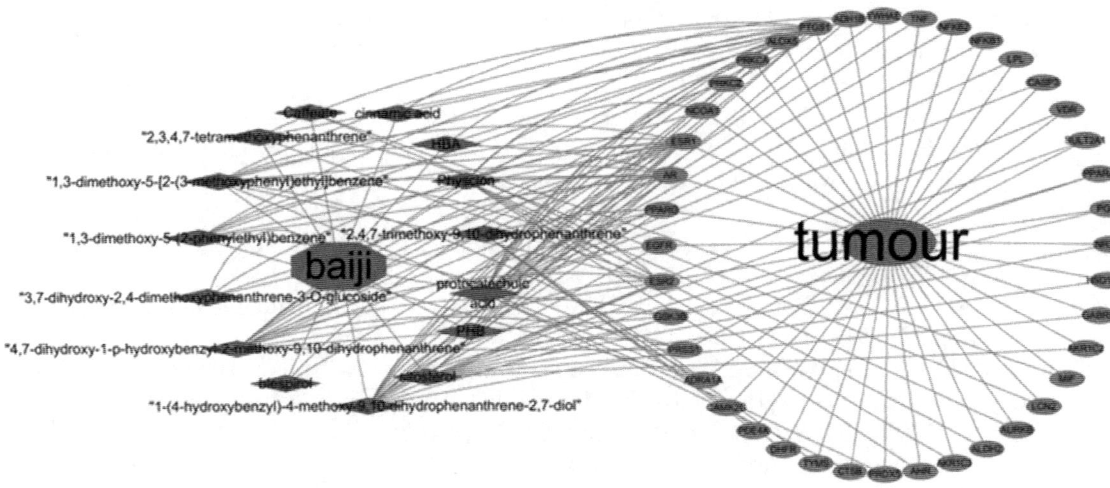


**Figure 4.** Component-target-disease interaction network.

## 4. Discussion

*Bletilla* Rhizoma has a good therapeutic effect against liver cancer, esophageal tumor, cardiac cancer and uterine fibroids [11], but the anti-tumor mechanism is still unclear. Pharmacology prediction shows that through the network, *Bletilla* Rhizoma main chemical composition are protocatechuic acid, physcion, cinnamic acid, 3,7 dihydroxy-2,4-dimethoxyphenanthrene -3-O-glucoside, these a few classes have anti-tumor effects. For example, physcion has a strong inhibitory effect on Hela cell growth of human cervical cancer [12]; cinnamic acid has obvious inhibitory effect on lung adenocarcinoma cancer cell proliferation is A-5491 lung adenocarcinoma cancer cell effective inhibitors [13]. These anti-tumor active components are mainly related to the genes of TNF, prkca, EGFR, CASP3 and ESR1 on proteoglycans in cancer, pathways in cancer and hepatitis C pathway.

In this study, the TNF gene Degree, Closeness Centrality, and Betweenness Centrality were larger in the key target screening, ranking first in the comprehensive ranking, and is one of the key targets for white and anti-tumor. TNF often binds to TNF receptors present on the surface of tumor cells to cause tumor cell death, or directly inhibits cell growth mediated by specific receptors. KEGG map showed that in the Proteoglycans in cancer pathway, the EGFR gene was activated by Decorin, which indirectly promoted the apoptosis of tumor cells through Casp3. In addition, Decorin (proteoglycan) can activate TNF gene after the expression of MAPK/Erk and p38 singaling, and finally make tumor cells Growth suppression. Therefore, *Bletilla* Rhizoma may play an anti-tumor effect by promoting tumor cell apoptosis or inhibiting its growth. This article provides relevant evidence for the later research on the anti-tumor mechanism of *Bletilla* Rhizoma, but the specific anti-tumor mechanism of *Bletilla* Rhizoma needs to be further studied and verified.

## Acknowledgments

This work was financially supported by Project of research initiation fund of Guizhou University of traditional Chinese medicine [2018] No.97; Science and technology plan project of Guizhou Province (qiankehe support [2018] 2772); Talent project of traditional Chinese medicine characteristic technology inheritance of State Administration of Traditional Chinese medicine (ZMC HR [2018] No. 204) and Major health science and technology cooperation project of the first affiliated hospital of guizhou university of traditional Chinese medicine, guiyang science and technology bureau (Zhuke contract [2019] 9-2-11).

## References

[1] National pharmacopoeia committee. Pharmacopoeia of the People's Republic of China (part I) [M]. China Science and Technology Press.,Beijing, 2015: 103.

[2] D. Vossoughi, D. Woitaschek, E. Oppermann et al. Combined transarterial chemoembolization and arterial administration of Bletilla striata in treatment of liver tumor in rats [J]. World J Gastroenterol, 2003, 9 (12): 2676-2680.

[3] Y.Z. Deng. Study on the active components and molecular mechanism of Bletilla striata on suppressing pulmonary fibrosis based on RAW264.7 macrophage inflammation modle [D]. Zhejiang Chinese Medical university, Hangzhou, 2017.

[4] X.B. LV, C.Q. Huang, Z.C. Wu, et al. The therapeutic effects of polysaccharides from Bletilla Striata on gastric ulcer rats [J]. Journal of Yunnan University of Traditional Chinese Medicine, 2012, 35 (1): 30-32, 45.

[5] X. Zou, M.Q. Chang, P.D. Tao, et al. Bletilla Striata in cancer treatment [J]. Central South Pharmacy, 2015, 13 (1): 58-60.

[6] S.H. Luo, C.S. Zheng, W.Y. Li, et al. Experimental study on anti-tumor effect of Bletilla Striatai polysaccharide in vitro [J]. Chinese Traditional Patent Medicine, 2014, 36 (1): 165-168.

[7] S.S. Chen, B.Wu, T.Tan, et al. Isolation, purification and structural characterization of Bletilla striata polysaccharides and its antitumor activity [J]. Chinese Traditional and Herbal Drugs, 2019, 50 (8): 1921-1926.

[8] A.J. Sun, S.Q. Pang, G.Q. Wang. Separation of Chemical Constituents from Bletilla striata and

Their Antitumor Activities [J]. Chinese Journal of Pharmaceuticals, 2016, 47 (1): 35-38.

[9] S.J. Liu, Y.L. Che. Experimental research on the explantive liver- cancer treated by Baiji [J]. Information on traditional Chineses medicine, 2008, 25 (1): 38-40

[10] Z.H. Sun, X.B. Sun. Network pharmacology: new opportunity for the modernization of traditional Chinese medicine [J]. Acta Pharmaceutica Sinica, 2012, 47 (6): 696-703.

[11] Q.X. Guan, D.D. Sun, G.Y. Zhang, et al. Docetaxel-loaded self-assembly stearic acid-modified Bletilla striata polysaccharide micelles and their anticancer effect:preparation, characterization, cellular uptake and in vitro evaluation [J]. Molecules, 2016, 21 (12): 1641.

[12] L.G. Ng, A.P. Sutherland, R. Newton, et al. B cell-activating fac-tor belonging to the TNF family (BAFF) -R is the principal BAFF receptor facilitating BAFF costimulation of circulating T and B cells [J]. J Immunol, 2004, 173 (2): 807-817.

[13] T. Wang, G. Jin, S.M. Wangi, et al. Induced dif ferentiation of human lung adenocarcinoma cells by cinnamic acid [J]. Chinese Journal of cancer, 2000, 19 (8): 782-785.

IOP Conf. Series: Materials Science and Engineering **914** (2020) 012010 doi:10.1088/1757-899X/914/1/012010

# Study on the mechanism of anti-cancer action of *rosa roxbughii tratt* based on network pharmacology

**Ying Mei[1, a], Xiang Pu[1], Xiangling Qu [2, *]**

[1]School of Guizhou University of Traditional Chinese Medicine, Guizhou, China
[2]The Second Affiliated Hospital of Guizhou University of Traditional Chinese Medicine, Guizhou, China

[*]Corresponding author e-mail: 125193321@qq.com, [a]37979960@qq.com

**Abstract**. The active chemical components and targets of *rosa roxbughii tratt* were screened by the method of network pharmacology, and the molecular mechanism of its action was discussed. By consulting relevant literature reports, we established drug-disease targets by using databases such as TCMSP, TCMIP, Genecards and OMIM. Using STRING database to construct the protein interaction network between *rosa roxbughii tratt* and cancer target; and Cytoscape software was used to construct the interactive network diagram of "active ingredient-target-disease". The function enrichment analysis of Gene Ontology (GO) and pathway enrichment analysis based on Kyoto Encyclopedia of genes and genomes (KEGG) were carried out by using biological information annotation database (David). We obtained 11 active components and 80 drug-disease targets were obtained, DAVID database was used to carry out GO function enrichment analysis and KEGG pathway enrichment analysis on 80 potential targets, and 303 biological processes and 114 signaling pathways were screened out to participate in the anti-cancer effect of *rosa roxbughii tratt*. The Pathway in cancer has the strongest correlation with cancer, which is related to the regulation of IL6, AR, CASP3 and other key genes. These key genes are related to myricetin, quercetin, kaempferol and other compounds in *rosa roxbughii tratt*. The anti-cancer compounds of *rosa roxbughii tratt* might were myricetin, quercetin, kaempferol and other compounds. The anti-cancer effect of *rosa roxbughii tratt* may regulation of key genes, indirectly affecting the sustained angiogenesis, developing apoptosis and promotion, so as to promote or inhibit the production of cancer.

## 1. Introduction

*Rosa roxbughii tratt* is a kind of perennial deciduous bush of Rosaceae, which is mainly distributed in southwest mountainous areas such as Yunnan, Guizhou, Sichuan, and also planting in Hubei, Guangdong and other places, now. It has the functions of relieving food, diarrhea and heatstroke. *Rosa roxbughii tratt* has rich nutritional value and broad prospects in the development of food, health products and medicines. Fresh *rosa roxbughii tratt* fruit is rich in vitamins, polyphenols, flavonoids, organic acids, polysaccharides, amino acids and trace elements [1]. Among them, the content of VC is the highest, ranking the first among many fruits, and a certain amount of VK1, VB2, VP and carotene [2]. Other chemical components are mainly flavonoids, and their aglycones are myricetin, quercetin and kaempferol their triterpenoids are pentacyclic triterpenoid ester glycoside, Euscaphic acid, *rosa roxbughii tratt* acid, *rosa roxbughii tratt* glycoside, etc[3]. These chemical components, such as myricetin, quercetin and kaempferol, have been reported to have anti-cancer activity.

Modern medical research showed that *rosa roxbughii tratt* has a variety of biological effects such as regulating immune function, detoxification, sedation, delaying senility, anti-atherosclerosis, and anti-cancer. In recent years, *rosa roxbughii tratt* has been shown to inhibit a variety of cancers, such as bladder [4], prostate [5]and liver cancer [6]. At present, CL1 from *rosa roxbughii tratt* can inhibit the growth of gastric cancer SGC-7901 cells in vitro, and does not significantly inhibit the proliferation and differentiation of hematopoietic stem/progenitor cells into granulocytes [7]. Although, many studies have shown that *rosa roxbughii tratt* plays an important role in anti-cancer, most of the studies are based on animal and in vitro cell experiments. However, the type and mechanism of cancer treatment are still unclear, which still needs further research and exploration.

Fortunately, the advent of network pharmacology provides a reference for the study of the mechanism of *rosa roxbughii tratt* in the treatment of cancer [8]. Due to the multi-component, multi-target and multi-level characteristics of traditional Chinese Medicine, Network pharmacology has the characteristics of wholeness, systematicness and focusing on drug interactions, which are consistent with the basic characteristics of traditional Chinese medicine [4]. Therefore, network pharmacology can be used to predict drug targets on the whole, to comprehensively and systematically study the law and mechanism of drug interaction with the body, and to provide theoretical basis for in depth experimental research on *rosa roxbughii tratt* and guidance of rational clinical drug use.

## 2. Materials and methods

### *2.1. Acquisition of active ingredients of rosa roxbughii tratt*

By referring to related literatures, the latest reported compounds of *rosa roxbughii tratt* were searched, and the active chemical components of *rosa roxbughii tratt* were screened through the screening of oral bioavailability (OB) $\geq$ 30% and drug-like property (DL) $\geq$0.18, or the compounds with the activity reported in literatures.

### *2.2. Prediction of rosa roxbughii tratt target and cancer target*

Using the function of TCMSP, TCMIP, BATMAN and other databases to predict the target of the effective components of *rosa roxbughii tratt*. In the human gene database Genecards and OMIM of Online Mendelian genetic platform, "cancer" is the key word to retrieve the related target genes of cancer, and map and compare with the target genes of active ingredients to screen the common target, which is the target of *rosa roxbughii tratt* active ingredients for cancer treatment.

### *2.3. Key target PPI network construction*

By searching the STRING database online, importing 80 common targets, defining species as "humans", and PPI network for cancer treatment of *rosa roxbughii tratt* was constructed. The results are exported in TSV format, and then the topological parameters of each target point in PPI network are obtained through the Network Analyzer plug-in of the software Cytoscape 3.6.1, such as Degree, Betweenness centrality and Closeness centrality, and the median of the three topological parameters was calculated. Selected the targets whose values of the above three topology parameters are greater than all median values, list them in the form of tables, and visualize the targets.

### *2.4. Go function enrichment analysis and KEGG pathway enrichment analysis*

The selected targets were analyzed by using David (https://david.ncifcrf.gov/) database for KEGG pathway analysis and Go (gene ontology) biological process analysis.

### *2.5. Labeling of KEGG signaling pathway*
Using the function of KEGG mapper in KEGG (https://www.genome.jp/kegg/) signal pathway database to mark the target points on the signal pathway closest to cancer, to verify that *rosa roxbughii tratt* plays an anti-cancer role through multiple targets and multiple channels.

## 3. Results

### *3.1. Screening of active ingredients*
By referencing to relevant literature, ADME parameters (OB≥ 30% and DL≥0.18) were used as criteria for the selection of compounds. Or refer to the existing literature and report that the compound has activity for screening, and a total of 11 active compounds are screened, the results are shown in Table 1.

**Table 1.** Active compounds and target numbers of *rosa roxbughii tratt* [9-10]

| Mol Id | Chemical compound | OB% | DL | Target number |
|---|---|---|---|---|
| MOL002008 | myricetin | 13.75 | 0.31 | 16 |
| MOL000098 | quercetin | 46.43 | 0.28 | 76 |
| MOL000422 | kaempferol | 41.88 | 0.24 | 35 |
| MOL001468 | MLT | 59.62 | 0.02 | 8 |
| MOL000131 | EIC | 41.9 | 0.14 | 6 |
| MOL005500 | linolenate | 45.01 | 0.15 | 2 |
| MOL001308 | oleic acid | 33.13 | 0.14 | 8 |
| MOL001002 | ellagic acid | 43.06 | 0.43 | 10 |
| MOL001641 | METHYL LINOLEATE | 41.93 | 0.17 | 2 |
| MOL002850 | butylated hydroxytoluene | 40.02 | 0.07 | 4 |
| MOL000635 | vanillin | 52 | 0.03 | 1 |

### *3.2. Prediction results of potential targets of rosa roxbughii tratt in the treatment of cancer*
In Genecards and OMIM databases, a total of 4862 cancer-related genes were retrieved with "Cancer" as the keyword, and Cancer genes were matched with *rosa roxbughii tratt* related targets and Venn diagram was drawn, as shown in figure 1.

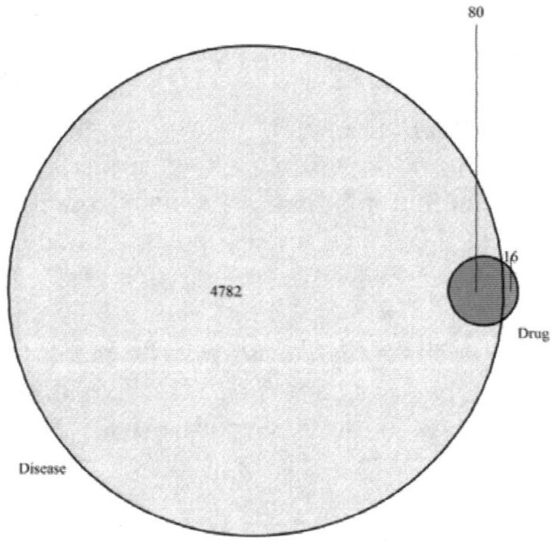


**Figure 1**. Match of cancer target gene and *rosa roxbughii tratt* target gene

### *3.3. PPI network analysis*

80 intersection targets were imported into the network diagram of the interaction relationship obtained in the STING database, the results were exported in TSV format, and the TSV file was imported into Cytoscape 3.6.1 software. The plug-in Network Analyzer is used to obtain the topological parameters of each target point, such as Degree, Betweenness centrality and Closeness centrality. The median of the three topological parameters is 15, 0.003825445 and 0.541523305, respectively. Degree and Betweenness centrality, Closeness centrality were greater than the median of targets， as shown in Table 2. According to the Degree value of the topological indicator, it is presented in terms of rank 1, 2-11, 12-31, and 32-80, as shown in Figure 2. It is suggested that these targets play an important role in PPI network, indicating that these targets play an important role in the treatment of cancer in *rosa roxbughii tratt*.

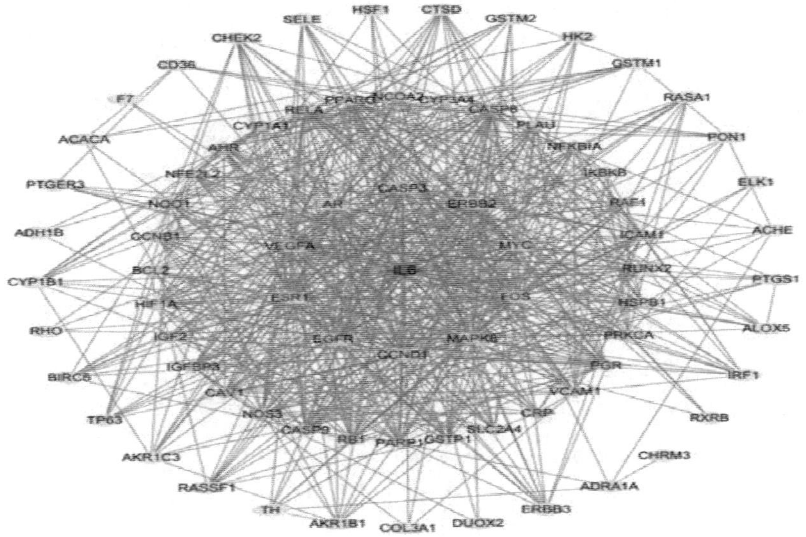


**Figure 2.** Gene association diagram (PPI network diagram)

**Table 2.** Analysis of gene topological parameters

| name | Degree | Betweenness Centrality | Closeness Centrality |
|---|---|---|---|
| IL6 | 52 | 0.1275 | 0.7500 |
| MAPK8 | 48 | 0.0794 | 0.7282 |
| CASP3 | 47 | 0.0554 | 0.7143 |
| VEGFA | 47 | 0.0605 | 0.7143 |
| EGFR | 44 | 0.0358 | 0.7009 |
| ESR1 | 42 | 0.0413 | 0.6881 |
| MYC | 42 | 0.0417 | 0.6881 |
| CCND1 | 38 | 0.0193 | 0.6579 |
| FOS | 36 | 0.0556 | 0.6522 |
| ERBB2 | 34 | 0.0173 | 0.6410 |
| AR | 33 | 0.0210 | 0.6356 |
| PPARG | 32 | 0.0455 | 0.6250 |
| RELA | 29 | 0.0100 | 0.6098 |
| CASP8 | 27 | 0.0123 | 0.5859 |
| PGR | 27 | 0.0108 | 0.6000 |

### *3.4. Enrichment analysis of Go biological function*

80 potential targets were mapped to the DAVID database for GO functional enrichment analysis, and 303 biological processes were obtained. The first 20 biological processes with P<0.01 and FDR< 0.05 were screened out, as shown in Table 3. It is shown as a bubble chart, and the result is shown in Figure 3. The results showed that the anti-cancer effect of *rosa roxbughii tratt* was related to the regulation of multiple biological processes, biological processes including positive regulation of transcription from RNA molecular ase II promoter, response to estradiol, negative regulation of apoptotic process, positive regulation of transcription, DNA-templated are involved. Among them, positive regulation of transcription from RNA polymerase II promoter is mainly a process of activating or increasing the frequency, rate or degree of transcription of RNA polymerase II promoter. Response to estradiol is any process by which the stimulation of estradiol (a C18 steroid hormone hydroxylated at C3 and C17 as an effective estrogen) results in a change in the state or activity of a cell or organism. These biological processes reflect the anti-cancer mechanism of *rosa roxbughii tratt*, which involves the abnormality of multiple biological processes in vivo, and also indicate that the active components of *rosa roxbughii tratt* may exert anti-cancer effect by regulating these biological processes.

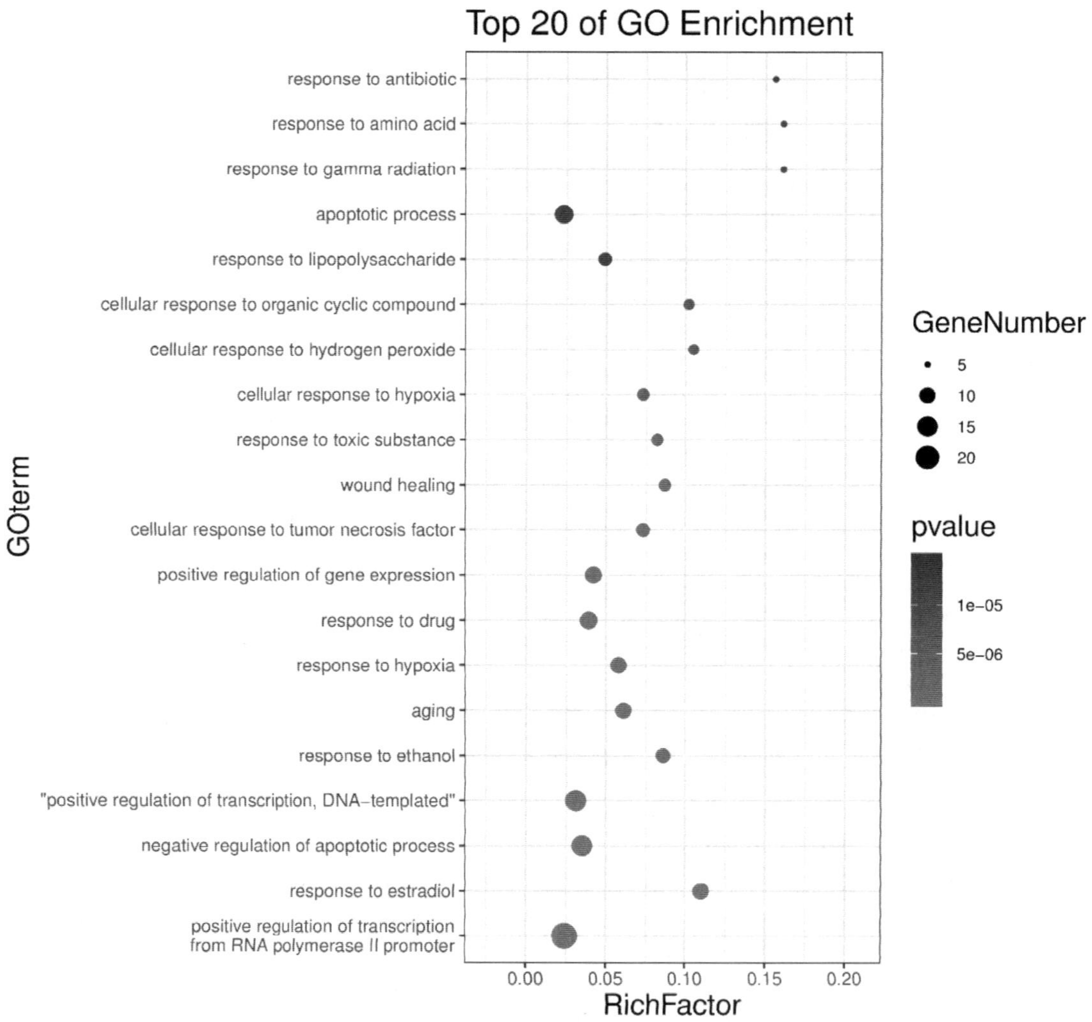


**Figure 3.** GO biological function enrichment analysis bubble chart

**Table 3.** GO biological function enrichment analysis results

| ID | Term | Cou nt | Count % | *P* Value | FDR |
|---|---|---|---|---|---|
| GO:00459 44 | positive regulation of transcription from RNA polymerase II promoter | 24 | 30 | 6.99296E- 11 | 1.16579E- 07 |
| GO:00323 55 | response to estradiol | 10 | 12.5 | 4.05895E- 10 | 6.76662E- 07 |
| GO:00430 66 | negative regulation of apoptotic process | 16 | 20 | 2.72842E- 09 | 4.54851E- 06 |
| GO:00458 93 | positive regulation of transcription, DNA-templated | 16 | 20 | 1.44413E- 08 | 2.40749E- 05 |
| GO:00454 71 | response to ethanol | 9 | 11.25 | 3.22447E- 08 | 5.37547E- 05 |
| GO:00075 68 | aging | 10 | 12.5 | 7.84009E- 08 | 0.0001307 01 |
| GO:00016 66 | response to hypoxia | 10 | 12.5 | 1.12009E- 07 | 0.0001867 29 |
| GO:00424 93 | response to drug | 12 | 15 | 1.72979E- 07 | 0.0002883 7 |
| GO:00106 28 | positive regulation of gene expression | 11 | 13.75 | 4.02959E- 07 | 0.0006717 66 |
| GO:00713 56 | cellular response to tumor necrosis factor | 8 | 10 | 8.39566E- 07 | 0.0013996 19 |
| GO:00420 60 | wound healing | 7 | 8.75 | 2.03722E- 06 | 0.0033961 64 |
| GO:00096 36 | response to toxic substance | 7 | 8.75 | 2.91027E- 06 | 0.0048515 64 |
| GO:00714 56 | cellular response to hypoxia | 7 | 8.75 | 5.92115E- 06 | 0.0098706 02 |
| GO:00703 01 | cellular response to hydrogen peroxide | 6 | 7.5 | 7.01059E- 06 | 0.0116866 17 |
| GO:00714 07 | cellular response to organic cyclic compound | 6 | 7.5 | 8.32105E- 06 | 0.0138709 97 |
| GO:00324 96 | response to lipopolysaccharide | 8 | 10 | 1.19777E- 05 | 0.0199660 42 |
| GO:00069 15 | apoptotic process | 13 | 16.25 | 1.27132E- 05 | 0.0211918 87 |
| GO:00103 32 | response to gamma radiation | 5 | 6.25 | 1.2962E- 05 | 0.0216065 29 |
| GO:00432 00 | response to amino acid | 5 | 6.25 | 1.2962E- 05 | 0.0216065 29 |
| GO:00466 77 | response to antibiotic | 5 | 6.25 | 1.47609E- 05 | 0.0246048 19 |

### *3.5. KEGG pathway analysis*

80 potential targets were mapped to the DAVID database and KEGG pathway enrichment analysis was performed. A total of 115 signal pathways were obtained. The first 20 signal pathways with $P<0.01$ and FDR$<$ 0.05 were screened out and displayed in a bubble chart, as shown in Figure 4. These pathways are closely related to the mechanism of anti-cancer of *rosa roxbughii tratt*, including Pathways in cancer,

Hepatitis B, Proteoglycans in cancer, Pancreatic cancer and other signaling pathways, as shown in Table 4. The corresponding diseases include liver cancer, pancreatic cancer, prostate cancer and bladder cancer. Using the KEGG mapper function in the KEGG signaling pathway database, 80 related target proteins were labeled on the most closely related signaling pathway with cancer. The results showed that 27 target proteins were involved in the regulation of pathways in cancer signaling pathway, as shown in Figure 5.

**Table 4.** Enrichment analysis results of KEGG pathway

| hsa ID | Term | Count | Count% | PValue | FDR |
|---|---|---|---|---|---|
| hsa05200 | Pathways in cancer | 27 | 33.75 | 1.62E-14 | 2.01E-11 |
| hsa05161 | Hepatitis B | 17 | 21.25 | 1.95E-12 | 2.42E-09 |
| hsa05205 | Proteoglycans in cancer | 15 | 18.75 | 2.30E-08 | 2.86E-05 |
| hsa05212 | Pancreatic cancer | 10 | 12.5 | 2.61E-08 | 3.23E-05 |
| hsa05215 | Prostate cancer | 11 | 13.75 | 2.93E-08 | 3.63E-05 |
| hsa05223 | Non-small cell lung cancer | 9 | 11.25 | 1.27E-07 | 1.58E-04 |
| hsa04668 | TNF signaling pathway | 11 | 13.75 | 1.93E-07 | 2.40E-04 |
| hsa05219 | Bladder cancer | 8 | 10 | 2.30E-07 | 2.85E-04 |
| hsa05210 | Colorectal cancer | 9 | 11.25 | 2.88E-07 | 3.57E-04 |
| hsa04066 | HIF-1 signaling pathway | 10 | 12.5 | 8.08E-07 | 0.001001554 |
| hsa05206 | MicroRNAs in cancer | 15 | 18.75 | 1.91E-06 | 0.00236837 |
| hsa05222 | Small cell lung cancer | 9 | 11.25 | 3.35E-06 | 0.00415496 |
| hsa04917 | Prolactin signaling pathway | 8 | 10 | 1.05E-05 | 0.013010193 |
| hsa05213 | Endometrial cancer | 7 | 8.75 | 1.88E-05 | 0.023262588 |
| hsa05145 | Toxoplasmosis | 9 | 11.25 | 2.29E-05 | 0.028444756 |
| hsa05134 | Legionellosis | 7 | 8.75 | 2.34E-05 | 0.029006305 |
| hsa04919 | Thyroid hormone signaling pathway | 9 | 11.25 | 3.17E-05 | 0.039312309 |
| hsa04012 | ErbB signaling pathway | 8 | 10 | 4.01E-05 | 0.049741114 |
| hsa04210 | Apoptosis | 7 | 8.75 | 5.20E-05 | 0.064433604 |
| hsa04510 | Focal adhesion | 11 | 13.75 | 7.07E-05 | 0.087548194 |

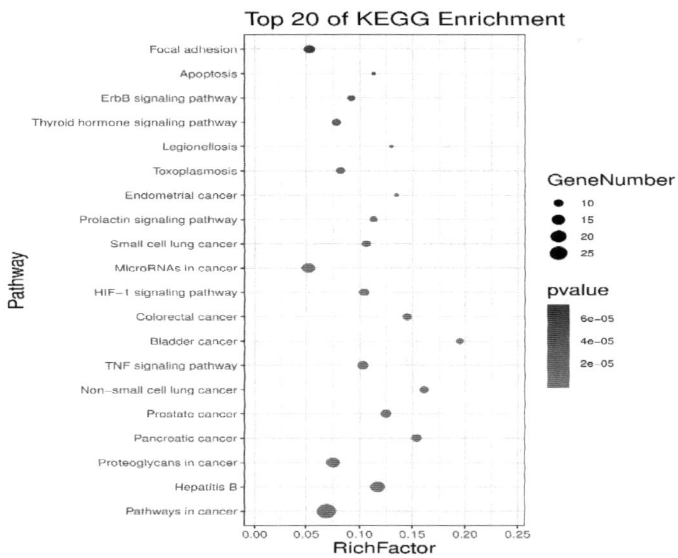


**Figure 4.** KEGG pathway enrichment analysis of bubble chart

IOP Conf. Series: Materials Science and Engineering **914** (2020) 012010 doi:10.1088/1757-899X/914/1/012010

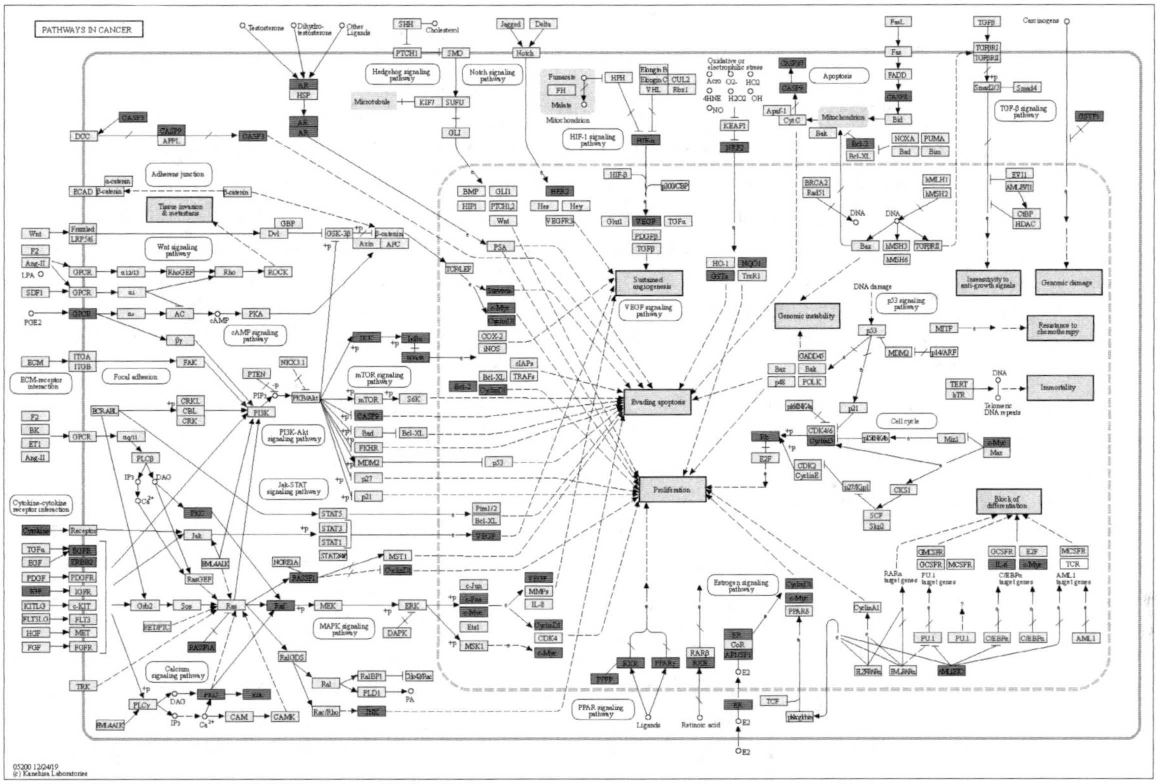


**Figure 5.** Labeling of potential targets on Pathways in cancer

By using the software of Cytoscape 3.6.1, the network visualization analysis of 80 common genes, diseases and chemical components were carried out, and the interaction network of anti-cancer of *rosa roxbughii tratt* was constructed. After selecting the corresponding interactive proteins and visualizing them with different colors and shapes, we can directly see the network relationship between the active chemical components and the target. 80 purple targets represent common targets, and the common targets in the purple inner circle are the top 20 with strong correlation. The red represent diseases, there are 11 triangles representing active ingredients of compounds, and the inverted triangles represent drugs. The larger the connectivity, the larger the shape, as shown in Figure 6.

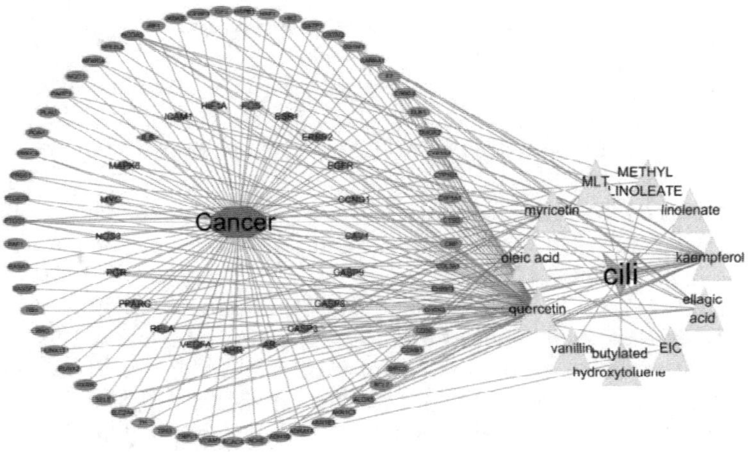


**Figure 6.** "component-target-disease" interaction network of *rosa roxbughii tratt* anti-cancer effect

## 4. Discussion

*Rosa roxbughii tratt* has a long history of medicine and food in China, as well as rich resources. At present, *rosa roxbughii tratt* has been developed into Chinese medicine preparation, such as compound Cili agent, in clinical treatment of gastric ulcer, insomnia and other diseases. It has been reported that *rosa roxbughii tratt* can inhibit or kill liver cancer cells, bladder cancer cells and pancreatic cancer cells in cell experiment or pharmacological experiment, but the mechanism of anti-cancer of *rosa roxbughii tratt* is still unclear. Through network pharmacology prediction, the main chemical components of *rosa roxbughii tratt* were myricetin, quercetin, kaempferol, etc., which all had anti-cancer effects. For example, myricetin can inhibit cancer by up-regulating the apoptosis pathway, and can also up-regulate the expression of proapoptotic gene BAX, so as to prompt mitochondria to produce cytochrome C and activate the cytochrome c-mediated apoptosis pathway [11]. Quercetin can significantly inhibit TGF-β1 induced EMT in HCC cells by increasing E-cadherin protein expression and decreasing the expression of N-cadherin and Vimentin proteins [12]. Kaempferol can inhibit SKOV-3 cell proliferation in vitro [13]. These anti-cancer Pathways are mainly related to Pathways in cancer, Hepatitis B, Proteoglycans in cancer and Pancreatic cancer, and their regulation is related to IL6, MAPK8, CASP3, VEGFA and EGFR.

In this study, IL6 gene in the key target screening, Degree, Closeness centrality and Betweenness centrality are all large, ranking the first in comprehensive ranking, which is one of the key targets of *rosa roxbughii tratt*. IL 6 protein is produced primarily at the site of acute and chronic inflammation, where it is secreted into the serum and induces transcriptional inflammatory responses through the interleukin-6 receptor A. It plays an important role in the final differentiation of B cells into immunoglobulin secreting cells, and is also involved in the differentiation of lymphocytes and monocytes, acting on T cells, liver cells, hematopoietic progenitor cells and cells of the central nervous system. According to KEGG map, Pathways in cancer are composed of VEGF signaling pathway, PPAR signaling pathway and P53 signaling pathway, etc. By activating or inhibiting IL6, MAPK8, CASP3 and VEGFA gene expression, and indirectly Sustained angiogenesis, Evading apoptosis and effort, promote or inhibit the occurrence of cancer. This article provides relevant evidence for the research of anti-cancer mechanism of *rosa roxbughii tratt*, but the specific mechanism of anti-cancer of *rosa roxbughii tratt* needs further research and verification.

**Acknowledgements**

This work was financially supported by the Guizhou Domestic First-Class Construction Project [(Chinese Materia Medica) (GNYL [2017] 008)]. The authors thank the government of China for their financial support.

**Reference:**

[1] C.N. Zhang, Y. Zhou, J.J. Wang. Progress in the studies of pharmacology of *rosa roxbughii tratt*[J]. J Med Postgrad, 2005,18(11):1049-1051.

[2] X.L. Zhang. Investingation on flavonoid from rose roxbughii tratt and its biological activity[D]. Shanghai:East China Normal University, 2005.

[3] G.Y. Liang. The isolation and structures of kaji-ichigoside and rosamultin[J]. Acta Botanica Sinica, 1988,30(4):409-413.

[4] Y.Q. Qing, X.Y. You, W. Zhao. Clinical observation on the preventive effect of "Rosa rose roxbughii tratt mixture" on bladder cancer and recurrence after operation[J]. J Labour Med, 1997,14(4):222-223.

[5] Z.Z. Chen, W.B. Cheng. Pharmacodynamic study and clinical application of Rosa roxburghii Tratt and its preparation[J]. Chin J Ethnomedicine Ethnopharmacy, 1994,03:20-23.

[6] J.E. Huang, J.Y. Jiang, Y. Luo, et al. Effect of Rosa roxburghii Tratt Triterpene on Proliferation of Human Hepatoma SMMC-7721 Cells[J]. Food Sci, 2013,13:275-279.

[7] Effects of Rosa roxburghii Tratt extract-CL1 on gastric carcinoma SGC-7901 cells growth and human umbilical cord CD34+ cells proliferation and differentiation[J]. Chin J Clini Pharma

Therapeut, 2006(07):829-832.
[8] J. Xie, S. Gao, L. Li, et al. Research progress and application strategy on network pharmacology in Chinese materia medica[J]. Chin Trad Herbal Drugs, 2019,50(10):2257-2265.
[9] T.T. Dai, X.S. Yang. Research progress in chemical constituents and pharmacological activities of Rosa roxburghii Tratt[J]. J Guiyang College of Trad Chin Med, 2015,37(04):93-97.
[10] Y.N. Shi, D.P. Wang, L. Ma, et al. Constituents and Antioxidant Activity of Fatty Acids in Seed Oil of Rosa roxburghii[J]. J Mountain Agric Biology, 2013,32(1):24-28.
[11] Boehning D, Patterson R L, Sedaghat L, et al.Cytochrome c binds to inositol (1, 4, 5) trisphosphate receptors, amplifying calciumdependent apoptosis[J].Nat Cell Biol, 2003, 5 (12) :1051-1061.
[12] Y, S, Zheng, W.D. Lei, M.L. Chen, et al. Effect of quercetin on the induction of EMT by TGF-β1 in SMMC-7721 human hepatoma cells[J]. J Toxicol, 2019,33(06):465-468.
[13] L.X. Yu, Y.H. Deng, B. Deng, et al. Effect of kaempferol on proliferation and apoptosis of ovarian cancer cells and its mechanism[J]. Chin J Hospital Pharma, 2019,20:2040-2043.

IOP Conf. Series: Materials Science and Engineering **914** (2020) 012011 doi:10.1088/1757-899X/914/1/012011

# Application and Research Based on Green Landscape Materials in Interior Design

**HuiJuan Li**

Wuhan software engineering Career Academy, School of Arts and media, HuBei 430000

disjill02@163.com

**Abstract.** Because indoor building materials are not environmentally friendly or the decoration methods are not reasonable enough, it poses certain harm to human health, and there are also problems such as waste of resources. With the proposal of the green economy and sustainable development strategy, the application of green building materials has gradually been valued. Nowadays, people pay more attention to environmental protection and health care. As a bamboo with low carbon and environmentally friendly materials, it is also the first choice for interior design materials. As an energy-saving and environmentally friendly material that can replace traditional building decoration materials, bamboo materials can not only improve the comprehensive utilization of resources and energy, but also reduce the pollution caused to the environment. Nowadays, bamboo fiber, bamboo charcoal and other health care materials have entered our lives, so the use of bamboo materials in interior design will have great potential for development. This paper analyzes bamboo functional materials as a case, and analyzes the advantages and disadvantages of bamboo artificial materials in modern interior design. From the perspective of design, it discusses how to exert the functional advantages of bamboo materials, and hopes to play the role of bamboo in the decoration industry in the future.

## 1. Introduction

The use of new high-quality energy-saving and environmentally-friendly materials in the interior design of construction projects not only can quickly improve the function of building buildings, but also improve the quality and comfort of the living environment. On the other hand, if materials with recycling characteristics are used in the interior design of construction projects, it can provide a platform for turning waste into treasure and recycling for various industrial solid wastes. In general, the promotion and application of various new energy-saving and environmentally-friendly materials in the interior design of construction projects can avoid the waste of energy and resources in essence, rapidly improve the utilization efficiency of energy and resources in China, and promote the construction field. Continue to develop the road[1].

The bamboo industry is an important part of the forestry industry. The development and utilization of bamboo resources and the production and trade of bamboo products have gradually increased the impact on forestry and local economic development, which has attracted more and more attention and recognition. Bamboo is widely distributed around the world and has a long history of use[2]. With the increasing environmental problems, the ecological concept is increasingly valued by people. Bamboo has re-entered people's vision with its good ecological adaptability. How to use this traditional material to rejuvenate it and become the current research. problem.

## 2. Green design requirements for decorative materials

### *2.1. The meaning of green design*

Green design is also known as "ecological design", "environmental design" or "environmental awareness design". Originated from the 1960s, an anti-consumer design movement in the United States led by environmental protection and sustainable development. The green design is essentially a design movement that protects the natural environment resources (as shown in Figure 1)[3] and prevents the pollution caused by industrial production from damaging the environmental balance. After the industrial revolution, people created an amazing material civilization, and it was followed that the living environment was also unimaginably destroyed[4]. Resources are heavily wasted and predatory, and the environment is heavily polluted. Therefore, the design thinking of many designers in the design world began to focus on environmental protection and ecological protection[5]. The green design reflects the reverse thinking of the contemporary civilization created by the human being on the natural environment and ecological destruction, and shows the rational thinking of the designers' professional ethics and social responsibility. At present, although the green design is still in its initial stage, it has pointed out an extremely important new trend for future development[6].

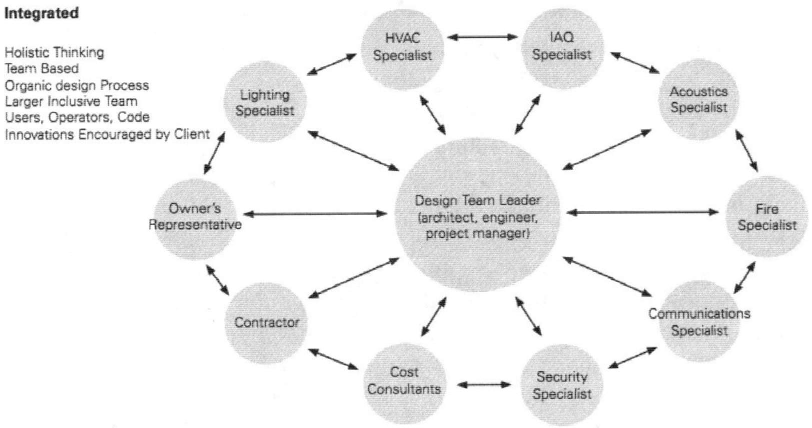


**Fig. 1** Green design process

### *2.2. Green design requirements for decorative materials*

*2.2.1. Non-toxic, harmless and non-polluting.* Green ecological design must adopt green ecological materials, that is, the materials used in the design can not cause any damage to human health and ecological environment, so that it is non-toxic, harmless, non-polluting, non-radioactive, no noise, and thus conducive to the ecological environment. Protection and human health. It is required that the decorative materials used in production will not produce or release toxic substances that can pollute the environment and destroy the ecological balance, and can reduce or effectively prevent adverse effects on the earth itself and the natural ecological environment, and even protect the environment

IOP Conf. Series: Materials Science and Engineering **914** (2020) 012011 doi:10.1088/1757-899X/914/1/012011

and the ecological balance[7]. It plays a certain positive sense and can build a beautiful living space for human beings that is warm, sweet, comfortable, safe and healthy.

*2.2.2. Recyclable.* In order to save more energy and protect the global environment, we must seek ways to reuse it. This includes land resources, water resources, mineral resources, and the recycling of various materials. In the initial stage of product design, the company fully considers the various parts of the product and the feasibility of material recovery, and the value of recycling. A series of conceivable problems related to the recovery and subsequent solution, such as the method of recycling and disposal, the process of recycling and processing, etc., to achieve the maximum effective use of parts and materials resources and energy, and to design a scheme with the lowest pollution environment. Ideas and approaches.

## 3. We take bamboo materials as an example to analyze the application of green materials in interior design

### *3.1. Characteristics of bamboo materials*

**(1) Natural features.** Bamboo is a one-time, one-time flowering plant that belongs to the family Poaceae. Bamboo has the characteristics of easy planting, easy harvesting, fast growth, short material production time and good output value. It is the fastest growing plant on the earth, and the slowest growing period can grow up to twenty centimeters per day and night. Thirty centimeters, the period of rapid growth can even grow from 150 to 200 centimeters per day and night.

**(2) Physical characteristics.** Bamboo has the unmatched performance of ordinary wood: the difference between the length difference after drying and the length before drying, with strong splitting, high-quality elasticity and toughness, and its good compressive strength and tensile strength. In addition, the absorption of moisture and moisture by bamboo and the absorption of external heat are far superior to wood, and it has the ability to resist the corrosive damage of surrounding media, the ability to avoid loss due to friction or use, and It is easy to produce material deformation ability, it is harder and denser than wood, and it has more advantages in compression resistance and flexural strength. Its surface texture can be elegant and elegant, and it can also adjust the moisture in the air spontaneously according to the space environment. It has high moisture resistance and high thermal conductivity, so it has excellent characteristics of warm winter and cool summer

**(3) Cultural characteristics.** Bamboo has been loved by people since ancient times in China's history. China has the history of the world's oldest bamboo research, cultivation and utilization. Reflecting on the profound significance of China’s long history and cultural development and the formation of spiritual civilization, it is not difficult to see the long history of cultural relations between bamboo and Chinese chess and painting and classical garden construction. 3) and the close connection between bamboo and people's daily lives. Therefore, China has been praised by the world as "a country with bamboo civilization." Among the civilizations passed down from generation to generation in China, bamboo symbolizes tenacious life, flexibility, health, harmony and spiritual guidance. Bamboo, delicate and elegant, graceful and graceful, delicate and beautiful, four seasons green, reflecting the youthful and young state; bamboo, natural and generous, upright and towering, beautiful and novel, unconventional, with ordinary and noble style; bamboo itself The structure is hollow inside, symbolizing humility, not boasting of your ability or value, and character can be self-control. The characteristics of bamboo are not bent, but they are constantly folded. The symbolic softness also contains the strong principle of doing things for others. The top-notch Lingyun, Lingshuang proud rain, and the shackles, the bamboo sticks are exposed, the bamboo branches are towering, and the metaphors and behaviors are high; The character is noble and not vulgar. The appearance of bamboo grows with bamboo knots, and it is also praised as a symbol of integrity.

**Fig. 2** Bamboo-themed paintings and paintings

*3.2. Application of Bamboo Materials in Interior Decoration*

**(1) Floor laying materials.** Bamboo flooring brings a new green and refreshing experience to the building materials market with its natural and natural characteristics and many good properties after processing into finished products. Bamboo flooring has the natural texture of bamboo itself, which is refreshing and elegant, giving users a different kind of natural, noble and elegant feeling. It has a lot of special features. Firstly, it replaces the traditional wooden floor with bamboo flooring. It not only has the original characteristics of wood, but also uses the quality of the national leading standard in the processing of bamboo flooring. Adhesive glue can effectively reduce the production of harmful substances such as formaldehyde and reduce the harm to the human body. In addition, the production equipment and technical means used in bamboo flooring are in the leading position in the world, and more than 20 processes are processed in the production process. The finished bamboo flooring obtained not only has the natural feeling of the log floor but also has the firmness and firmness of the ceramic floor tiles.

**(2) Wall materials.** The bamboo raw material is processed into bamboo pulp and then processed. Bamboo fiber, such as bamboo, bamboo, and bamboo, is used as the main raw material, and the bamboo fiber is processed and the shape and length are between wood and grass fiber. Easy to apply glue, bamboo pulp is a kind of pulp with suitable fiber length, fine and has good softness. The bamboo pulp is diastolic and thick, has high tear strength, and has low burst strength and tensile strength. Has a high mechanical strength. The surface of the wallpaper made of bamboo pulp is elastic, and has sound insulation, heat insulation, good heat preservation performance, and soft and comfortable hand feeling. The biggest feature of bamboo wallpaper is that it does not release poison gas, no harm, no special smell, good ventilation performance, and paper type stability. It can be scrubbed at any time after being contaminated, and the service life is more than twice that of ordinary wallpaper. Therefore, this new type of wallpaper is more and more popular in the market and consumers, and will become the "dominant series" of the decorative materials market in the near future.

**(3) Structure and decorative materials.** The screen is a form of furniture used to block the wind in ancient buildings. The so-called "screen is also windy." Screens are an important part of traditional furniture and have been around for a long time. The screen is an important piece of furniture and decorations in the ancient Chinese home interior. Its production form, surface pattern texture and text contain a lot of cultural information, which can not only express the elegant sentiment of the literati, but also contain people praying for happiness. The far-reaching significance. The various forms of screens also condense the wisdom and skill of the craftsmen to gather ideas. The screen is generally placed in an obvious position in the room, which serves as a function of separating space, beautifying the environment, shielding the wind and wind, and coordinating the environment. It echoes with traditional Chinese furniture, interacts with each other, and is inseparable. It becomes an indispensable part of home decoration, and presents a harmonious and beautiful beauty, quiet and peaceful beauty

IOP Conf. Series: Materials Science and Engineering **914** (2020) 012011 doi:10.1088/1757-899X/914/1/012011

The word "screen" gives people a feeling of poetry and painting, which creates a quiet space that seems to be divided into non-points and breaks.

**(4) Decorative materials.** Bamboo is used as a raw material for handicrafts, and then various paintings, sculptures, and other artistic themes are performed on the surface, and bamboo crafts are obtained. Compared with ordinary wood crafts, the bamboo craft has a much longer shelf life than wooden crafts. This is determined by the special processing techniques such as cooking and drying during the processing of bamboo. Also, due to the strong and unyielding characteristics of bamboo in its natural form, it has been praised by people who know the book and can be noble in character since ancient times (as shown in Figure 3), so with the art of wood. The comparison of bamboo art is more accepted by people and is more popular.

**Fig. 3** Bamboo jug and bamboo carving incense burner

## 4. Cases of bamboo in interior design

### *4.1. Shanghai World Expo*

The biggest "bamboo" highlight of the Expo Vanke Pavilion lies in the materials used on the ground. The floor materials used in the entire pavilion are made of high-resistance bamboo flooring. The high-resistance bamboo flooring is the heavy bamboo flooring or silk bamboo flooring that we often say. It has firmness and hardness. It has the advantages of high density, wear resistance and no insects. The bamboo floor of the Expo Vanke Pavilion uses an unusual arrangement of curved and radial equipment in the paving method. The distinctive materials and the designer's ingenious laying method make the entire venue visually visual. It has been one of the bright. In addition, the materials selected in the theme of the venue are the building materials made of wheat mustard that can be recycled and reused. The combination of two new green materials reflects a very strong environmental awareness.

### *4.2. Vietnamese Bamboo Wings*

The entire bamboo wing is an open space of twelve meters long. Its wing-shaped structure is not designed with a column support (as shown in Figure 4), creating a wide, unobstructed space. In this space, weddings, social gatherings, live concerts and many other events can be held. The reason why bamboo is selected as the main building material is because bamboo is a new type of ecological building material, which has low cost, high tensile strength and elasticity which cannot be compared with steel. The bamboo used after scientific and technical means can reach the service life. About thirty years. In addition, Vietnam is located in the south of the Tropic of Cancer. It has a tropical monsoon climate, with hot climate and abundant rain. The bamboo winged roof also makes the whole building extremely ventilated, which can effectively reduce the frequency of air conditioning and reduce the consumption of natural materials. Bamboo Wing is a successful building in terms of building bamboo as a building material. In addition, it also expressed another message to all countries with bamboo planting area, that is, bamboo will be used as a new type of ecological material in the construction field. It is gradually replacing traditional materials as a popular development direction.

**Fig. 4** Vietnamese bamboo wing

## 5. Conclusion

Due to the very close relationship between the use of building decoration materials and the ultimate energy-saving effect and the immediate interests of the occupants, the use of energy-saving and environmentally-friendly materials for building decoration has become a hot topic and has received extensive attention from consumers. Decorative materials occupy a large proportion in the environmental labeling products formulated by the China Environmental Labeling Products Certification Committee. Before carrying out the interior design work of architectural engineering, it is necessary to fully grasp the basic characteristics of various decorative materials and the energy saving and environmental protection effects. Under the premise of aesthetic safety requirements, the materials with the best energy-saving and environmental protection effects are selected to meet the potential environmental and health needs of residents.

## References

[1] Elgayar, A., Beheiry, S. M., Jabbar, A., & Ansari, H. A. A benchmarking approach to the progress of green materials and systems' use in the uae construction industry. World Journal of Science Technology & Sustainable Development,Vol. 9 (2016) No. 23, p.315-327.

[2] Paço, A. D., Ferreira, J. M., Raposo, M., Rodrigues, R. G., & Dinis, A. Entrepreneurial intentions: is education enough?. International Entrepreneurship & Management Journal,Vol. 7 (2016) No. 12, p.57-75.

[3] Celik, K., Meral, C., Gursel, A. P., Mehta, P. K., Horvath, A., & Monteiro, P. J. M. Mechanical properties, durability, and life-cycle assessment of self-consolidating concrete mixtures made with blended portland cements containing fly ash and limestone powder. Cement & Concrete Composites,Vol. 9 (2017) No. 17, p.59-72.

[4] Meraj, G., Romshoo, S. A., Yousuf, A. R., Altaf, S., & Altaf, F. Assessing the influence of watershed characteristics on the flood vulnerability of jhelum basin in kashmir himalaya. Natural Hazards,Vol. 3 (2016) No.23, p.153-175.

[5] Dubey, S. R., Singh, S. K., & Singh, R. K. A multi-channel based illumination compensation mechanism for brightness invariant image retrieval. Multimedia Tools & Applications,Vol. 1 (2017) No. 14, p.112-116.

[6] Liu, J., Heidarinejad, M., Gracik, S., Srebric, J., & Yu, N. An indirect validation of convective heat transfer coefficients (chtcs) for external building surfaces in an actual urban environment. Building Simulation,Vol. 5 (2017) No.16, p.337-352.

[7] Wei, Y., Xu, Y., Wang, W., & Kanopolous, P. Dependence of dynamic tensile strength of longyou sandstone on heat-treatment temperature and loading rate. Rock Mechanics & Rock Engineering,Vol. 12 (2016) No. 19, p.129-131.

IOP Conf. Series: Materials Science and Engineering **914** (2020) 012012 doi:10.1088/1757-899X/914/1/012012

# Heterologous Expression and Enzymatic Properties of Exocellulase CelA

**Ming Sui[1], Jiachao Rong[2], Yang Zhang[2], Peng Qu[3], Yusong Zhu[3]**

[1]Department of Wine and Food Engineering, Sichuan Technology & Business College, Dujiangyan, Sichuan Province, China
[2]Department of economics and management, Tongren Polytechnic College, Gui Zhou, Tongren, 554300, China
[3]Xinjiang Oilfield Company

**Abstract.** The gene of glycosyl hydrolase CeIA is derived from Bacillus licheniformis and the GenBank accession number is AAU40776.1. The sequence analysis of this gene was carried out by online software, and the celA gene was cloned from the genome of the strain by homologous amplification. The gene was ligated into the pMAL-c2x vector, transformed into E. coli Transetta for heterologous expression, and the catalytic properties of the recombinant enzyme CeIA were studied.

## 1. Introduction

In recent years, the intensification of the resource crisis has led to a sharp increase in the demand for bioenergy. Lignocellulose is the most abundant biomass resource, and its biotransformation research has attracted much attention. Cellulase, especially progressive exocellulase, is considered to be an important component of the biotransformase system and has an important influence on the rate of transformation. In this study, the GH48 family exocellulase CeIA derived from Bacillus licheniformis was cloned and heterologously expressed, and the enzymatic properties and related functional residues of the enzyme were studied. The CBM domain was also explored. Effect on the catalytic properties of CeIA.

## 2. Medium and buffer

(a). SOXTI}E buffer (50 times agarose gel running buffer): Weigh 242 g of Tris dissolved in 800 mL of deionized water, add 57.1 mL of acetic acid, stir and mix, and dilute to 1 L with deionized water. Adjust the pH of the buffer to 8.5 and store at room temperature.

(b). Agarose gel: Weigh agarose in a concentration of 0.8% to 1.5% (W/V), add it to 1*TAE buffer, dissolve it in a microwave oven, cool it to a certain temperature at room temperature, add the nucleic acid dye, mix and quickly pour Into the glue tank, solidified at room temperature.

(c). LB medium: 10 g of trypsin, Sg yeast extract and 10 g of NaCl were dissolved in 1 L of deionized water to adjust the pH to 7.0. 1.8% agar was added to the solid medium, and all liquid and solid media were sterilized in a sterilizer at 115 ° C for 30 min.

(d). Ampicillin antibiotics: Acetomycin was weighed at a concentration of 100 mg/mL and dissolved in sterile water. Dissolved in 1.5 mL sterile EP tubes and stored at -20 °C.

(e). Isopropylthio-galactose (IPTG): IPTG was weighed in sterile water at a concentration of 1 mol/L, filtered and sterilized, and dispensed in 1.5 mL of sterile EP tube and stored at -20 °C.

**Table 1.** Formulation of citric acid-sodium citrate buffer

| Buffer pH | Citric acid (mL) | Sodium citrate (mL) |
|---|---|---|
| 3.0 | 18.6 | 1.4 |
| 3.0 | 13.1 | 6.9 |
| 5.0 | 8.2 | 11.8 |
| 6.0 | 3.8 | 16.2 |

**Table 2.** Formulation of Tris-HCL Buffer

| Buffer pH | 0.1 mol/L Tris(mL) | 0.1 mol/L HCL(mL) | H2O (mL) |
|---|---|---|---|
| 8.0 | 50 | 29.2 | 21.8 |
| 9.0 | 50 | 7.0 | 43.0 |

## 3. Sequence analysis and homology modeling

The gene sequence of exo-glucanase CeIA was obtained from NCBI database, and the amino acid sequence was analyzed by NCBI BLAST program. The domain of the enzyme and its family were predicted. The online software Prot Param was used to predict CeIA. The physicochemical properties and amino acid composition of the enzyme were analyzed; the signal peptide sequence of the enzyme was predicted using the online software SignalP-5.0 Server; the transmembrane region of the recombinant protein was predicted using the online software TMHMM Server v.2.0. Homology modeling was performed using SWISS-MODEL online protein structure prediction software.

The Bacillus lichen rmisATCC14580 strain was inoculated into Sml LB liquid medium, shaken overnight at 37 ° C, the cells were collected by centrifugation, and the whole genome DNA of the strain was extracted using the bacterial whole genome DNA extraction kit, and stored in a refrigerator at -80 ° C. .

PCR amplification was carried out using the strain Bacillus lichen such as rmisATCC14580 whole genome DNA as template. The PCR reaction conditions are shown in Table 3:

**Table 3.** Reaction conditions

| | | |
|---|---|---|
| 94℃ | 5 min | 1 time |
| 94℃ | 30 s | 30 times |
| 56℃ | 30 s | 30 times |
| 72℃ | 2 min 30 s | 30 times |
| 72℃ | 10 min | 1 time |

The PCR product was detected by agarose gel electrophoresis, and the target band was recovered using a gel recovery kit.

## 4. Expression and purification of recombinant enzyme

The sequence-verified recombinant plasmid was transformed into the expression host E. cohTransetta (DE3) for heterologous expression according to the method of 2.3.2.4.

(a). Randomly pick a single colony to prepare the seed solution, inoculate the seed solution in a 1 / 40 ratio into an Erlenmeyer flask containing 200 ml of LB liquid medium, shake culture at 37 ° C, 190 rpm, and add IPTG at a final concentration of 1 mM. The culture temperature was adjusted to 30 ° C, the shaking speed was adjusted to about 165 rpm, and the culture was continued for about 15 hours.

(b). The fermentation broth obtained in the previous step was centrifuged at 4 ° C in a 10,000 rpm sub-supersonic cryogenic centrifuge for 20 min. The supernatant was decanted, and the precipitate was washed by adding Tris-HCL buffer, and then centrifuged to collect a precipitate.

(c). 8 ml of pMAL column buffer was added to the precipitate, suspended and mixed, placed in an ice bath, and the cells were disrupted by an ultrasonic cell pulverizer.

(d). The liquid to be crushed was cleared, and it was dispensed into a 2 mL EP tube, and centrifuged at 4 ° C, 10000 rpm high-speed low-temperature centrifuge for 20 min, and the supernatant and the precipitate were separated and placed in a refrigerator at 4 ° C for use.

Purification by amylose resin affinity chromatography: 2 mL of Amylose Resin injection column stored in 20% ethanol, and washed with sterile water to add pMAL Column Buffe: balance column material; Add the crushed supernatant, make the enzyme solution fully contact with Amylose Resin, mix at low temperature for 2h, and gently flip it during the period; fix the column, let Amylose Resin settle to the bottom naturally, install the sieve plate and add the elution buffer sequentially. Elute and collect. Each part of the eluate was stored at -20 ° C for use.

## 5. Determination of protein concentration

The BCA working kit was used to determine the protein concentration. The bovine serum albumin (BSA) was used as a standard. The 20 ul system was added with 200 ul of BCA working solution, placed at 37 ° C for 35 min, and the absorbance at 570 nm was measured by a microplate reader to map the protein. standard curve line. After the sample to be tested is diluted to a suitable concentration, the same operation is performed, and the protein concentration of the sample is calculated from the standard curve and the dilution.

Take RAC, methyl cellulose, cotton wool, microcrystalline cellulose, corn stover, cellulose powder CF11 as substrates, add appropriate amount of enzyme, react at 50 ° C for 30 min, use DNS method to measure the yield of reducing sugar, calculate reorganization The enzymatic activity of the enzyme on different substrates to characterize the substrate specificity of the enzyme.

The pure enzyme components were separately added to each pH buffer and allowed to stand at 4 ° C for 24 h. The enzymatic solution and the RAC substrate were thoroughly mixed under optimal conditions for enzymatic reaction, and the cellulase activity was determined according to the DNS method. The highest enzyme activity was 100, and the relationship between the relative enzyme activity and the incubation pH was plotted. .

Add methanol, ethanol, isopropanol, glycerol, Triton-X 100 at a final concentration of 1% and 10%, respectively, to the enzyme reaction system, and add EDTA, PMSF, and urea at a final concentration of 1 mM and 10 mM, respectively. After the reaction under the reaction conditions, the cellulase activity was measured according to the DNS method, and the relative enzyme activity of the enzyme in the reaction system was calculated by using the enzyme activity of 100% without any reagent.

The maltose-removing recombinase is mixed with 5 mg/mL of cellotriose, cellotetraose and celloose in a certain ratio, and reacted under the optimum reaction conditions for a period of time, and the boiling water bath is used for 10 min to terminate the reaction and centrifuge. The supernatant was taken and the product analysis was carried out by the TLC method.

## 6. Sequence analysis and homology modeling

According to the NCBI database, the sequence of the CeIA gene derived from Bacillus lichen rmis ATCC14580 is 2115 bp long and encodes a protein sequence of 704 amino acids. The protein has only one functional domain belonging to the GH48 family. The amino acid sequence of CeIA was analyzed by the online software ProParam, and the molecular weight of the protein was predicted to be 79463.69, and the theoretical pI was 4.92. This indicates that the protein is structurally stable. The half-life in E. coli is estimated to be >10 hours and greater than 20 hours in yeast. The total number of negatively charged residues of the amino acid of the enzyme was 82, and the total number of positively charged residues was 51. The aliphatic index was 57.19 and the average hydrophilicity was -0.670. The enzyme signal surface was predicted by the online software SignalP 4.1 Server. The predicted result is shown in Figure 1. CeIA has no obvious signal peptide sequence and is a non-secreted protein.

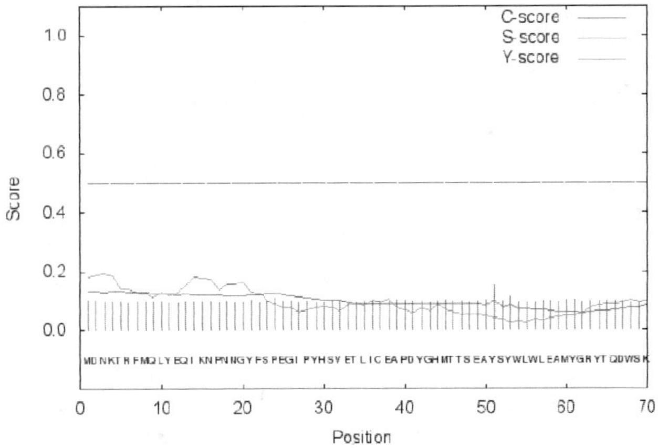


**Figure 1.** Signal peptide prediction of CeIA

The transmembrane region of the enzyme was analyzed using the online software TMHMMServer v. 2.0. The predicted results are shown in Figure 2. CeIA is an extramembrane protein and there is no transmembrane region.

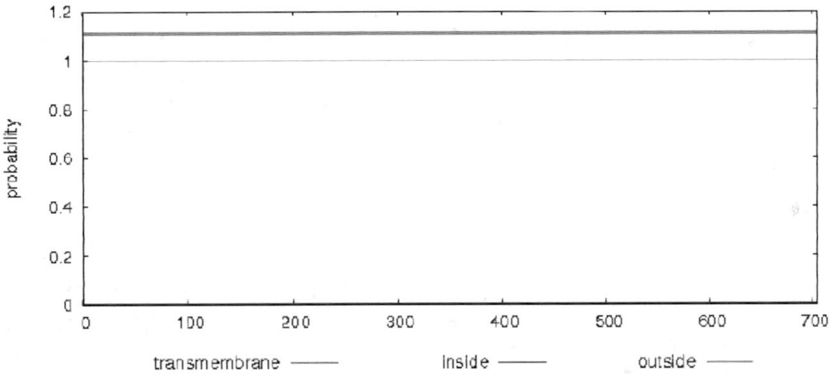


**Figure 2.** CeIA transmembrane channel protein prediction results

The structure determines the function, and the three-dimensional structure of the protein helps to understand the function of the protein. With the development of bioinformatics technology and the analysis of more and more protein structures, one can use the 3D model of protein homology with high homology as a template to model the structure of unknown enzyme molecules to understand its structure. feature. This paper uses the online software SWISS-MODEL to model the homology of CeIA. The template is the three-dimensional structure of GH48 derived from Bacillus pumilus. The sequence similarity is 73.43%, the GMQE score is 0.94, and the QMEAN score is 0.23. The structure of CeIA is shown in Figure 3. It has the same barrel folding as other enzymes of the GH48 family, which is a typical structural feature of the GH48 family of enzymes. At the same time, the molded structure also showed that there is a long closed substrate at the N-terminus of the enzyme molecule in combination with the tunnel, and the tunnel is followed by an open crack, which may be the outlet of the hydrolyzate.

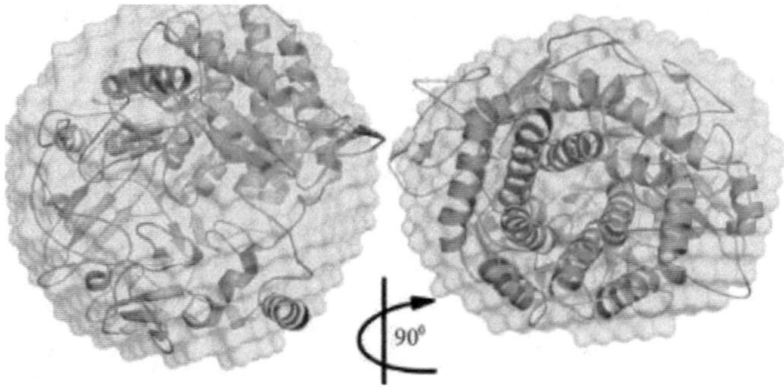


**Figure 3.** CelA model

## 7. Conclusion

In this paper, the exo-cellulase derived from Bacillus lichenniformisATCC14580 was cloned and heterologously expressed, and the properties of the recombinant enzyme were studied. The results obtained were: CelA enzyme gene was 2115 bp, encoding one containing 704 amino acids. A sequence of proteins that have only one functional domain that is part of the GH48 family. Homology modeling shows that CelA has barrel-like folding like other enzymes of the GH48 family, which is a typical structural feature of the GH48 family of enzymes. The celA gene was cloned by homologous amplification and ligated into pMAL-c2X vector, heterologously expressed in the host, and the expressed product was isolated and purified to obtain a recombinant protein with exocellulase activity.

## References

[1] Das KC, Haque N, Baruah KK, et al. Comparative nutrient utilization, growth, and rumen enzyme profile of mithun (Bas frontalis) and Tho-tho cattle (Bos indicus) fed on tree-leaves-based ration [J] ]. Tropical animal health and production, 2016, 43 (1): 209-214.

[2] Ma G, Chang H, Li S, et al. Phylogenetic relationships and status quo of colonies for gayal based on analysis of cytochrome 6 gene partial sequences[J]. Journal of genetics and genomics, 2007, 34 (5): 413-419.

[3] Ye Daiyong, Huang Hong, Fu Heqing. Progress in cellulose chemistry research [J]. Journal of Chemical Industry and Engineering (China),2016, (8):1782-1791.

[4] Voon L K, Pang S C, Chin S F. Regeneration of cello-oligomers via oelective oepolymerization of cellulose fibers derived from printed paper pastes [J]. Carbohydrate polymers, 2016, 142:31-37.

[5] Ye D, Farriol X. Improving accessibility and reactivity of celluloses of annual plants for the synthesis of methylcellulose [J]. Cellulose, 2015, 12 (5): 507-515.

[6] Zugenmaier P. Conformation and packing of various crystalline cellulose fibers [J]. Progress in polymer science, 2011, 26 (9): 1341-1417.

IOP Conf. Series: Materials Science and Engineering **914** (2020) 012013 doi:10.1088/1757-899X/914/1/012013

# Study on the Variation of Microbial Diversity during the Fermentation of Potato Liquor Based on Biolog ECO Technology

**Liuwei Yang[a], Qiao Lin[*], Yuxin Yang[b], Shudong Zhao[c], Yanling Ren[d], Yao Chen[e] and Bangzhou Luo[f]**

School of Xichang University, Sichuan, China

*Corresponding author e-mail: 66656736@qq.com, [a]1606285355@qq.com, [b]982488891@qq.com, [c]1104419850@qq.com, [d]1010322800, [e]719739236@qq.com, [f]375031607@qq.com

**Abstract**. There are many kinds of microorganisms in the fermentation process of potato liquor. Using the technology of Biolog Eco, the changes of microbial diversity in different fermentation periods were studied. The results showed that the AWCD value of microorganism in the fermentation process of potato liquor was increasing, and the AWCD value of the sample on the sixth day of fermentation was higher than that of other fermentation time samples. The results showed that the values of dominance index and diversity index of the samples on the sixth day of fermentation were higher than those of the samples in other fermentation periods, but the overall difference was not significant; the value of richness index of the samples before loading was the highest, and the value of richness index decreased after the start of fermentation, and the difference in the later stage of fermentation was small. The utilization of various carbon sources by microorganisms showed that in the process of potato liquor fermentation, microorganisms mainly used carbohydrate, polymer and amino acid carbon sources, but less used carboxylic acid, amine and phenolic acid carbon sources. The results of principal component analysis of microbial community showed that the highest carbon source was D-xylose / pentanose, and the lowest carbon source was D, l-α-Glycerin phosphate.

## 1. Introduction

Potato is a kind of potato plant, its tuber contains a lot of starch, and has a series of unique properties. There is a long history of using potato to brew liquor in foreign countries. Using potato brewed liquor, in its fermentation process involves many physical and chemical changes, these changes directly affect the product quality and safety, accurate understanding of the changes in microbial fermentation process is particularly important for the formation of good quality and good flavor products [1], the core and the potato liquor fermentation is the role of microorganisms in the fermentation process, therefore in the process of potato liquor fermentation the change rule of microbial diversity research, is of great significance for optimizing the potato liquor fermentation process [2]. With the continuous development of microbiology and neighborhood, at present a lot of methods to explore microbial diversity, because of Biolog ECO technology simple operation, good sensitivity, wide detection range, according to microbial utilization of different carbon source, the changes of microbial community metabolism and function display features [3], has become an important tool to study the microbial functional diversity, and is widely applied in microbiology, agriculture, food science, medicine and other fields.

Therefore, this paper takes the microorganisms in the process of potato liquor fermentation as the research object, studies the microbial flora dynamics in the process of potato liquor fermentation, objectively and accurately grasps the change law of microorganisms in the process of potato liquor fermentation, and lays a theoretical basis for further analysis of the change law of microbial diversity in the process of potato liquor fermentation.

## 2. Materials and Methods

### *2.1. Sample collection*
According to the standard of DB 34 / T 2264-2014, the fermented grains in the process of potato liquor fermentation are sampled with sterile sampling bags at regular intervals every day. After sampling, the samples are stored in a refrigerator at -20°C for standby.

### *2.2. Main instruments and equipment*
EnSpire automatic enzyme labeling instrument (Perkin Elmer Enterprise Management Co., Ltd.), WE-2 dual function water bath thermostatic oscillator (Beijing Haitian Youcheng Technology Co., Ltd.), LRH-250F biochemical incubator (Wuxi marrett Technology Co., Ltd.), YXQ-LS-18SI portable pressure steam sterilization pot (medical equipment factory of Shanghai bosun Industry Co., Ltd.), BIOLOG ECO board (American Biolog company).

### *2.3. Determination method*

*2.3.1. Weigh 10.0g fermented grains accurately, pour them into conical flask, add 90ml (0.85% NaCl, w/V) of sterile physiological saline, and seal them with sterile cotton plug.*

*2.3.2. Put the conical flask with fermented grains into the water bath constant temperature oscillator, shake and mix for 30 minutes, take out and stand for 15 minutes, use the pipette gun to transfer 10ml of supernatant on the sterile operation platform to another empty conical flask, and then add 90ml of sterile physiological saline.*

*2.3.3. Dilute the fermented grains to 10-3g/ml according to the operation method of 2.3.2.*

*2.3.4. Suck 150μl of 10-3g/ml fermented grains suspension on the sterile operation platform with a pipette gun, and inoculate it into each micropore of the Biolog ECO plate in turn.*

*2.3.5. Cover the inoculated Biolog ECO plate and put it into a biochemical incubator at 25°C for 6 days.*

*2.3.6. The absorbance value of each micropore was measured at 590nm and 750nm every 24h with the full-automatic enzyme labeling instrument, and the data was recorded.*

### *2.4. Data processing*
The difference between the OD value of 590nm and that of 750nm represents the metabolic activity of microorganisms. When the difference is less than 0.06, the correction result is 0 [4]. The average activity of microorganisms was evaluated by AWCD of Biolog ECO plate [5], and the calculation formula was as follows:

$$AWCD=\Sigma(Ci-R)/n \quad (1)$$

In the formula, Ci and R are the difference between the reaction pores and the control pores $OD_{590}$ and $OD_{750}$ of the eco plate respectively. If the Ci-R value is less than zero, then zero treatment is used for calculation; n is the type of carbon source [6].

The richness index Shannon (H), dominance index Simpson (D), diversity index Mclntosh (U) are used to analyze the functional diversity of microbial community [7], and the calculation is based on the data of 96h culture [8], and the formula is shown in Table 1.

**Table 1.** Calculation formula of microbial community diversity index.

| Diversity index | purpose | formula | Remarks |
|---|---|---|---|
| Shannon index | Assess richness and average | $H=-\sum pi\ln pi$ | pi is the ratio of the relative absorptivity (C-R) of i holes to the relative absorption value of the whole plate |
| Simposn index | Evaluate the dominance index of some of the most common species | $D=\sum\frac{(ni(ni-1))}{(N(N-1))}$ | ni is the relative absorption value (C-R) of the i hole; N is the sum of the relative absorption values; Simpson index is expressed by 1 / D value |
| Mclntosh index | Evaluation of diversity index based on eucidian distance in multi-dimensional space of community species | $U=\sqrt{(\sum ni^2)}$ | ni is the relative absorption value (C-R) of the i pore |

SPSS statistics 23 software was used for data statistical analysis.

## 3. Results and Analysis

### *3.1. Analysis of microbial metabolic activity in potato liquor fermentation*

The change of AWCD value of microorganisms in fermented grains at different fermentation time is shown in Figure 1. It can be seen from Figure 1 that the AWCD value of each sample is on the whole increasing trend, and the AWCD value of each sample increases the fastest in the 48h-96h range; the AWCD value of the sample on the sixth day of fermentation is significantly higher than that of other fermentation time. It shows that during the culture period of samples, all kinds of microorganisms use microplate for rapid growth and propagation, and the activity of microorganisms is the most frequent in 48h-96h, and the overall metabolic activity of microorganisms in the sixth day of fermentation is higher than that in other fermentation times [9].

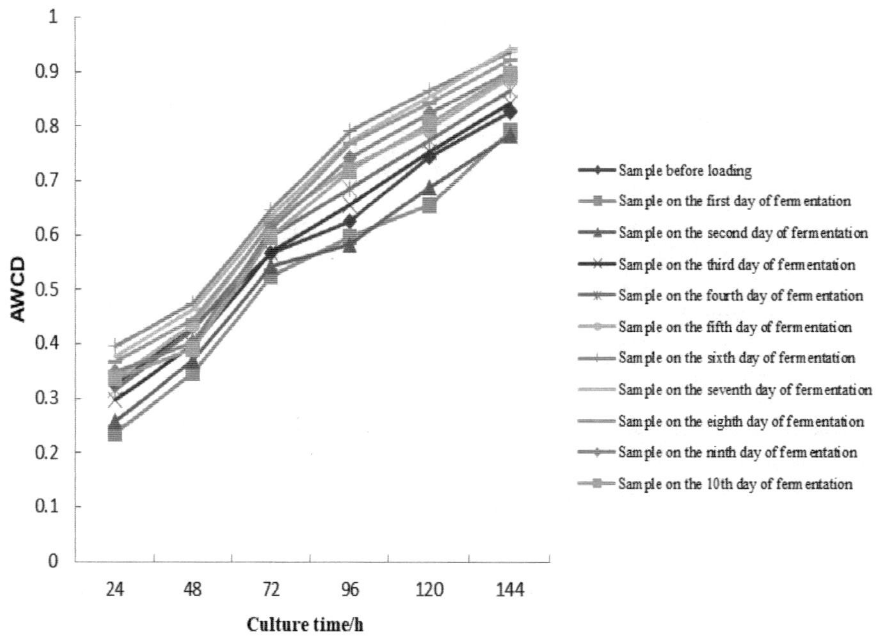


**Figure 1.** AWCD value curve of different samples.

It can be seen from Figure 1 that the total difference of AWCD value among different samples cultured for 96h is the largest, so we use the data of cultured for 96h to explore the overall characteristics [10], the results are shown in Figure 2, the AWCD value of samples fermented for 0-2 days shows a downward trend, the AWCD value starts to increase after the second day, the AWCD value reaches the maximum value on the sixth day, and the AWCD value starts to slowly decline after the sixth day. In the process of potato liquor fermentation, within 2 days after the start of fermentation, the microorganisms that do not adapt to the fermentation environment begin to die or the environment has inhibited the growth of microorganisms [11]. After 2 days, the dominant population begins to grow and propagate rapidly and carry out various life activities. On the sixth day, the microorganisms are most active. After 6 days of fermentation, the increase of acidity and alcohol concentration in the fermentation tank inhibited the growth and reproduction of microorganisms, and the overall metabolic activity of microorganisms began to decline [12].

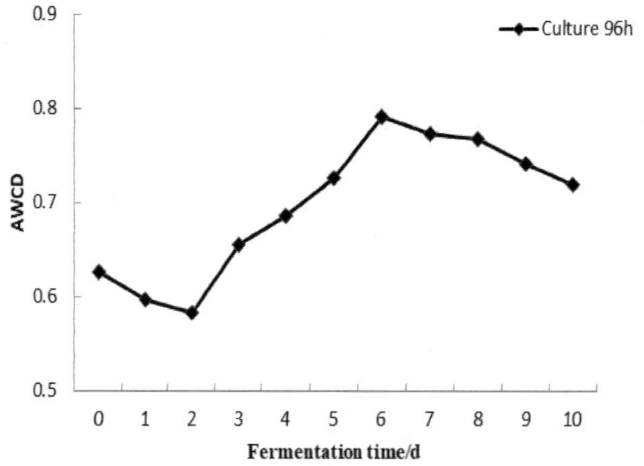


**Figure 2.** AWCD curve of each sample cultured for 96h.

*3.2. Utilization of various carbon sources by microorganisms in potato liquor fermentation*

There are 31 different carbon sources in the Biolog ECO panel, which can be divided into 6 categories: amino acids, polymers, sugars, phenolic acids, carboxylic acids and amines. The utilization of carbon sources by different microorganisms will be different. The AWCD value of each type of carbon source will be calculated in the experiment, and the experimental data will be drawn as shown in Figure 3 below. It can be seen from Figure 3 that during the fermentation process of potato liquor, microorganisms are relatively beneficial the largest utilization rate is carbohydrate carbon source, followed by polymer carbon source; the relative utilization rate of carboxylic acid carbon source increases gradually, amino acid carbon source decreases gradually, and the relative utilization rate of amine and phenolic acid carbon source is less, and the daily relative utilization rate change is not significant [13].

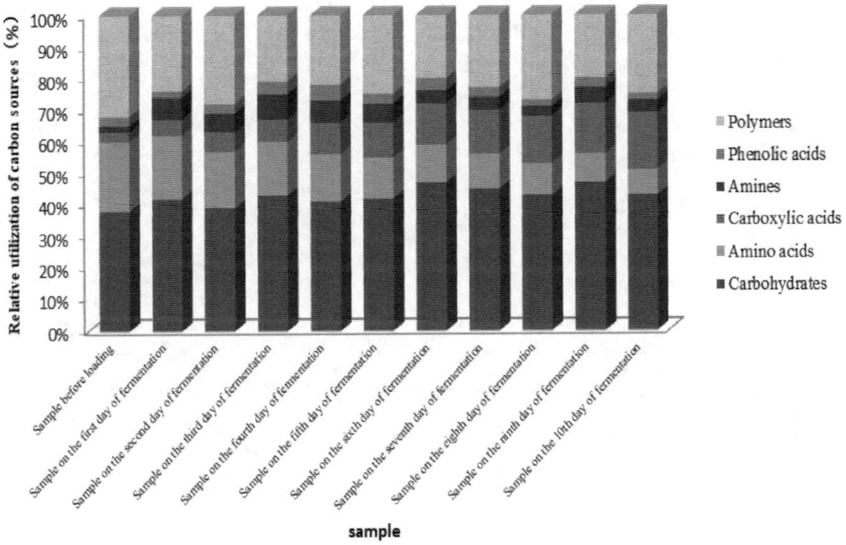


**Figure 3.** Distribution of utilization ratio of samples to different carbon sources.

According to the specific situation of microorganism using different kinds of carbon sources, carry out comparative analysis, select the best three kinds of carbon sources used in each fermentation stage, and the results are shown in Table 2.

**Table 2.** Best carbon source of microbial community of each sample.

| sample | carbon source |
|---|---|
| Before the altar | Glycogen, D-xylose / pentanose, L-arginine |
| Fermentation day 1 | D-xylose / pentanose, Glycogen, D-glucosamine |
| Fermentation day 2 | Glycogen, β-methyl-D-glucoside, D-xylose / pentanose |
| Fermentation day 3 | D-xylose / pentanose, L-phenylalanine, putrescine |
| Fermentation day 4 | D-xylose / pentanose, Glycyl-L-glutamate, Glycogen |
| Fermentation day 5 | D-xylose / pentanose, L-asparagine, Tween 80 |
| Fermentation day 6 | N-acetyl-D-glucosamine, D-xylose / pentanose, 2-hydroxybenzoic acid |
| Fermentation day 7 | D-xylose / pentanose, Itaconic acid, N-acetyl-D-glucosamine |
| Fermentation day 8 | D-xylose / pentanose, N-acetyl-D-glucosamine, Methyl pyruvate |
| Fermentation day 9 | D-xylose / pentanose, L-asparagine, Glycogen |
| Fermentation day 10 | D-xylose / pentanose, D-galacturonic acid, α-butanone acid |

IOP Conf. Series: Materials Science and Engineering **914** (2020) 012013 doi:10.1088/1757-899X/914/1/012013

It can be seen from table 2 that D-xylose / pentanose of sugar is the most widely used carbon source in the whole fermentation process of potato liquor, followed by liver sugar of polymer, and N-acetyl-D-glucosamine of sugar.

### *3.3. Analysis of microbial diversity in potato liquor fermentation*

According to the calculation method of Yang Yonghua et al. [14], calculate the dominance index Simpson (D), richness index Shannon (H), diversity index Mclntosh (U) of each sample, and the calculation results are shown in Table 3.

**Table 3.** Analysis of microbial community diversity index of each sample.

| Sample | Simpson (D) | Shannon (H) | Mclntosh (U) |
|---|---|---|---|
| Before the altar | 0.9332 | 3.2325 | 0.9636 |
| Fermentation day 1 | 0.9235 | 3.0144 | 0.9742 |
| Fermentation day 2 | 0.9225 | 2.8586 | 0.9635 |
| Fermentation day 3 | 0.9375 | 2.8385 | 0.9785 |
| Fermentation day 4 | 0.9459 | 2.8432 | 0.9743 |
| Fermentation day 5 | 0.9632 | 2.8687 | 0.9815 |
| Fermentation day 6 | 0.9875 | 2.9623 | 0.9963 |
| Fermentation day 7 | 0.9844 | 2.8897 | 0.9887 |
| Fermentation day 8 | 0.9765 | 2.8541 | 0.9756 |
| Fermentation day 9 | 0.9631 | 2.8456 | 0.9653 |
| Fermentation day 10 | 0.9653 | 2.8494 | 0.9765 |

It can be seen from table 3 that the difference between the dominance index value and diversity index value of each sample is small, and the dominance index value and diversity index value of the sample on the sixth day of fermentation are the largest compared with other fermentation time samples; the comparison between the sample before loading and other sample data shows that the richness index value is the highest, followed by the richness index of the sample on the first day of fermentation, and the richness index of other samples the difference is not obvious.

### *3.4. Principal component analysis of functional diversity of microbial community in potato liquor fermentation*

Based on the data of 96h culture, standardized transformation was carried out, and the utilization of different carbon sources by microorganisms in the fermentation process of potato liquor was further described by principal component analysis. It can be seen from Figure 4 that the characteristics of microbial community metabolism can be directly reflected after dimensionality reduction treatment. The contribution rate of principal component 1 (PCA1) is 69.119%, and that of principal component 2 (PCA2) is 20.945%. According to the distribution characteristics in the figure, six types of carbon sources can be divided into three parts: the first part is amino acids, sugars, carboxylic acids, amines, the second part is polymers, and the third part is phenolic acids Thus, we can know the overall distribution of carbon sources used in the fermentation process of potato liquor.

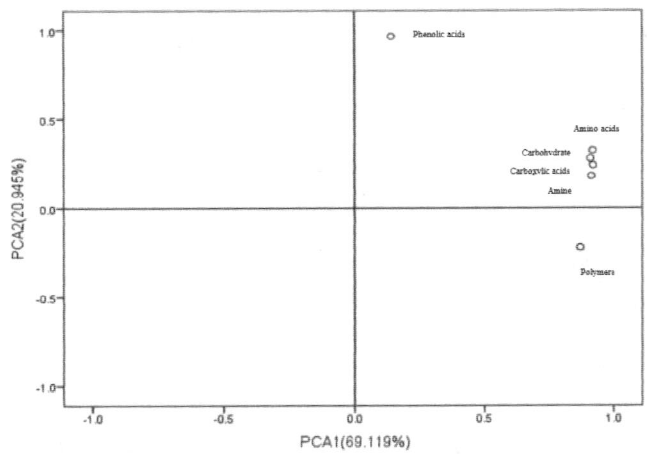


**Figure 4.** The first and the second main composition and dispersion point diagram of six types of carbon sources.

In order to further analyze the utilization of 31 kinds of carbon sources in the whole process of potato liquor fermentation, the characteristic chart of microbial community utilization of carbon sources was drawn. After standardizing the sample data of 96h culture, principal component analysis was carried out the contribution rate of principal component 1 (PCA1) is 46.623%, the contribution rate of principal component 2 (PCA2) is 21.248%, and the contribution rate of principal component 3 (PCA3) is 11.243%. There are differences in the spatial distribution of different carbon sources. Through statistical analysis of the contribution rate and correlation of the three principal components, the relative ranking of the utilization rate of carbon sources in the whole process of potato liquor fermentation is:

D-xylose / pentanose > N-acetyl-D-glucosamine > Glycogen > L-asparagine > Tween 80 > D-galacturonic acid > D-glucosamine acid > Tween 40 > Methyl pyruvate > 4-hydroxybenzoic acid > Glycyl-L-glutamate > D-mannitol > i-erythritol > γ-hydroxybutyric acid > 1-glucose phosphate > α-cyclodextrin > Putrescine > L-serine > 2-hydroxybenzoic acid > α-butanone acid > L-Phenylalanine > β-methyl-D-glucoside > D-malic acid > L-threonine > Phenylethylamine > L-arginine > D-cellobiose > α-D-lactose > D-galactoseγ-lactone > Itaconic acid > D, L-α-glycerin phosphate.

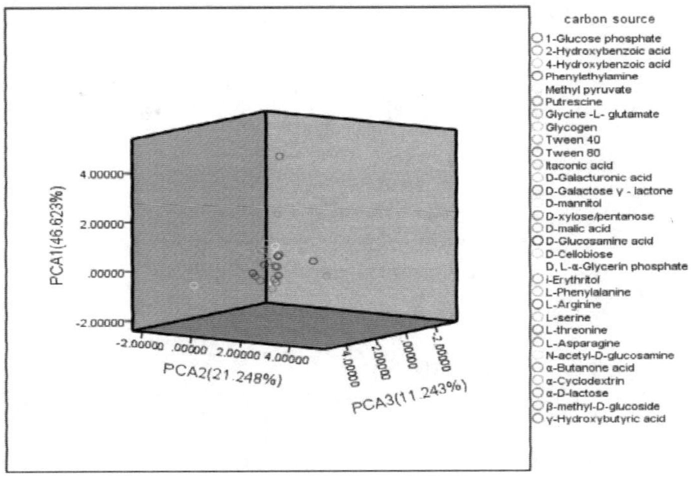


**Figure 5.** Principal component analysis of utilization ratio of 31 carbon sources by microbial community.

## 4. Conclusion

In this study, Biolog ECO technology was used to analyze the diversity of microorganisms in the fermentation process of potato liquor. The average absorbance value (AWCD) can reflect the overall metabolic activity and carbon source utilization capacity of microorganisms, and the diversity index can reflect the changes of micro organisms in the fermentation process of potato liquor [15].

According to the analysis results, the AWCD value of each fermentation sample showed an overall increasing trend, indicating that during the sample culture period, microorganisms were making more and more use of carbon sources in the ECO board. The AWCD value of the samples on the sixth day of fermentation was the largest compared with other samples, indicating that the metabolic activity of microorganisms on the sixth day was the strongest. The results of Simpson (D) and Mclntosh (U) also indicated that the microbial growth and reproduction capacity of the fermentation samples on the sixth day was the strongest. The richness index Shannon (H) and the AWCD curve of 96h of sample culture showed the overall trend of the microbial quantity in the fermentation process of potato liquor. After the fermentation of fermented grains, the microbial quantity in the fermented grains first began to decrease, then began to increase after the second day of fermentation, and reached the highest level on the sixth day. Potato liquor fermentation microorganisms during the process of utilization of carbon sources, sugar carbon source is the microbial use most of the carbon source [16], based on the whole potato liquor fermentation process of carbon source utilization rate of relative principal component analysis data show that microbial communities throughout the fermentation process of D-xylose / pentanose relative efficiency is the highest, further suggests that D-xylose / pentanose is the best carbon source in the potato liquor fermentation process.

The experimental results laid a foundation for more effectively grasping the rule of microbial changes in the process of potato liquor fermentation, provided theoretical reference for improving the yield of potato liquor and other optimization processes, and further demonstrated the reliability and simplicity of Biolog ECO technology in the field of food.

**Acknowledgments**

This work was financially supported by the following funds. (1) Applied Basic Research Project of Science and Technology Office of Sichuan Province, "Research on microbial community of potato fermentation products -- taking Tartary buckwheat wine as an example" (2018JY0396); (2) Key Scientific Research Project of Education Office of Sichuan Province, "Research on key technology of solid-state liquor brewing" (17ZA351); (3) Xichang University "Two High" Talent Scientific Research Support Project, "Based on high-throughput sequencing of the potato liquor fermentation process microbial community" (LGZ201917); (4) Scientific Research Project of Education and Science Bureau of Xichang City, "Research and application of the development mode of potato brewing cycle ecological industry chain" (19YYJS14); (5) Xichang University Undergraduate Innovation and Entrepreneurship Training Program, "Study on the variation of microbial diversity during the fermentation of potato liquor based on Biolog ECO technology" (201810628021).

## References

[1] Zou Huifang, Qu Chang, Wu Hao, Tian Haijiao, Yang Hongyan. Research progress of microbial diversity in the fermentation process of pickled vegetables [J]. Chinese condiments, 2013, 38 (11): 107-112

[2] Qiu Zengyu, Wang Yaping, Li Boyi, Xie Biao, Liu Guofeng, Xiao Dongguang. The effect of temperature on the microflora structure of high temperature Daqu liquid culture process [J]. Chinese brewing, 2017, 36 (05): 30-35

[3] Wang Haijing. Effect of nitrogen fertilizer on soil microbial diversity [D]. Jilin Agricultural University, 2011

[4] Chen Yuzhu, Tang Qiyong, Gu Meiying, Zhu Jing, Liu Xiaojing, Cui Weidong, Zhang Zhidong. Distribution characteristics of microorganisms in the root of claw and the effect of salt concentration on carbon source metabolism [J / OL]. Microbiological Bulletin: 1-9 [2019-

10-10].
[5] Ji Yanyan, Zhang Guilong, Zhang Rui, Liu Yusheng, Yang Dianlin, Wang Cailing. Effects of farming methods on functional diversity of soil microorganisms in farmland [J]. China agronomy bulletin, 2013, 29 (06): 117-123.
[6] Sun Wei, Qian Xun, Fu Qingxia, Hu Ting, Gu Jie, Wang Xiaojuan, Gao Hua. Effect of bio organic fertilizer on soil microbial community and enzyme activity of walnut garden in Qinba Mountain Area [J]. Journal of plant nutrition and fertilizer, 2013, 19 (05): 1224-1233.
[7] Zhang Bianhua, Jin Dongsheng, Gao Chunhua, Gao Yajing, Li Jianhua. Comparison of microbial diversity in rhizosphere of different plants in industrial and mining reclamation area by different methods [J]. Jiangsu agricultural science, 2019, 47 (04): 223-226.
[8] Zhou Qingwu, Li Hongge, Li Anjun, Liu Guoying, Tang Youhong, he Hongkui, Li Xiaohuan. Biolog eco analysis of functional diversity characteristics of Daqu microbial community in different habitats [J]. Food and fermentation technology, 2014, 50 (03): 53-56.
[9] Shang Yuling. Study on yeast flocculation in advance caused by malt of beer barley [D]. Jiangnan University, 2015.
[10] He Hongkui, Li Lan, Tang Youhong, Li Xiaohuan, Chen Bin. Study on the relationship between sensory evaluation of gujinggong Daqu and its microorganism, physical and chemical indexes [J]. Brewing, 2017, 44 (01): 58-63.
[11] Liu yunya. Microbial community structure in Shaoxing rice wine fermentation and its influence on flavor substances [D]. Jiangnan University, 2015.
[12] Lu Lei. Analysis and Research on aroma production of key microorganisms in sesame flavor liquor [D]. Qilu University of technology, 2014.
[13] Lu Ze, Hu Xiaomin, an Jing, Wei Wei. The effect of compound pollution of musk and cadmium on the function and abundance of soil microbial community [J]. Journal of agricultural environmental science, 2017, 36 (01): 66-75.
[14] Yang Yonghua, Yao Jian, Hua Xiaomei. Effects of pesticide pollution on functional diversity of soil microbial community [J]. Journal of Microbiology, 2000 (02): 23-25 + 47.
[15] Li Hao, Luo Huibo, Hou Hua, Zhen pan. Microbial diversity of Luzhou flavor Daqu by Biolog microplate technology [J]. Brewing, 2009, 36 (01): 54-56.
[16] Zhang Yongmei. Research on the principle of effective recovery and enhanced denitrification of kitchen waste carbon source [D]. Xi'an University of architecture and technology, 2016..

IOP Conf. Series: Materials Science and Engineering **914** (2020) 012014 doi:10.1088/1757-899X/914/1/012014

# Study on the preparation technology of Sanqi hemostatic spongia for percutaneous absorption

**Shijun Liu[1, 2, 3, *], Lin Li[1,2,3], Zhishu Tang[1, 2, 3], Yu Zhang[1, 2, 3], Jing Sun[1, 2, 3], Zhongxing Song[1, 2, 3], Hongbo Xu[1 ,2, 3], Chunli Cui[1, 2, 3], Hongbo Liu[1, 2, 3]**

[1]Shaanxi University of Chinese Medicine/Shaanxi Collaborative Innovation Center of Chinese Medicinal Resource Industrialization, Xianyang, Shaanxi, 712083, China
[2]State Key Laboratory of Research & Development of Characteristic Qin Medicine Resources (Cultivation), Xianyang, Shaanxi, 712083, China
[3]Shaanxi Innovative Drug Research Center, Xianyang, Shaanxi,712083, China

*Corresponding author e-mail: 1181618@126.com

**Abstract.** Objective: To optimize the preparation technology and matrix ratio of Sanqi hemostatic spongia. Methods: The effects of Twain 80, formaldehyde and twelve sodium alkyl sulfate on the matrix formability of the spongia were investigated by the $L_9$ ($3^3$) orthogonal test, and the optimum ratio of the matrix was optimized through the water absorption, the thickness uniformity, the softness and the color degree. The orthogonal test results showed that the matrix ratio was Twain 80 1.4g, formaldehyde 1.6 mL, sodium dodecyl sulfate 1.0 g, and Sanqi hemostatic spongia was superior when the proportion of other substrates was fixed. Conclusion: Sanqi hemostatic spongia was produced by orthogonal test and matrix ratio is stable, uniform in thickness and good in quality.

## 1. Introduction

Traditional Chinese medicine spongia is a transdermal drug system (TDDS) [1]-[2], It is a kind of sponge solid sterilization preparation with strong water absorption, which is made of hydrophilic colloid solution and medicine through foaming, hardening, freezing and drying [3]. It is easy to use, high safety, especially suitable for patients with massive hemorrhage. However, there is no relevant report about Sanqi hemostatic spongia, so the research group chose to study it to provide basis for clinical medication.

## 2. Methods and results

### *2.1. Extraction and separation of dencichine*

50 g of crude powder of *Panax notoginseng* was extracted twice with 10 times of methanol at room temperature, combine the extract and store it. The residue was dried and extracted with 10 times of water for three times, then the residue was filtered to get a green extract, which was concentrated to a certain amount. The concentration of alcohol was adjusted to 75% with 95% ethanol. Filter out the precipitate, and the filtrate was extracted with equal volume n-butanol. The purified solution was obtained by combining the aqueous phase, and the purified solution was passed through cation

exchange resin. First wash the purified solution to the effluent to colorless, then eluted with 0.1mol/L ammonia water, collected in stages. The collected solution is developed under n-butanol: glacial acetic acid: anhydrous ethanol: water (4:1:1:2). Ninhydrin solution is used as the developer on the silica gel G thin-layer plate for identification, and the collected solution with only one purple red spot is combined and the same height as the control of dencichine, which proves that this fraction is dencichine extract [4].

### *2.2. Orthogonal test design.*

On the basis of pre-test, $L_9$ ($3^3$) test is used to design the level table of orthogonal factors. The levels and factors of orthogonal test were shown Table 1.

**Table 1.** Factors and levels of orthogonal test

| Levels | Factors | | |
|---|---|---|---|
| | Twain 80 (g) A | Formaldehyde (mL)B | Sodium dodecyl sulfate (g) C |
| 1 | 1.4 | 1.2 | 0.6 |
| 2 | 1.6 | 1.4 | 0.8 |
| 3 | 1.8 | 1.6 | 1.0 |

### *2.3. Orthogonal test procedure*

Take 12g gelatin, soak it in distilled water for 1h, heat it in water bath to make it completely dissolved, add a proper amount of Tween 80, formaldehyde and sodium dodecyl sulfate (according to the prescription designed in the orthogonal test design table) under high-speed stirring, continuously stir it to produce a large number of uniform and delicate bubbles, then separately pack it in the mold, put it in the refrigerator at -20 ℃ for 24h, take it out and dry it in the freeze dryer for 3 days, Take it out and put it into the dryer for storage [5].

### *2.4. Scoring standards [3] and results of orthogonal test*

**Table 2.** Quality evaluation standard of Sanqi hemostatic spongia

| Assessment items | Quality evaluation standard | Score |
|---|---|---|
| Thickness uniformity | The more uniform the thickness, the higher the score | 25 |
| Softness | The better the rebound after extrusion, the higher the score | 25 |
| Water absorption | The greater the water absorption, the higher the score | 25 |
| Color | The more uniform the color, the higher the score | 25 |
| Total score | | 100 |

**Table 3.** Results of orthogonal test

| Serial number | A | B | C | Total score |
|---|---|---|---|---|
| 1 | 1.4 | 1.2 | 0.6 | 75 |
| 2 | 1.4 | 1.4 | 0.8 | 78 |
| 3 | 1.4 | 1.6 | 1 | 94 |
| 4 | 1.6 | 1.2 | 0.8 | 70 |
| 5 | 1.6 | 1.4 | 1 | 88 |
| 6 | 1.6 | 1.6 | 0.6 | 87 |
| 7 | 1.8 | 1.2 | 1 | 86 |
| 8 | 1.8 | 1.4 | 0.6 | 79 |
| 9 | 1.8 | 1.6 | 0.8 | 84 |
| $K_1$ | 247 | 231 | 241 | G=741 CT=61009 |
| $K_2$ | 245 | 245 | 232 | |
| $K_3$ | 249 | 265 | 268 | |
| $k_1$ | 82.33 | 77.00 | 80.33 | |
| $k_2$ | 81.67 | 81.67 | 77.33 | |
| $k_3$ | 83 | 88.33 | 89.33 | |
| R | 1.33 | 11.33 | 12 | |

*2.5. Analysis of variance results*

**Table 4.** Analysis of variance results

| Source of variance | Sum of squares from mean deviation | Degree of freedom | Variance | F | P |
|---|---|---|---|---|---|
| A | 2.67 | 2 | 1.335 | | |
| B | 194.67 | 2 | 97.335 | 72.91 | <0.05 |
| C | 234.00 | 2 | 117.00 | 87.64 | <0.05 |
| Error(e=A) | | | | | |

Note: $F_{0.05}(2, 2)$ =19.00 $F_{0.01}(2, 2)$ =99.00
The results of orthogonal test and analysis of variance show that the amount of factor B and C, that is, formaldehyde and sodium dodecyl sulfate, has a significant effect on the forming of sponge agent, and factor A, that is, the amount of Tween 80, has no significant effect on the forming of sponge agent. Based on cost saving, the level of factor A is selected, which not only saves cost but does not affect the forming of sponge agent, and coincides with the results of the third group with the highest score. To sum up, the best horizontal combination of factors is $A_1B_3C_3$, that is, Tween 80 1.4g, formaldehyde 1.6g mL, sodium dodecyl sulfate 1.0g.

## 3. Verification test

Take 12g of gelatin and soak it in distilled water for 1h, then heat it in water bath to make it completely dissolved, take 2ml of Sanqi extract and add it into gelatin solution, and add appropriate amount of auxiliary materials with the best ratio under high-speed agitation for three batches of repeated verification test. The results show that the optimal ratio is reasonable and feasible, and the thickness uniformity, softness, water absorption, color of the prepared Sanqi hemostatic spongia is the best.

## 4. Conclusion

Foaming is the key operation to ensure the quality of sponge agent, which directly affects the appearance, looseness, water absorption and softness of the finished product. Formaldehyde solution is a curing agent, in addition to its dosage should be strictly controlled during preparation, and attention should be paid to slowly add formaldehyde solution during high-speed mixing, which can make gelatin crosslinked and solidified evenly. Stirring speed and stirring time have influence on foaming performance, cooling temperature and cooling time have certain influence on curing, so it is also important to optimize stirring speed, stirring time, cooling temperature, cooling time and other process parameters. In general, the solidified foam should be cooled rapidly at about 18 ℃ below zero, and must be frozen thoroughly for at least 24 hours, in order to ensure that the gelatin network structure is fixed and will not be deformed during the drying process.

**Acknowledgments**
This work was supported by Shaanxi Project (Grant no. 2015KTCL03-14, 15JF001, ZYMS 025, 2018TD-005, 2018SF-285) and Shaanxi Education Finance ((2013)171); Key disciplines of traditional Chinese Medicine Pharmaceutical Engineering of Shaanxi Provincial Administration of traditional Chinese Medicine (2018).

**References**

[1] Li CJ, Xiao XC, Huang XJ, et al. Progress in percutaneous absorption preparation [J]. Technological Development of Enterprise, 2009, 28 (12): 24-25, 28.
[2] Zhang ZW. Monograph on traditional Chinese medicine pharmaceutics [M]. Beijing. People's Health Publishing House,2009:329-330.
[3] Jiang PC, Li J, Jiang FD, et al. Optimization of preparation technology of sponge [J]. China Pharmacist, 2010,13 (5): 658- 659.
[4] Xie GX, Qiu MF, Zhao AH, et al. Separation, purification and structural analysis of dencichine from *Panax notoginseng* [J]. Nat Prod Res Dev, 2007,19 (6):1059-1061.
[5] Zhang LJ. Study on the therapeutic material basis of *Rehmannia Glutinosa* Libosch and the best pepartion technology of foam-rubber [D]. Shaanxi University of Chinese Medicine, 2011.

IOP Conf. Series: Materials Science and Engineering **914** (2020) 012015 doi:10.1088/1757-899X/914/1/012015

# Potato wine Brewing Process Optimization by Response Surface Analysis

**Qiao Lin[1], Ting Qiu[1], Xiaoyan Liu[1], Junhua Liu[2], Cheng Li[3]**

[1]Xichang University,Xichang 615000, China;
[2]Liangshan Tianmeiyi Agricultural Science Co., Ltd, Xichang 615000, China
[3]New Hope Group Xichang Feed Company, Xichang 615000, China.

**Abstract.** The volatile aroma components in potato wine were detected by static headspace-gc-ms with potato wine as raw material. The effects of equilibrium temperature, equilibrium time, sampling amount and NACL addition on the number of effective peaks of aroma components in potato wine were studied by single factor experiment. Then the detection conditions were optimized by Orthogonal test. The results showed that under the conditions of 0.2 g NaCl addition, 6 mL sampling, 35 min equilibrium time and 40 °c equilibrium temperature, the number of effective peaks was 15, which was more than 5 aroma components before optimization. It laid a solid foundation for the establishment of a rapid and simple method for the determination of aroma components in wine.

## 1. Introduction

Potato wine has a long history of development, originated from Poland, and is biased towards faint scent wine. It smells fragrant and tastes delicious. Potato wine is rich in nutrition and contains a variety of vitamins and minerals. The production of potato wine is also very simple, first saccharifying potatoes, then adding distiller's yeast for fermentation, and finally distilling to obtain the finished product of potato wine. In view of its high nutritional value and simple fermentation method, many people began to study potato wine.

Aroma components in potato wine are the key to evaluate the quality of potato wine. Different contents of aroma components make wine show different qualities and styles. In order to make the industrialization of potato wine develop more smoothly, it is particularly important to study the aroma components of potato wine. Chromatographic technology is now widely used in the analysis of wine aroma components. There are many methods for extraction of wine aroma components, including headspace solid phase microextraction, steam distillation, purge and trap, static headspace technology, and liquid extraction technology [1]. Static headspace technology does not need liquid-liquid extraction of aroma components in potato wine [2], which can reduce the pretreatment time of samples and make aroma components easier to be detected. Therefore, static headspace technology was used in this experiment [3]. Because GC-MS can effectively separate complex aroma components from volatile wine, and mass spectrometer has high sensitivity, fast analysis speed and strong identification ability, GC-MS technology is adopted to detect aroma components in potato wine [4].

In this experiment, the chromatogram obtained by GC-MS was compared with NIST standard library by MS3000 quick retrieval program, and the characteristic peaks with similarity greater than

65% were screened out, and then the aroma components were further determined [5]. In general, by optimizing the detection conditions, five more target substances can be detected after optimization than before. It provides a theoretical basis for the establishment of a rapid and simple method for the determination of aroma components in potato wine in the future.

## 2. Materials and Methods

### *2.1. Test material*
Potato wine

### *2.2. Test reagent*
Sodium chloride, 51 kinds of liquor mixed standard, ethanol

### *2.3. Instrumentation and equipments*
GC-MS3000 desktop chromatography-mass spectrometry
AHS-6890A automatic headspace sampler.
DB-WAX 57 CB (50m×0.25mm×0.20μm)
FA2204B electronic balance
Pipetting gun, top empty bottle, gland clamp

### *2.4. Method*

*2.4.1. Experiment on the effect of the amount of NaCl on the number of effective peaks.* In order to explore the effect of NaCl addition on the number of effective peaks, under the conditions of 6mL of sample, 40℃ of equilibrium temperature and 30min of equilibrium time, the effect of NaCl addition on the number of effective peaks was investigated and the optimal addition of NaCl was determined 0.2g

*2.4.2. Effect of sampling amount of potato wine on effective peak number.* In order to study the effect of sampling amount on the number of effective peaks, under the conditions of NaCl addition of 0.2g, equilibrium temperature of 40℃ and equilibrium time of 30min, the effect on the number of effective peaks when the sampling amounts of potato wine were 4mL, 6 mL, 8mL and 10 mL respectively was investigated, and the optimal sampling amount of potato wine was determined.

*2.4.3. Experiment on the effect of equilibrium temperature on the number of effective peaks.* In order to study the effect of equilibrium temperature on the number of effective peaks, when the addition of NaCl is 0.2g, the equilibrium time is 30min and the sampling amount is 6mL, the effect on the number of effective peaks when the equilibrium temperatures are 20℃, 30℃, 40℃ and 50℃ respectively is investigated, and the best equilibrium temperature is determined.

*2.4.4. Experiment on the effect of equilibrium time on the number of effective peaks.* In order to study the effect of equilibrium time on the number of effective peaks, when the addition of NaCl is 0.2g, the equilibrium temperature is 40℃, and the sampling amount is 6mL, the effect on the number of effective peaks when the equilibrium time is 10min, 20min, 30min and 40min respectively is investigated, and the best equilibrium time is determined.

*2.4.5. Orthogonal test.* The horizontal range of the four factors of NaCl addition amount, sampling amount, equilibrium time and equilibrium temperature was determined through single factor test results. The number of effective peaks with matching degree >65% was taken as the evaluation index, and the best combination was analyzed through $L_{16}(4^4)$ orthogonal test. The orthogonal test results are shown in Table 1.

**Table 1.** Potato wine aroma ingredient test conditions optimization orthogonal test factors and level

| Level | (A)NaCl addition /g | (B)Sampling quantity | (C) Equilibrium time /min | (D) Equilibrium temperature/℃ |
|---|---|---|---|---|
| 1 | 0 | 5 | 20 | 30 |
| 2 | 0.1 | 6 | 25 | 35 |
| 3 | 0.2 | 7 | 30 | 40 |
| 4 | 0.3 | 8 | 35 | 45 |

*2.4.6. Detection method*

*2.4.6.1 Sample pretreatment.*
Take out 6mL of aged potato wine and 0.1gNaCl, add them into a 20mL top empty bottle, seal them, and put them into the sample chamber of the headspace sampler for testing.

*2.4.6.2 Instrument parameters.*
Headspace sampler parameters [6] [8]: sample volume 1mL, pressurization time 0.2min, sample injection time 1min, equilibrium temperature 40℃, equilibrium time 30min.

Gas chromatography conditions: sample inlet temperature 230℃, programmed temperature rise, initial temperature 45℃, retention time 6min, rise to 150℃ at 5℃/min, rise to 210℃ at 10℃/min, retention time 5min, column flow rate 1mL/min, split ratio 30: 1, purge flow rate 3mL/min, carrier gas pressure 60kpa, [9] carrier gas 99.99% helium.

Mass spectrum conditions [10],[12]: ion source temperature 250℃, interface temperature 230℃, column temperature 40℃, gasification chamber temperature 230℃, solvent peak time 2min, scanning mode: full scanning, multiplier high pressure 1000V v.

*2.4.6.3 Qualitative method.*
The chromatogram obtained in full scanning mode is compared with the standard chromatogram in NIST, and the target substance with similarity > 65% is screened out, and then quantitative analysis is carried out [13]

*2.4.6.4 Quantitative method.*
The relative content of aroma components was normalized by peak area method [14], and then the relative area of aroma components was brought into the standard curve of 51 detected internal standard substances [15], and the concentration of characteristic aroma components was calculated.

*2.4.6.5 Data processing.*
The experimental data were analyzed by variance analysis using IBM SPSS Statistics 19.0.

IOP Conf. Series: Materials Science and Engineering **914** (2020) 012015 doi:10.1088/1757-899X/914/1/012015

## 3. Results and Analysis

### *3.1. Effect of NaCl content on the number of effective peaks of aroma components in potato wine*

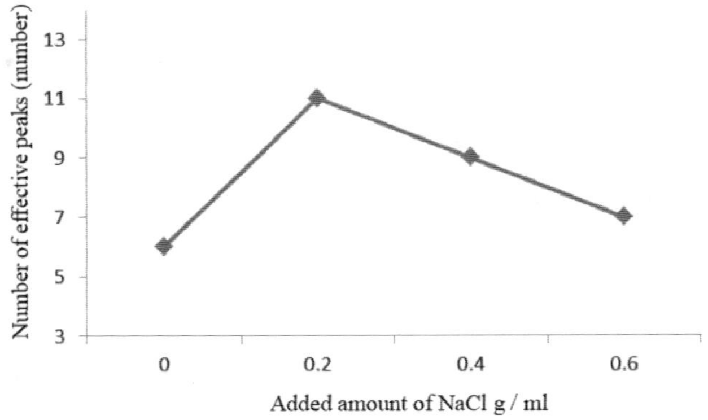


**Fig. 1** Effect of the addition of NaCl on the number of effective peaks of the aroma composition of potato wine

As can be seen from fig. 1, the number of effective peaks increases with the increase of NaCl addition. at 0.2g, the maximum number of effective peaks reaches 11. Salt is added to reduce the solubility of aroma components, eliminate matrix effect and precipitate volatile aroma components. However, the increase of salt will affect the viscosity of the wine sample. The higher the viscosity, the lower the diffusion rate of the analyte [16], and the aroma components are not easy to be detected. Therefore, according to the test results, 0.2g NaCl is selected as the optimal addition amount.

### *3.2. The effect of sampling quantity of potato wine on the number of effective peaks of aroma components in potato wine*

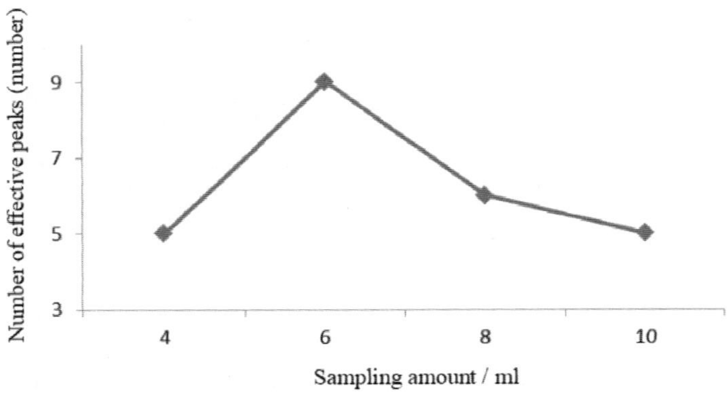


**Fig. 2** Effect of sampling volume on the number of active peaks of potato wine aroma composition

It can be seen from Figure 2 that the number of effective peaks increases with the increase of potato wine sampling amount. The amount of potato wine sample will affect the vapor-liquid distribution balance in the headspace bottle. The volume of gas in the headspace bottle will be smaller and the pressure will be larger. In order to re-establish the balance, the aroma components will overflow. When the sampling amount is more than 6ml, the space in the headspace bottle becomes smaller, which is not conducive to the overflow of aroma components. Therefore, the best sampling amount of potato wine is 6ml.

*3.3. Effect of equilibrium time on the number of effective peaks of aroma components in potato wine*

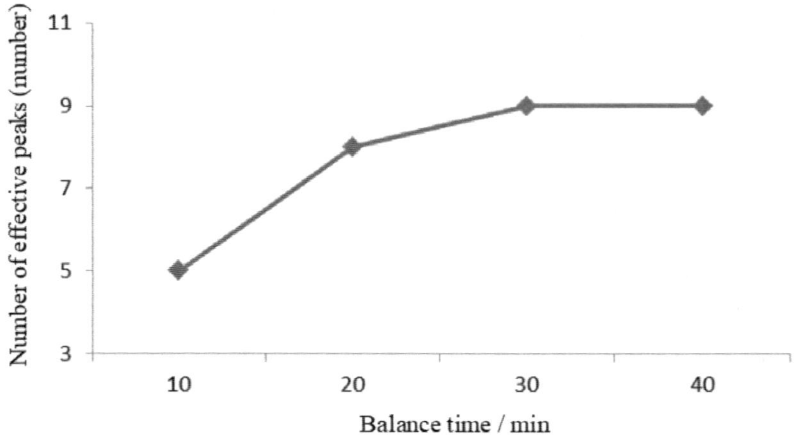


**Fig. 3** Effect of balance time on the number of effective peaks of potato wine aroma composition

It can be seen from fig. 3 that the increase of equilibrium time increases the number of effective peaks to a certain extent, but it will remain unchanged after reaching 30min. Because after the sample is heated, volatile and semi-volatile substances in the sample are released and adsorbed [17], and heat balance is achieved in the gas and liquid phases, so that more volatilization of aroma components of potato wine can be detected more easily. However, the equilibrium time depends on the diffusion speed of the aroma components in the wine. The larger the molecular diffusion coefficient, the faster the diffusion speed of the aroma components, and the shorter the equilibrium time required for the sample. If the equilibrium time is too long, the volatile components will not change much, and the number of effective peaks will even decrease. Therefore, 30min is the best balance time.

*3.4. Effect of equilibrium temperature on the number of effective peaks of aroma components in potato wine*

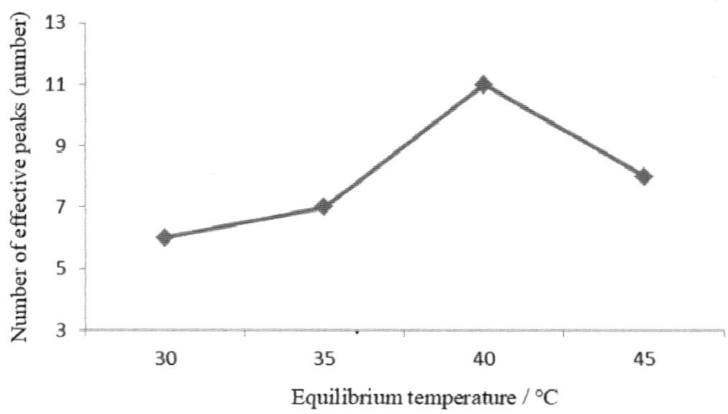


**Fig. 4** Effect of equilibrium temperature on the number of active peaks of the aroma composition of potato wine

As can be seen from fig. 4, as the equilibrium temperature increases, the aroma components are decomposed, the faster the movement speed is, the easier it is to be detected, and the number of effective peaks is also increasing. When the equilibrium temperature is 40℃, the maximum number of effective peaks is 11, and then it will remain unchanged. Because of different temperatures, the distribution coefficient of aroma substances in the top-empty bottle is different [18]. With the increase of temperature, the higher the vapor pressure, the shorter the time to reach equilibrium, and the higher the sensitivity of analysis, while the decrease of distribution coefficient is not conducive to the detection of aroma components. Choose a lower equilibrium temperature, because too high a temperature will affect the decomposition of aroma components in potato wine, making the top empty bottle pressure too high. Therefore, 40℃ is chosen as the best equilibrium temperature.

### *3.5. Orthogonal test results*

The horizontal range of NaCl addition amount, sample amount, equilibrium temperature and equilibrium time was determined through the results of single factor test. The optimal combination of conditions for detecting aroma components was determined through orthogonal test with the number of effective peaks as evaluation index.

**Table 2.** Results and analysis of positive test results for the detection conditions of the aroma ingredients of potato wine

| Test number | A | B | C | D | Number of effective peaks (number) |
|---|---|---|---|---|---|
| | NaCl addition | Sampling quantity | Equilibrium time | Equilibrium temperature | |
| 1 | 1 | 1 | 1 | 1 | 6 |
| 2 | 1 | 2 | 2 | 2 | 8 |
| 3 | 1 | 3 | 3 | 3 | 13 |
| 4 | 1 | 4 | 4 | 4 | 9 |
| 5 | 2 | 1 | 2 | 3 | 10 |
| 6 | 2 | 2 | 1 | 4 | 8 |
| 7 | 2 | 3 | 4 | 1 | 13 |
| 8 | 2 | 4 | 3 | 2 | 10 |
| 9 | 3 | 1 | 3 | 4 | 5 |
| 10 | 3 | 2 | 4 | 3 | 15 |
| 11 | 3 | 3 | 1 | 2 | 9 |
| 12 | 3 | 4 | 2 | 1 | 10 |
| 13 | 4 | 1 | 4 | 2 | 6 |
| 14 | 4 | 2 | 3 | 1 | 11 |
| 15 | 4 | 3 | 2 | 4 | 6 |
| 16 | 4 | 4 | 1 | 3 | 9 |
| K1 | 36 | 27 | 32 | 40 | |
| K2 | 41 | 42 | 34 | 33 | |
| K3 | 39 | 41 | 39 | 47 | |
| K4 | 32 | 38 | 43 | 28 | |
| K1j | 9 | 6.75 | 8 | 10 | |
| K2j | 10.25 | 10.5 | 8.5 | 8.25 | |
| K3j | 9.75 | 10.25 | 9.75 | 11.75 | |
| K4j | 8 | 9.5 | 10.75 | 7 | |
| R | 2.25 | 3.75 | 2.75 | 4.75 | |
| Excellent combination | $A_3B_2C_4D_3$ | | | | |

**Table 3.** Analysis of variance of orthogonal experimental results

| Source of variation | Sum of squares | Variance | Mean square | Variance ratio | Significant |
|---|---|---|---|---|---|
| A | 11.5 | 3 | 3.833 | 5.75 | 0.092 |
| B | 35.5 | 3 | 11.833 | 17.75 | 0.021* |
| C | 18.5 | 3 | 6.167 | 9.25 | 0.050 |
| D | 51.5 | 3 | 17.167 | 25.75 | 0.012* |
| Error | 2 | 3 | 0.667 | | |
| Total variation | 1488 | 16 | | | |

Note: "**" means that the effect on the results is extremely significant when $P < 0.01$, "*" means that the effect on the results is significant when $P < 0.05$

According to Table 2, it can be seen that the range R of four factors A, B, C and D and the equilibrium temperature are the most important factors, followed by the sampling amount, equilibrium time and NaCl addition amount. The optimal level group of optimized detection conditions is $A_3B_2C_4D_3$., while the optimal result of single factor test is $A_3B_2C_3D_3$, in which the number of effective peaks is 15 and 10 respectively, with no significant difference in equilibrium time. therefore, it can be determined that the optimal combination for detection of aroma components in potato wine is $A_3B_2C_4D_3$.

Table 3 shows that the difference in equilibrium temperature is the most significant, followed by sampling amount. The difference in equilibrium time and NaCl addition amount is not significant, which is consistent with the result of range analysis.

To sum up, the best conditions for static headspace-air mass spectrometer to detect aroma components in potato wine are NaCl addition of 0.2g, sampling amount of 6mL, equilibrium time of 35min, and equilibrium temperature of 40℃.

### *3.6. Comparison before and after optimization of detection conditions for aroma components in potato wine*

According to the orthogonal test, the optimal conditions for detecting aroma components in potato wine were optimized, and the obtained spectra were compared with NIST standard spectra library.

There are 10 kinds of aroma components before optimization, which are ethyl acetate, 2-ethylhexanol, propanol, acetic acid, isopropyl acetate, acetal, isobutanol, pentanol, butyl lactate, ethyl hexanoate.

After optimization, 15 aroma components were detected, including ethyl acetate, propanol, β -phenylethanol, acetic acid, isopropyl acetate, ethyl decanoate, isobutanol, acetal, 2-ethylhexanol, butyl lactate, propylene glycol, methyl 3-methylhexanoate, butyl 5-methylpropionate, pentanol and ethyl hexanoate.

IOP Conf. Series: Materials Science and Engineering **914** (2020) 012015 doi:10.1088/1757-899X/914/1/012015

### *3.7. Quantitative analysis of aroma components in potato wine*

**Table 4.** Comparison of aroma composition before and after potato wine test method optimization

| Retention time | Compound | Molecular formula | Before optimization | Postoptimality |
|---|---|---|---|---|
| | | | Relative content/% | |
| 3.32 | Ethyl acetate | C4H8O2 | 28.64% | 31.19% |
| 4.11 | Propyl alcohol | C3H8O | 2.17 | 3.15% |
| 4.15 | Acetic acid | C2H4O2 | 0.07% | 0.33% |
| 4.33 | Isopropyl acetate | C5H10O2 | 1.51% | 1.78% |
| 5.38 | Isobutanol | C4H10O | 0.46% | 2.58% |
| 5.53 | Acetal | C6H14O2 | 0.66% | 2.99% |
| 5.99 | Butyl lactate | C7H14O3 | 0.45% | 1.41% |
| 7.29 | Propylene glycol | C3H8O2 | - | 1.16% |
| 10.94 | Methyl 3-methylhexanoate | C8H16O2 | - | 1.4% |
| 11.17 | Butyl 5-methylpropionate | C8H16O2 | - | 0.13% |
| 17.06 | Amyl alcohol | C5H12O | 0.49% | 0.87% |
| 20.95 | Ethyl hexanoate | C8H16O2 | 11.02% | 18.3% |
| 26.26 | 2-ethylhexanol | C8H18O | 0.51% | 2.64% |
| 28.93 | β - phenylethanol | C8H10O | - | 0.29% |
| 33.06 | Ethyl decanoate | C12H24O2 | - | 0.72% |

"-"means that the above components are not detected and the sum is not 100, and column loss is inevitable.

After the optimization of the static headspace-gas chromatography-mass spectrometry detection conditions, there are 5 kinds more than before, namely propylene glycol, methyl 3- methyl acetate, butyl 5- methyl propionate, β-phenylethanol and ethyl decanoate. The peak area normalization method was used for quantitative analysis of each aroma component [20].

## 4. Conclusion and Discussion

This experiment is to study the detection conditions of aroma components in potato wine and make optimization. The test results are as follows:

(1) The instruments used in this test are mainly static headspace sampler and gas chromatography-mass spectrometry [22]. Firstly, headspace sampling is carried out through single factor test, and then the optimal levels of four factors, namely, equilibrium time, equilibrium temperature, NaCl addition amount and potato wine sampling amount, are respectively determined according to the number of effective peaks. The orthogonal test of $L_{16}(4^4)$ was further used to determine the best combination. The optimal combination of orthogonal test was the optimized detection conditions. The effective peaks before and after optimization were compared. It was found that the optimized detection conditions of static headspace-gas chromatography-mass spectrometry instrument could have 5 more aroma components than before optimization, namely propylene glycol, methyl 3- methyl acetate, butyl 5-methyl propionate, β-phenylethanol and ethyl decanoate.

(2) In the process of detecting potato aroma components by static headspace-gas chromatography-mass spectrometry [23], the equilibrium temperature has the greatest influence on the detection results, followed by the sampling amount of wine, the equilibrium time should not be too long, and the addition amount of NaCl has no significant influence on the detection of potato aroma components.

(3) The best conditions for static headspace-gas chromatography-mass spectrometry to detect the aroma components of potato wine in wine-making laboratory are: sample volume 6 mL, NaCl addition 0.2 g, equilibrium temperature 40℃, equilibrium time 35 min.

This experiment studied the detection of aroma components in potato wine under the use of static headspace-gas chromatography-mass spectrometry. The effective peak number was used as the evaluation basis. Compared with liquid-liquid extraction, headspace solid-phase microextraction and purge-and-trap technologies [24], sample pretreatment still needs to be improved. However, static headspace sampler can be used for simple and rapid test and pre-test [25]. Therefore, static headspace-GC-MS technology can lay a solid foundation for the subsequent establishment of a fast and simple detection method for wine aroma components.

**Acknowledgments**
2018JY0396 Applied Basic Research Project of Sichuan Provincial Science and Technology Department, Study on Microbial Community of Potato Fermented Products-Taking Potato Tartary Buckwheat Wine as an Example (above); 17ZA351 Key Research Project of Sichuan Provincial Education Department, Research on Key Technology of Making Potato Liquor by Solid State Method; LGZ201917 Xichang College's "Two Highs" Talent Research Support Program Project, Based on Microbial Communities in Potato Liquor Fermentation Process in high-throughput sequencing; 19YYJS14 Research Project of Xichang Municipal Bureau of Education and Science Research Research and Application of Development Model of Potato Brewing Circular Eco-industrial Chain.

**References**
[1] Li Dengkun, Liu Xiangping, Zha Hexia, etc. Determination of volatile compounds in wine by static headspace-gas chromatography [J]. Chinese Journal of Hygiene Inspection, Vol. 25 (2015) No. 11, p. 1730-1732.
[2] Ji Houwei, Yang Jingguo, Wang Fang et al. Application progress of static headspace-gas chromatography in the tobacco industry [J]. Physical and Chemical Inspection (Chemical Volume), Vol. 50 (2014) No. 09, p. 1188-1192.
[3] Mo Yanxia, Yin Juyi, Gu Xiaojun, etc. Research status and application prospects of wine organic acids [J]. Science and Technology of Food Industry, Vol. 36 (2015) No. 06, p. 380-384.
[4] Gong Hongsheng, Liao Liewen, Hu Wenbin et al. Research progress of the application of GC-MS in the detection of food pesticide residues [J] .Food and Machinery, Vol. 29 (2013) No. 05, p. 245-248.
[5] You Hongmei. Analysis of plasticizers and aroma components in wine by gas chromatography-mass spectrometry [D]. Harbin University of Science and Technology, 2014.
[6] Zhang Qian, Liu Weilun, Lu Yanan et al. Application progress of headspace gas chromatography-mass spectrometry [J] .Chinese Journal of Chromatography, Vol. 36 (2018) No. 10, p. 962-971.
[7] DARTIGENAVEC, JEANDETP, MAUJEAN A. Study of the contri-bution of the major organic acids of wine to the buffering capacity of wine in model solutions[J]. Am J Enol Viticul, Vol. 51(2000) No. 4, p. 352-356.
[8] MORATAA,GOMEZ-CORDOVCSMC.SUBERVIOLAJet,al.Ad-sorptionofanthocyanins by yeast cell walls during the fermentation of red wines[J]. J Agr Food Chem, Vol. 51(2003) No. 14, p. 4084-4089.
[9] Dun Zhuciren, Zhan Xiuxiu, Labazaxi et al. Study on volatile components in gardenia based on headspace-mass spectrometry usage [J]. Tibet Science and Technology, (2018) No. 12, p. 70-73 + 77.
[10] Zhou Wenjie, Zhang Fang, Wang Peng et al. Study on the characteristic aroma components of Korla fragrant pear wine based on GC-MS / GC-O combined with chemometrics [J]. Food Science, Vol. 39 ( 2018) No. 10, p. 222-227.
[11] REBIERE L, CLARKAC, SCHMIDTKEL M, et al. A robust method for quantification of volatile compounds within and between vintages using headspace-solidphase micro-extraction coupled with GC-MS application on Semillon wines[J]. Analytica chimica acta,

Vol. 660 (2010) No. 1/2, p. 149-157.
[12] CABREDO-PINILLOSS, CEDRON-FERNANDEZT,SAENZ-BARRIOC. Comparison of different extraction methods applied to volatile compounds in wine samples previous to the determination by gas chromatography[J]. Analytical letters, Vol. 37 (2004) No. 14, p. 3063-308.
[13] Lu Ning, Wan Xiaochun, Pan Dong. Preliminary study on the relationship between aroma components and quality of jasmine tea [J]. Food Science, 2004 (06): 93-97. [11] GB / T 10345-2007, Baijiu Analysis method [S].
[14] Shen Haiyue. Research on aroma substances of sauce-flavor liquor [D]. Jiangnan University, 2010.
[15] Zhang Ting, Chen Xiaowei, Zhang Shasha et al. GC-MS analysis of aroma components of cassava wine produced by fermentation of different koji [J]. Food Research and Development, Vol. 40 (2019) No. 02, p. 125-131.
[16] Wu Zengsheng. Research on Analysis Methods of Volatile Flavor Substances in Beer [D]. Northeastern University, 2008.
[17] Chai Chuan, Yu Sheng, Cui Xiaobing, et al. Static headspace-mass spectrometry analysis of volatile components in light tempeh [J]. Food Research and Development, Vol. 34 (2013) No. 14, p. 81-85.
[18] Huang Huizhen, Liang Hui, Liu Xiucai, etc. Analysis of volatile aroma components in Fujian special tea by trapped headspace-mass spectrometry [J]. China Test, Vol. 43 (2017) No. 11, p. 59-63.
[19] Wang Duliu, Ye Wenbin, Zhang Shaofei et al. Analysis of aroma components of "Mingliuzi Wine" by gas chromatography-mass spectrometry [J]. Brewing Science and Technology, (2017) No. 08, p. 132-135.
[20] Li Weiqing, Wang Hao, Jiang Liting et al. Determination of volatile components in litchi fruit vinegar by headspace solid phase microextraction coupled with gas chromatography-mass spectrometry [J]. China Brewing, (2011) No. 12, p. 160-162.
[21] Kong Chengshi. GC-MS study on volatile flavor compounds of dry wine [D]. Qingdao University of Science and Technology, 2013.
[22] Qi Xinchun, Zhang Huining, Wei Leipeng et al. Optimization of conditions for detecting aroma components in wine by static headspace-mass spectrometry [J]. China Brewing, Vol. 36 (2017) No. 09, p. 168-172. Kong Chengshi.
[23] Wang Yang, Wang Jie, Liu Yaqiong et al. Determination of aroma components in apple pomace fermentation distilled wine by headspace solid phase microextraction-gas chromatography-mass spectrometry [J]. Food Science, Vol. 33 (2012) No. 12, p. 205-209.
[24] You Hongmei, Li Fengqin, Yu Zhigang. Analysis of aroma components of Bei Okura liquor by gas chromatography-mass spectrometry [J]. Chemical Research and Application, Vol. 26 (2014) No. 11, p. 1820-1824.
[25] Müller A, Düchting P, Weiler E W. A multiplex GC-MS/MS technique for the sensitive and quantitative single-run analysis of acidic phytohormones and related compounds, and its application to Arabidopsis thaliana [J]. Planta, Vol. 216 (2002) No. 1, p. 44-56.

IOP Conf. Series: Materials Science and Engineering **914** (2020) 012016 doi:10.1088/1757-899X/914/1/012016

# Fatigue and Recovery of Wushu Athletes Based on Fatigue Damage Model

**Xue Yu**

Department of Sports Education and Research, Northwestern University, ShaanXi, Xian 710069, china

xueyu8528@163.com

**Abstract.** The problem of sports injury and rehabilitation has always been a hot topic in competitive sports. With the continuous development of martial arts requirements and competition system, competitions are increasingly fierce and special technical requirements are more stringent. Exercise-induced fatigue can not only cause the decline of athletes' body ability, but also lead to the occurrence of sports diseases and sports injuries. Only by understanding the mechanism of fatigue, understanding the body's recovery process, recovery methods. Only in this way can we effectively improve our athletic performance and prolong our athletic life. The increase of training difficulty and intensity leads to the occurrence of sports injury more and more frequently, so the rehabilitation and recovery of athletes after injury is particularly important. According to the characteristics of sports fatigue of Wushu Athletes in sports training competition, this paper puts forward the means to eliminate the physical and psychological fatigue of Wushu athletes. In order to improve martial arts training, improve the competitive ability of athletes and improve the health of athletes to provide help.

## 1. Introduction

Martial arts is a kind of competitive sports mainly based on dynamic exercises. With the vigorous development of sports in recent years and the success of Beijing's 2008 Olympic bid, Wushu has gradually been recognized and valued by more and more people. With the continuous improvement of martial arts athletes' technical level, in order to further strengthen the training effect and achieve excellent results, martial arts athletes have been bearing a large load for a long time [1]. Sports fatigue, as a physiological and psychological phenomenon, has always accompanied the development of martial arts. Sports fatigue is an important factor that leads to the decline of athletes' functional ability, and it is also the main cause of athletes' injuries. It restricts the improvement of athletes' competitive ability [2]. On the basis of understanding the mechanism of martial arts athletes' sports fatigue, it has a positive effect to scientifically use physiological and biochemical indexes to evaluate the degree of sports fatigue in martial arts training, reasonably arrange sports training load and prevent overtraining and sports injury [3]. Sports fatigue can not only cause athletes' physical decline, but also cause sports diseases and sports injuries [4]. Affect the improvement of athletes' technical level and training effect, and end their training career prematurely.

IOP Conf. Series: Materials Science and Engineering **914** (2020) 012016 doi:10.1088/1757-899X/914/1/012016

A certain degree of fatigue in the athlete's body during martial arts training is the biological basis for improving the level of exercise, and it is the basic measure to excavate and improve the athlete's physical function and enhance the athlete's physical fitness [5]. Sports fatigue restricts the improvement of athletes' athletic ability. Martial arts is a power endurance project that needs to overcome its own weight. It has high requirements for athletes' strength, speed, dexterity, balance and stability. Martial arts training is a complex and scientific diversified project. Fatigue and recovery determine the success or failure of training [6]. Only by understanding the mechanism of fatigue, the recovery process and recovery method of the body can we effectively improve the performance and extend the life of the exercise. The improvement of athletes' performance level is caused by moderate exercise fatigue caused by training, and reasonable recovery is applied, so that the human body's functional status can obtain adaptive results at a new level [7]. The process of fatigue and recovery is an indispensable part of training, and reasonable functional recovery methods are an important guarantee for improving the level of training and physical function. Only by understanding the mechanism of fatigue, the recovery process and recovery methods of the body, and mastering the recovery methods of fatigue can we effectively improve the athletic performance and prolong the athlete's sports life. According to the characteristics of martial arts athletes producing sports fatigue in sports training competitions, this article proposes a method to eliminate physical and psychological fatigue of martial arts athletes.

## 2. The Mechanism of Wushu Athletes' Sports Fatigue

Exercise-induced fatigue refers to the temporary decline of human body's exercise ability due to physiological and biochemical changes during exercise. It is a physiological phenomenon occurring in the process of human body movement and a sign of human body functional activities. During training, a large number of exciting impulses are transmitted to corresponding nerve cells in cerebral cortex, and the long-term exciting impulses of nerve cells lead to increased energy and material consumption. When it is consumed to a certain extent, the corresponding nerve cells will produce protective inhibition, resulting in central nerve innervation imbalance. At the same time, due to large energy loss, lactic acid increases, while lactic acid increases and glycogen decreases will cause central fatigue. The unique form of martial arts makes a large number of exciting impulses transmitted to the corresponding nerve cells in the cerebral cortex. In order to avoid excessive energy consumption, when consumed to a certain extent, the corresponding nerve cells will produce protective inhibition. There is imbalance of central nerve control, decreased motor ability and fatigue. When martial arts athletes are training, lactic acid production lowers the pH value of muscles, inhibits the activity of phosphofructokinase, a key enzyme in sugar metabolism, and hinders the energy supply of glycolysis metabolism. At the same time, ammonia production in blood and tissues affects muscle working ability, and muscle spasm occurs when ammonia production increases in brain tissues. In martial arts training, excessive fat in the body is mobilized and free fatty acid in the blood rises greatly.

Functional exercise for injury prevention refers to the special protective exercise for the parts of the body that are prone to injury in order to prevent the sports injury in sports during the special training process. The system structure of martial arts special sports injury prevention method is shown in Figure 1.

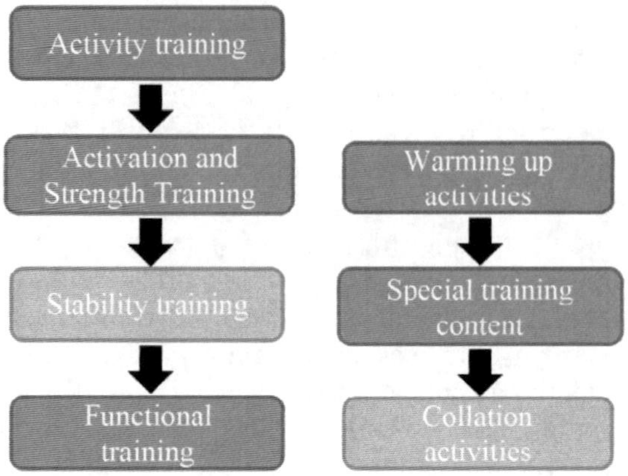


**Figure 1.** System structure of Wushu special sports injury prevention method

The energy and materials consumed during training are gradually recovered after the exercise, and the recovery is good or even excessive. During the training, the internal environment and the balance of nervous, immune and endocrine systems were destroyed. During the recovery period, it should be recovered as soon as possible and as well as possible to improve the body's adjustment ability. For long fist training, it is mainly a high-intensity exercise with anaerobic glycolysis as the main energy supply, and the direct energy comes from ATP in skeletal muscle [8]. In the type of sugar supplement, glucose and fructose are suitable. Use as little sucrose as possible, fructose mainly increases glycogen storage in liver. Glucose mainly increases the content of myoglycogen. As an excellent martial arts coach, one should make timely and correct judgment on the physical condition of athletes according to various situations and information. According to the nature of the load to determine the intermittent time and mode.

## 3. Recovery of Wushu Athletes' Sports Fatigue

In the process of martial arts training, the application of biochemical indexes is an important means to scientifically formulate training plans, master appropriate intensity, evaluate training effects and diagnose functional fatigue. Individual's low or high achievement needs may be related to their different understanding of success and failure. During the training, the contents of the training can be changed in time and scientifically, the training load can be adjusted, and the movement parts can be changed to rest. This can accelerate the recovery of fatigue. One of the causes of exercise fatigue is the consumption of energy materials in the body. Its recovery method is to make reasonable nutrition arrangements according to the nutritional characteristics of different sports items. The main purpose of recovery is to eliminate fatigue, and its progress is related to athletes' physical condition and athletic ability. A regular living system can ensure the elimination of fatigue after exercise. Martial arts athletes each have their own characteristics, so in the training class, we should choose comprehensive training methods and methods suitable for individual characteristics. In the course of training, interval time should be arranged reasonably and training methods and contents should be changed frequently. Athletes should strictly abide by the prescribed work and rest system, pay attention to food hygiene, create good sleep conditions to improve sleep quality, and overcome bad habits such as smoking and drinking.

Martial arts athletes do not rest to relieve fatigue after they are tired. But to change the content of martial arts training in a timely and scientific manner, adjust the training load and change the movement parts as rest. In order to achieve the goal of eliminating fatigue. The traditional definition of time loss includes the classification of martial arts injuries and the changes of injury states. The change of damage state is shown in Figure 2.

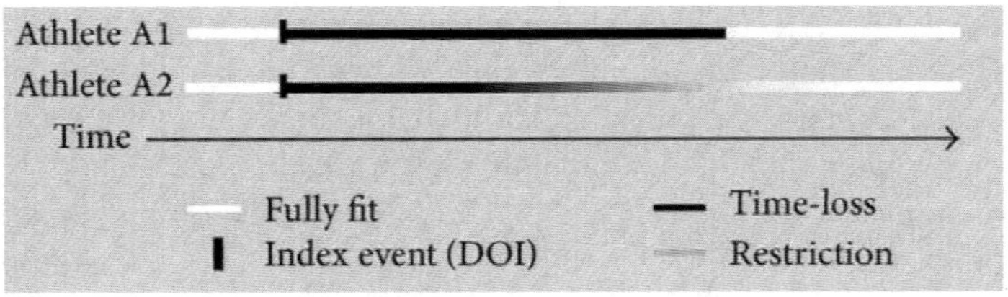


**Figure 2.** Change of damage status

Maximum strength and fast strength are important qualities that reflect muscle strength. The maximum force is the highest force value displayed when the will of the neuromuscular system is exerted to the greatest extent against resistance. Statistics are made on the test results of various mechanical indexes before and after training. The experimental results are shown in Figure 3.

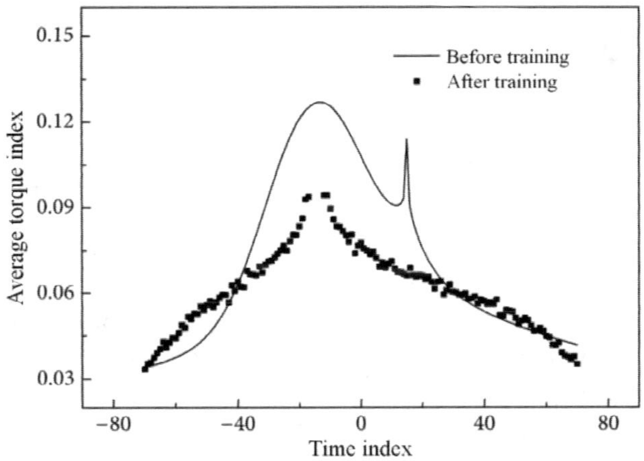


**Figure 3.** Comparison of average muscle strength at different time periods of knee extension before and after training.

From the perspective of martial arts training, basic skills account for a large proportion in the training. Every skill of athletes requires tens of millions of repetitions to finalize the design and establish a good proprioceptive feeling. There are a lot of running and jumping exercises in martial arts, so before, during and after the exercise, adding more carbohydrate and starch with high energy content is undoubtedly a promoting factor to improve martial arts athletes' sports ability. In order to speed up the recovery of athletes' sports fatigue, meat and protein should be added to the diet. Athletes should eat high protein and low fat food, and eat more food containing vitamin B1 and vitamin C. Martial arts training is an anaerobic exercise with sub-maximum intensity. The energy in the training process mainly comes from the aerobic metabolism or anaerobic metabolism of sugar, fat and protein [9]. Bathing therapy and hot compress can accelerate blood circulation and stimulate blood vessel expansion. Promote metabolism, accelerate the excretion of metabolites, is conducive to the absorption of nutrients in the body, at the same time is conducive to relax muscles, promote the recovery of fatigue. By allowing athletes to participate in beneficial social activities appropriately, they can meet the needs of the news media and the society. They can also deeply feel the satisfaction brought by hard training, so that athletes are full of hope for future training and have better psychological expectations.

## 4. Summary

With the development of modern competitive martial arts, the number of matches is becoming more and more frequent, and the difficulty and height of movements are constantly increasing and improving. Wushu athletes have to bear more and more load in sports training, and the probability of sports fatigue is also higher. Sports fatigue in Wushu athletes' training includes muscle fatigue, nerve fatigue and psychological fatigue, which is determined by the characteristics of Wushu sports and training. This article reviews the mechanism, diagnosis and recovery measures of martial arts athletes' sports fatigue, so that coaches and athletes can understand the symptoms of fatigue at any time during training, and provide good help for improving the physical health of martial arts athletes and improving their competitive level. Martial arts athletes' sports fatigue and recovery are a unified whole. They influence and interact with each other and jointly promote the deepening and development of training. In the specific practical operation process, according to the specific physiological function and training intensity of martial arts athletes, several appropriate measures and means are selected to cooperate with each other, which can accelerate the elimination of fatigue and the recovery of physical strength, so as to ensure the healthy development of athletes' body and sports training and improve the sports ability of martial arts athletes as soon as possible.

**References**

[1] Ma Fei. Pros and Cons of Premature Specialized Training for Young Wushu Routine Athletes [J]. Wushu Research, 2019 (7): 55-57.

[2] Ji, MinJoon. Analysis of injury types for mixed martial arts athletes [J]. Journal of Physical Therapy Science, 2016, 28 (5):1544-1546.

[3] James L P, Haff G G, Kelly V G, et al. Towards a Determination of the Physiological Characteristics Distinguishing Successful Mixed Martial Arts Athletes: A Systematic Review of Combat Sport Literature [J]. Sports Medicine, 2016, 46 (10):1525-1551.

[4] Structural differences in basal ganglia of elite running versus martial arts athletes: a diffusion tensor imaging study [J]. Experimental Brain Research, 2015, 233 (7):2239-2248.

[5] Tavares F, Smith T B, Driller M. Fatigue and Recovery in Rugby: A Review [J]. Sports Medicine, 2017, 47 (8):1515-1530.

[6] Schiffer T, MLlinger A, Sperlich B, et al. Kinesio Taping and Jump Performance in Elite Female Track and Field Athletes and Jump Performance in Elite Female Track and Field Athletes [J]. Journal of Sport Rehabilitation, 2015, 24 (1):47-50.

[7] Mascarenhas V V, Rego P, Dantas P, et al. Imaging Prevalence of Femoroacetabular Impingement in Symptomatic patients, Athletes, and Asymptomatic Individuals: a Systematic Review [J]. European Journal of Radiology, 2016, 85 (1):73-95.

[8] Lenetsky S, Nates R J, Brughelli M, et al. Is effective mass in combat sports punching above its weight [J]. Human Movement Science, 2015, 40:89-97.

[9] Lystad R P, Graham P L, Poulos R G. Epidemiology of training injuries in amateur taekwondo athletes:a retrospective cohort study [J]. Biology of Sport, 2015, 32 (3):213-218.

IOP Conf. Series: Materials Science and Engineering **914** (2020) 012017 doi:10.1088/1757-899X/914/1/012017

# Adsorption Characteristics of Ethyl Paraben from Aqueous Solution Using Rice Husk Biochar

**Meng Guo[1, 2, 3], Wenjing Chen[1, 2, 4], Han Ru[1, 2, 4], Lili Yang[1, 2, 4] and Yu Zhang[1, 2, *]**

[1]Shaanxi University of Chinese Medicine, Xianyang 712083, China
[2]Shaanxi Collaborative Innovation Center of Chinese Medicinal Resource Industrialization, Xianyang 712083, China
[3]The first clinical medical college of Shaanxi University of Chinese Medicine, Xianyang 712083, China
[4]College of pharmacy of Shaanxi University of Chinese Medicine 712083, China

*Corresponding author e-mail: zhangyu19432@163.com

**Abstract**. The rice husk was used as raw material to prepare the biochar adsorbent at 300°C, 400°C and 500°C, which were named BC300, BC400 and BC500. Ethyl paraben was subject to batch adsorption in aqueous medium by the prepared rice husk biochar. The results of adsorption experiments show that the preparation temperature of biochar significantly affects its adsorption effect on ethyl paraben. The adsorption capacity of ethyl paraben from three kinds of biochar was BC500 > BC400 > BC300. In addition, the initial concentration of ethyl paraben in waste water, the temperature and time of adsorption can also affect the adsorption effect. The reaction reached adsorption equilibrium after 270min. The results showed a theoretical basis for the removal of organic pollutants such as ethyl paraben.

## 1. Introduction

Rice husk, as an important renewable biomass resource, has attracted more and more attention [1-2]. Due to the decomposition of non carbon elements in the process of carbonization, biochar has formed a loose porous structure. As a useful adsorption material, biochar attracts more and more attention from researchers in adsorption of organic pollutants and improvement of soil environment [3-4]. Ethyl paraben (EP) is mainly used as antiseptic for food, cosmetics and medicine. In this study, rice husk was used as raw material to prepare biochar at different pyrolysis temperatures to adsorb EP in wastewater. The effects of pyrolysis temperature, initial concentration of EP, adsorption temperature and adsorption time on the adsorption efficiency of EP were studied. The adsorption characteristics of rice husk biochar on EP were revealed, which provided a reference for the resource utilization of rice husk.

## 2. Materials and Methods

### 2.1. Biochar preparation

The rice husk was taken from a rural area of Hanzhong City, Shaanxi Province. After cleaning and drying, it was dried in an oven at 60°C for 72h. After crushing, it was placed in a porcelain crucible, covered, and put into a muffle furnace. The temperature was raised to the target temperature (300°C, 400°C, 500°C) at a rate of 2°C/min, and then kept for 2h. Take it out after cooling, grind it through 60 mesh sieve, and store it in the dryer for standby. Marked as BC300, BC400, BC500.

### *2.2. Experimental design*

The removal efficiency of rice husk biochar at different pyrolysis temperature was studied by setting EP initial concentration gradients (20, 30, 40, 50, 60, 70, 80 mg/L).

The adsorption of rice husk biochar was studied at 25°C, 35°C and 45°C, respectively.

### *2.3. Determination of EP content*

The content of EP was detected by UV spectrophotometry. The maximum adsorption wavelength of EP was 247nm when the background solution was blank and scanned in the wavelength of 200-400 nm. Three parallel tests were carried out in each group, and the average value was taken. Accurately weigh 0.125g EP, put it into a 250ml volumetric flask, add distilled water to dilute it to the scale, and obtain 500mg/L EP stock solution. The solution was diluted to 10, 20, 30, 40 and 50mg/L with distilled water, and the adsorption values were measured with ultraviolet spectrophotometer (Shimadzu uv-2600, Japan) at 247nm. The regression equation was y = 0.0709x + 0.0127, $R^2$ = 0.9999 (n = 5).

### *2.4. Adsorption of EP by rice husk biochar*

Accurately weigh 0.1g BC300, BC400 and BC500 respectively, add 25ml EP solution with concentration of 20, 30, 40, 50, 60, 70 and 80mg/L respectively, place it in constant temperature oscillator, shake it for 4h under the temperature of 25, 35, 45°C and rotation speed of 150r/m respectively, and then pass through 0.45μm filter membrane, and use ultraviolet visible spectrophotometry (λ = 247 nm) to determine the concentration of EP after adsorption, three parallel experiments were carried out in each group, and the average value was taken. The amount of adsorption $q_e$ and removal rate $\eta$ of EP after reaching adsorption equilibrium are calculated by the following formula:

$$q_e = \frac{(C_0 - C_e)}{W} \cdot V \tag{1}$$

$$\eta = \frac{C_0 - C_e}{C_0} \times 100\% \tag{2}$$

Formula: $q_e$ is the adsorption capacity of equilibrium, mg/g, $C_0$ and $C_e$ are content of phenol before adsorption and after adsorption, mg/L, V is volume of solution, L, W is adsorbent dosage, g, respectively.

### *2.5. Adsorption changes with time*

Accurately weigh 0.2g BC300, BC400 and BC500 respectively, add 50ml EP solution with concentration of 50mg/L, place it in constant temperature oscillator, and fully oscillate at 30°C and 150r/m respectively at 5, 10, 30, 60, 90, 120, 150, 180, 210, 240, 270min. The residual EP concentration was measured, and the adsorption of EP by biochar was studied.

## 3. Results and Discuss

### *3.1. Adsorption of EP by rice husk biochar at different pyrolysis temperatures*

The adsorption efficiency of rice husk biochar to EP is different at different pyrolysis temperature. In the three selected temperature ranges, the removal efficiency increases with the increase of pyrolysis temperature which is shown in figure 1.

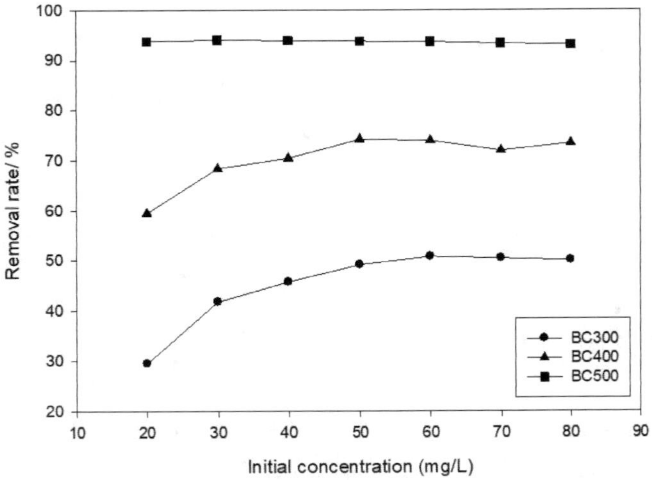


**Figure 1.** The removal rate of EP with different initial concentrations by rice husk biochar.

### *3.2. Effect of temperature on adsorption of EP by rice husk biochar*

The EP removal rates of BC300 after adsorption at 25, 35 and 45°C for 4 hours are shown in Figure 2. The removal rate of EP by rice husk biochar increased with the increase of temperature. The temperature has an obvious effect on the adsorption of EP.

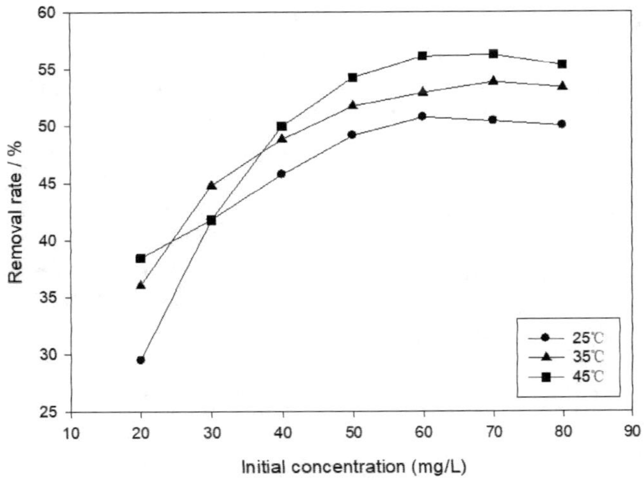


**Figure 2.** Effect of temperature on adsorption of different concentrations of EP.

### *3.3. Changes of adsorption rate of rice husk biochar to EP with time*

In the same adsorption time, the adsorption capacity of three biochar to EP is BC500 > BC400 > BC300. When the adsorption equilibrium is reached, the adsorption rate of BC300 is nearly 50%, that of BC400 is nearly 70%, and that of BC500 is almost 90%.

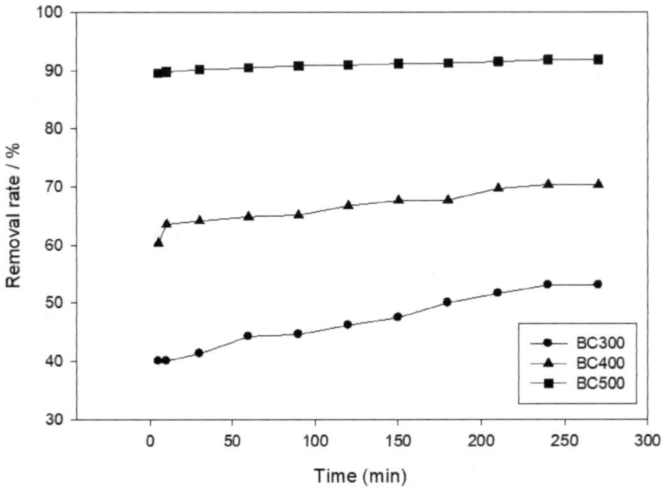


**Figure 3.** Relationship between adsorption rate and adsorption time of EP by rice hull biochar.

## 4. Conclusion

Rice husk biochar is a good absorbent, which has a broad application prospect in the field of sewage treatment. As a new type of absorbent, the raw material is easy to obtain and the preparation method is simple, which has high development and application value.

## Acknowledgments

This work was financially supported by the Innovation and entrepreneurship training program for College Students of Shaanxi University of Chinese Medicine (201805076).

## References

[1] Lingling Hu, Zhen He and Shipeng Zhang. Sustainable use of rice husk ash in cement-based materials: Environmental evaluation and performance improvement [J]. Journal of Cleaner Production, DOI: https://doi.org/10.1016/j.jclepro.2020.121744.

[2] Zahra Shamsollahi, Ali Partovinia. Recent advances on pollutants removal by rice husk as a bio-based adsorbent: A critical review [J]. Journal of Environmental Management, 2019, 246: 314-323.

[3] Deepshikha Pandey, Achlesh Daverey and Kusum Arunachalam. Biochar: Production, properties and emerging role as a support for enzyme immobilization [J]. Journal of Cleaner Production, DOI: https://doi.org/10.1016/j.jclepro.2020.120267.

[4] Mohammad Z. Rahman, Tomas Edvinsson and Philip Kwong. Biochar for Electrochemical Applications [J]. Current Opinion in Green and Sustainable Chemistry, DOI: https://doi.org/10.1016/j.cogsc.2020.04.007.

IOP Conf. Series: Materials Science and Engineering **914** (2020) 012018 doi:10.1088/1757-899X/914/1/012018

# Research on the Evaluation Model of Multi-objective Optimal Allocation of Water Resources in Guizhou Province

**Huan Wang[1], Zaimei Mei[1]**

[1]Guizhou Normal University, Guiyang 550025, China

**Abstract**. Aiming at the characteristics of multiple water sources and multiple users in Guizhou's urban water system, this paper establishes a multi-objective allocation model of urban water resources. The paper uses multi-objective programming to solve the model, and analyzes the impact of different target weights on the water resources allocation results of Guizhou Province in the fuzzy multi-objective decision model. The results show that the optimization model and fuzzy decision are reasonable and feasible.

## 1. Introduction

Generally speaking, multi-objective optimal allocation of water resources is to scientifically optimize the realization of water resources from the perspective of present and future, in order to facilitate the sustainable development, utilization, protection and management of water resources, so that multiple objectives can be optimized as best as possible in a given area at the same time. Guizhou Province is in the east of Yunnan Guizhou Plateau. Like most cities in China, many cities in Guizhou Province are close to medium and above rivers. There are 14 rivers managed by the provincial water administrative department (or above) in Guizhou Province, including Wujiang River, Liuchong River, Sancha River, Furong River, Qingshui River, Chishui River, Qingshui River and Wuyang River in the Yangtze River Basin, Huangni River, Duliu River, Beipan River, Hongshui River, Mengjiang River and Nanpan River in the Pearl River Basin. The current water supply sources of 21 cities (counties) are mainly the nearby big rivers, small and medium rivers, reservoirs, groundwater (Springs), etc. there are 10 cities directly taking water from the main stream of the big rivers, accounting for 47.6% [1]. The main contradictions and problems of the current water sources are as follows: among the current water sources, 7 cities have groundwater or spring water the proportion is 33.3%. Almost all these cities occupy the environment water. Every time there is a drought, the downstream section of the spring almost stops flowing, seriously damaging the ecological environment. In the current water supply sources, there are 11 cities with reservoir water sources, accounting for 52.4%. Almost all these cities are occupied with irrigation water and power generation water, resulting in the original irrigation surface cannot be fully irrigated, and the power generation efficiency of hydropower station is poor [2].

## 2. Relationship between water resources allocation system and water resource circulation system

The water resources system is a complex large system. Before human activities did not interfere with it, it was a natural system. The functions of precipitation replenishment, runoff, confluence, runoff, and conversion of surface water and groundwater were all performed in accordance with natural laws.

The water resource circulation system currently is a natural water circulation process. However, under the influence of human activities, the original water resources circulation system (including the structure of the water resources system, the runoff process, and the mechanism of action) has been artificially changed, which makes the original water resources system more complicated. Figure 1 below is a diagram of the relationship between the water resources allocation system and the water resource circulation system. In fact, the two cross each other and complement each other. There is no strict limit. This figure is only for the purpose of illustrating the relationship between the two.

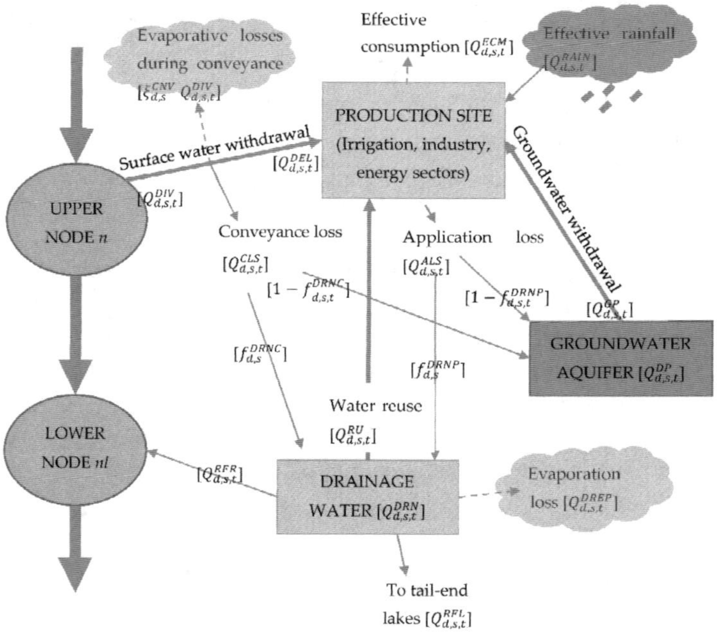


**Figure 1.** Relationship between water resources allocation system and water resource circulation system

The optimal allocation of water resources is to use systems engineering theory to optimize the allocation of regional or river basin water resources among various districts and water use departments. An optimization model with objective function and constraint conditions must be established.

First, it is necessary to divide the sub-regions, determine the water source route, and the water sector. Let the study area be divided into K sub-areas, k = 1, 2, ..., K; k sub-area has I (k) independent water sources and J (k) water use departments. There are M public water sources in the study area, c = 1, 2, ..., M. The amount of water allocated by the public water source c to the k sub-region is represented by $D_c^k$. Its water volume, like other independent water sources, needs to be distributed among water users. Therefore, for the k sub-region, it is the optimal allocation of water resources for I (k) + M water sources and J (k) water users. Second, you need to determine the model goals. The optimal allocation model of water resources for sustainable development pursues the greatest comprehensive benefits of society, economy and environment. According to the different methods of establishing objective function, it can be divided into multi-objective model and single-objective model. Finally, enumerate all constraints of the model.

## 3. Technical roadmap for regional water resources optimal allocation

In the study of the optimal allocation of regional water resources, it is necessary to seriously study the coordination between water resources management, socio-economic development and ecological environmental protection. In today's rapid social and economic development and the acceleration of urbanization, there are two main ways to solve the shortage of water resources: First, we must find our own reasons. In terms of water conservancy construction and water use, we must strengthen water conservancy projects Raising the effective utilization coefficient of canal water, improving the level of farmland irrigation technology, and improving the industrial technology level are beneficial measures for "throttling"; the other is to adopt "open source" measures. The general technical roadmap of regional water resources optimal allocation research is as follows [3]:

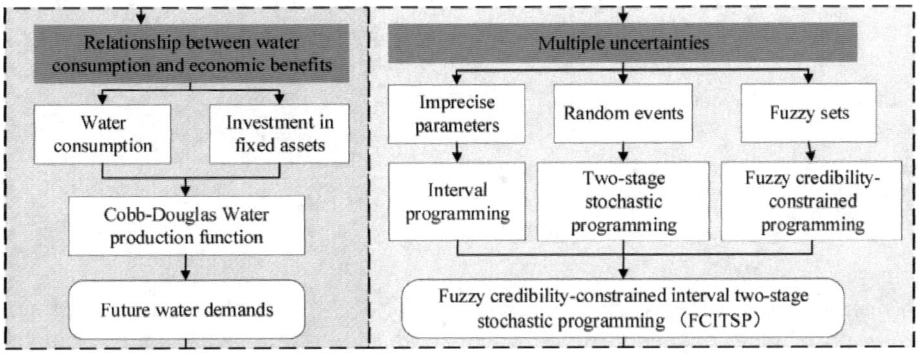


**Figure 2.** Technical route of regional water resources optimal allocation research

## 4. Establishment of optimal allocation model of water resources in Guizhou Province

The urban water resources system is a complex large system with multiple water sources and multiple water users. Its water sources include local fresh water resources, external water transfer, seawater, rainwater resources and recycled water. Water users include urban residents' domestic water, industry and ecology. The optimal allocation model of water resources, like the general optimization model, should consist of an objective function and constraints. The general form is as follows:

$$\max(\min)\left[f_1(x),f_2(x),f_3(x)\right] \quad (1)$$
$$s.t.G(x)\leq 0 \quad x\geq 0$$

In the above formula: x is the decision variable f1 (x), f2 (x), and f3 (x) are the economic benefits, environmental benefits, and social benefits ; G (x) is the constraint set, which indicates the water resource carrying capacity and environment Capacity, land resources, other social constraints and subsystem state equations. The key to establishing the above model is to determine the objective function and constraints [4].

### *4.1. Objective function*

Social benefits: Because social benefits are not easy to measure, they can be reflected indirectly by minimizing the total regional water shortage. Because the amount or degree of regional water shortage directly affects social development and stability, which is a side reflection of social benefits:

$$\min f_1(x)=\sum_{k=1}^{K}\sum_{j=1}^{J}D_j^k-\sum_{k=1}^{K}\sum_{j=1}^{J}\sum_{i=1}^{I}x_{ij}^k \quad (2)$$

In the above formula: $D_j^k$ is the water demand of the user j in the k sub-zone (10,000 $m^3$), and $x_{ij}^k$ is the water supply quantity of the user j in the k sub-zone i (10,000 $m^3$).

Economic benefits: quantified by GDP maximum:

$$\max f_2(x)=\sum_{k=1}^{K}\sum_{j=1}^{J} gdp_j^k \sum_{i=1}^{I} x_{ij}^k \tag{3}$$

In the above formula: $gdp_j^k$ is the unilateral aquatic production value coefficient of the production sector of sub-k of j, which is represented by unilateral water GDP.

Environmental benefits: Environmental issues directly related to the use of water resources are quantified by the maximum guarantee rate of water supply for ecological environment users:

$$\max f_3(x)=\sum_{k=1}^{K}\sum_{i=1}^{sI}\sum_{j=1}^{sJ} x_{ij} / \sum_{k=1}^{K}\sum_{j}^{sJ} S_j \tag{4}$$

In the above formula: $S_j$ is the water demand of the ecological user j (10,000 m$^3$).

### 4.2. Constraints

In order to determine the most reasonable water supply for each water source, the water supply should be as close as possible to the configuration target, that is, the objective function value of the optimization model is the largest or the smallest, and the water supply of each water source must meet certain conditions, which is reflected in the set of constraints of the optimization model. on.

Constraints on the amount of water available from the source are as follows. Where $W_i$ is the water supply of the i water source (10,000 m$^3$).

$$\sum_{j}^{J} x_{ji} \le W_i \tag{5}$$

The project water delivery volume is as follows. In the above formula: $Q_{ij}$ is the water delivery capacity (10,000 m$^3$) of water source i to user j.

$$x_{ij} \le Q_{ij} \tag{6}$$

The water sector's water demand constraints are as follows. In the above formula: $D_{j,\min}^k$ , $D_{j,\max}^k$ , respectively, are the lower and upper limits of the water demand change for users in k sub-zone j.

$$D_{j,\min}^k \le \sum_{i}^{I} x_{ij} \le D_{j,\max}^k \tag{7}$$

Non-negative constraints on decision variables:

$$x_{ij} \ge 0 \tag{8}$$

### 4.3. Solving the Multi-Objective Model

Because the objective functions in multi-objective optimization problems are often unfair, there is no unique solution. In contrast, there are only valid solutions, that is, there is no better solution than the effective solution within the allowed range. There are many solutions to multi-objective optimization problems. The weights and methods are introduced below. This method transforms the objective function problem into a scalar problem of the sum of all objective weights, namely: $\min_{x\in\Omega} f(x)=\sum_{t=1}^{m}\omega_t * F_t(x)^2$. Among them, there are many methods for selecting the weighting factor

$\omega_t$, including expert scoring method, $\alpha$ method, tolerance method and weight factor decomposition method. This paper uses MATLAB programming to solve the multi-objective model [5].

## 5. Optimization results and analysis

Starting from the principle of coordinating economic, social and environmental benefits, priority is given to domestic water use, followed by ecological and industrial water, and finally the principle of agricultural water. Different initial values are assigned to various variables. MATLAB optimization functions are used to solve Multi-objective programming problem, can get the optimal configuration scheme with different initial values. After comparison and optimization of various parameters, the results of the configuration scheme determined by the final calculation are shown in Table 1 and Table 2.

**Table 1.** Optimized water resources allocation plan (2010) (10,000 $m^3$)

| Water resources | Surface water | groundwater | Sewage treatment | Outflow water diversion | Shallow brackish water | total |
|---|---|---|---|---|---|---|
| Thermal power industry | 940 | 230 | 20 | 180 | 4410 | 5780 |
| General industry | 2140 | 3220 | 30 | 540 | 8010 | 13940 |
| Rural industry | 4140 | 8050 | 50 | 1040 | 15540 | 28820 |
| Life of urban residents | - | 10940 | - | 1570 | - | 12510 |
| Urban utilities | 1340 | 2890 | 20 | 310 | 4020 | 8580 |
| Rural farmland irrigation | 39900 | 116480 | 240 | 11790 | - | 168410 |
| Rural forestry, animal husbandry and fishery | 810 | 1340 | 20 | 250 | 2090 | 4510 |
| Life of rural residents | - | 12820 | - | 1110 | - | 13930 |
| Rural livestock life | - | 2950 | - | 710 | - | 3660 |
| ecosystem | 5340 | 13550 | 60 | 110 | 20950 | 40010 |
| total | 54610 | 172470 | 440 | 17610 | 55020 | 300150 |

**Table 2.** Optimized allocation plan of water resources (2015) (10,000 $m^3$)

| Water resources | Surface water | groundwater | Sewage treatment | Outflow water diversion | Shallow brackish water | total |
|---|---|---|---|---|---|---|
| Thermal power industry | 930 | 730 | 20 | 330 | 4520 | 6530 |
| General industry | 2120 | 4230 | 30 | 740 | 9020 | 16140 |
| Rural industry | 4110 | 10060 | 50 | 1430 | 15440 | 31090 |
| Life of urban residents | - | 9450 | - | 3650 | - | 13100 |
| Urban utilities | 1330 | 2690 | 20 | 370 | 5020 | 9430 |
| Rural farmland irrigation | 39530 | 111480 | 240 | 10680 | - | 161930 |
| Rural forestry, animal husbandry and fishery | 800 | 1940 | 20 | 280 | 1010 | 4050 |
| Life of rural residents | - | 13840 | - | 700 | - | 14540 |
| Rural livestock life | - | 3550 | - | 340 | - | 3890 |
| ecosystem | 5300 | 14550 | 60 | 1850 | 20050 | 41810 |
| total | 54120 | 172520 | 440 | 20370 | 55060 | 302510 |

IOP Conf. Series: Materials Science and Engineering **914** (2020) 012018 doi:10.1088/1757-899X/914/1/012018

The overall benefit function of the allocation plan for each planning year is based on first meeting the premise of domestic water use. The final allocation plan shows that in the 2010 water allocation plan, industrial water consumption is 485.4 million $m^3$, domestic water consumption is 31 million $m^3$, and public utility water consumption. 85.8 million $m^3$, urban public utilities use 85.8 million $m^3$, ecological water consumption is 40.1 million $m^3$, forest, animal husbandry and fishery water consumption is 45.1 million $m^3$, agricultural irrigation water consumption is 168410 $m^3$; industrial water consumption in the 2015 water allocation plan is 537.6 million $m^3$, domestic water consumption is 315.3 million $m^3$, public utility water consumption is 94.3 million $m^3$, ecological water consumption is 418.1 million $m^3$, forest, animal husbandry and fishery water consumption is 45.1 million $m^3$, agricultural irrigation water consumption is 16.193 million $m^3$; industrial water consumption and ecology Water consumption continues to increase. Due to water-saving measures and adjustment of planting structure, agricultural irrigation water continues to decrease [6].

## 6. Conclusion

The problem of multi-objective optimization of water resources allocation has been increasingly valued by water resources management and decision makers. Many scholars have also used various algorithms to solve some practical multi-objective optimization problems, but each algorithm has its own Defects of its own. Therefore, with the continuous development of various multi-objective optimization algorithms, it is necessary to further study more effective methods to solve the multi-objective optimization problem of water resources allocation.

## References

[1] Wang, R, Li, Y, & Tan, Q. A review of inexact optimization modelling and its application to integrated water resources management. Frontiers of Earth Science, 9 (1) (2015) 51-64.

[2] Binglong, Wang, Yanpeng, Xinâ An, Yin, & Qian, et al. Integrated approach of system dynamics, orthogonal experimental design and inexact optimization for supporting water resources management under uncertainty., 31 (3) (2018) 1-30.

[3] Jiang, Q., Zhou, Z., Wang, Z., Fu, Q., Wang, T., & Zhao, Y. Risk assessment and optimization of water resources shortage based on water and land resources coupling., 33 (12) (2017) 136-143.

[4] Liu Yuan, Jianzhong Zhou, Zijun Mai, & Yuanzheng Li. Random fuzzy optimization model for short-term hydropower scheduling considering uncertainty of power load. Water Resources Management an International Journal Published for the European Water Resources Association, 31 (9) (2017) 2713-2728.

[5] Mojtaba Shourian, & S. Jamshid Mousavi. Performance assessment of a coupled particle swarm optimization and network flow programming model for optimum water allocation. Water Resources Management, 31(12) (2017)1-19.

[6] Maryam Soleimani-Alyar, Alireza Ghaffari-Hadigheh, & Fatemeh Sadeghi. Controlling floods by optimization methods. Water Resources Management, 30 (12) (2016) 1-10.

IOP Conf. Series: Materials Science and Engineering **914** (2020) 012019 doi:10.1088/1757-899X/914/1/012019

# Design and implementation of code system for measuring instruments

**Chang Hu, Yanan Wu, Ying Zheng, Xiao Zhang**

Xi'an Institute of Measurement and Testing Technology, Xi'an, Shanxi, 710068, China

**Abstract**. This article introduces the preparatory preparation, structural principles, development process, application effects and existing problems of the mandatory verification measuring instrument code management system, and comprehensively expounds the birth, development and mature process of the first mandatory verification measuring instrument code management system in China.

## 1. Introduction
In November 2015, China's first compulsory verification measuring instrument management system went online in Xi'an. The compulsory verification of measuring instruments in Xi'an has been fully initiated. Since then, every strong measuring instrument in Xi'an has its own identification number. In the past three years, the code assignment work has fundamentally solved the long-term, unclear foundation and weak supervision of the mandatory inspection.

## 2. Reason for starting code management
The strong inspection system has been implemented in China for more than 30 years, and there have been serious problems such as "unclear foundation", "weak supervision", "concealment, omission of reports, non-reportation, rejection of inspections and over-cycle use". The main reasons for these problems are:

a). The uniqueness of the equipment information cannot be guaranteed. The same equipment is repeatedly built and cannot be accurately tracked and located. There is no way to talk about the strong inspection base, coverage rate and re-inspection rate.

b). The census, filing, updating, and supervision work was concentrated in the district (county) quality supervision branch bureaus, creating a management bottleneck. As a result, the quality of strong inspection data collection is not high, and a large amount of dirty data is formed.

c). The filing and updating procedures are cumbersome, and the existing data of the verification agency cannot be used. Increasing the burden on enterprises and low cooperation among enterprises.

d). Supervision data cannot be used by the public, and there is a lack of social supervision.

## 3. Preparing for code management work

### *3.1. Overall thinking*
The fundamental method to solve the above problem is to let each appliance have its own identity code, that is, to implement the tracking management of the entire inspection process and life cycle of the appliance by assigning a code. And build "network intelligent archives", "measurement verification data sharing and real-time updates", "real-time dynamic supervision of grassroots quality supervision

bureaus", "online supervision of metrology management departments", "user online management appliances", "system automatic warning "Six in one new dynamic management model.

### *3.2. basic research*
The "Catalogue of Compulsory Verification Work Measurement Apparatus" shall be detailed into the scientific name of each appliance in the applicable scope of the verification regulations. And establish the "instrument scientific name" and its verification rules, the verification cycle (longest), the use of appliances, inspection methods, units of measurement, accuracy levels, main parameters.

### *3.3. Coding scheme*
Take FRID as the carrier. It adopts the coding method of "user organization code, three-in-one code, personal ID card + branch code + user assigned code sequence number". At the same time, each user has established his own library of built-in appliances.

### *3.4. Choice of handheld terminal RF instrument*
By scanning the two-dimensional code on the measuring instrument, the handheld terminal radio frequency instrument can observe all the relevant information of the instrument, namely the basic information of the instrument, the calibration information of the instrument, etc. The supervisory department can check whether the device is the device by observing the basic information of the device by grasping this information, and further determine whether the measuring device meets the regulatory requirements through the verification information. As shown in Figure 1.

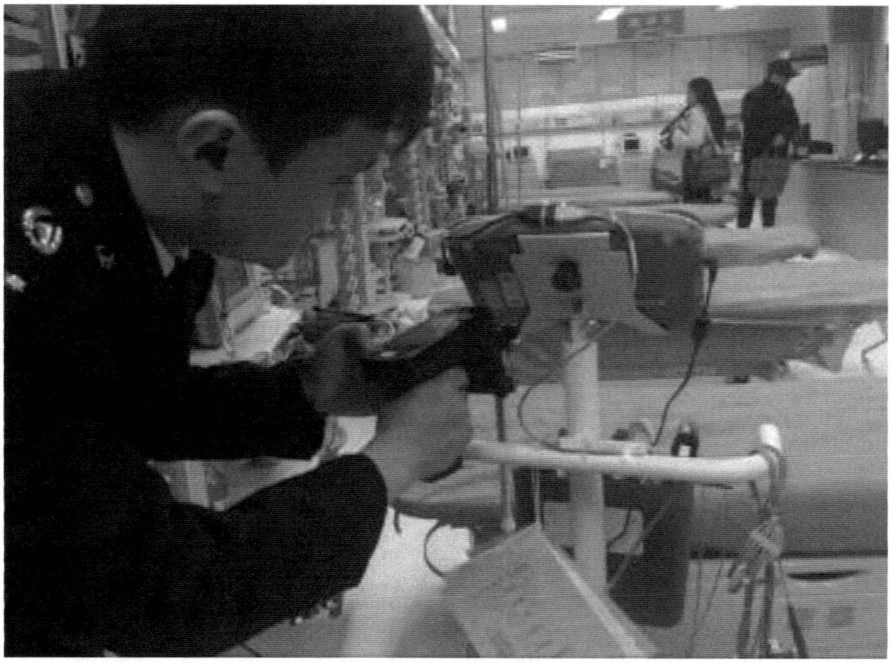

**Figure 1.** Field information collection map

## 4. Code Assignment Management System Design

### *4.1. Process Design*
Through repeated demonstration of the code assignment process, a strong detection code assignment process as shown in FIG. 2 is finally determined.

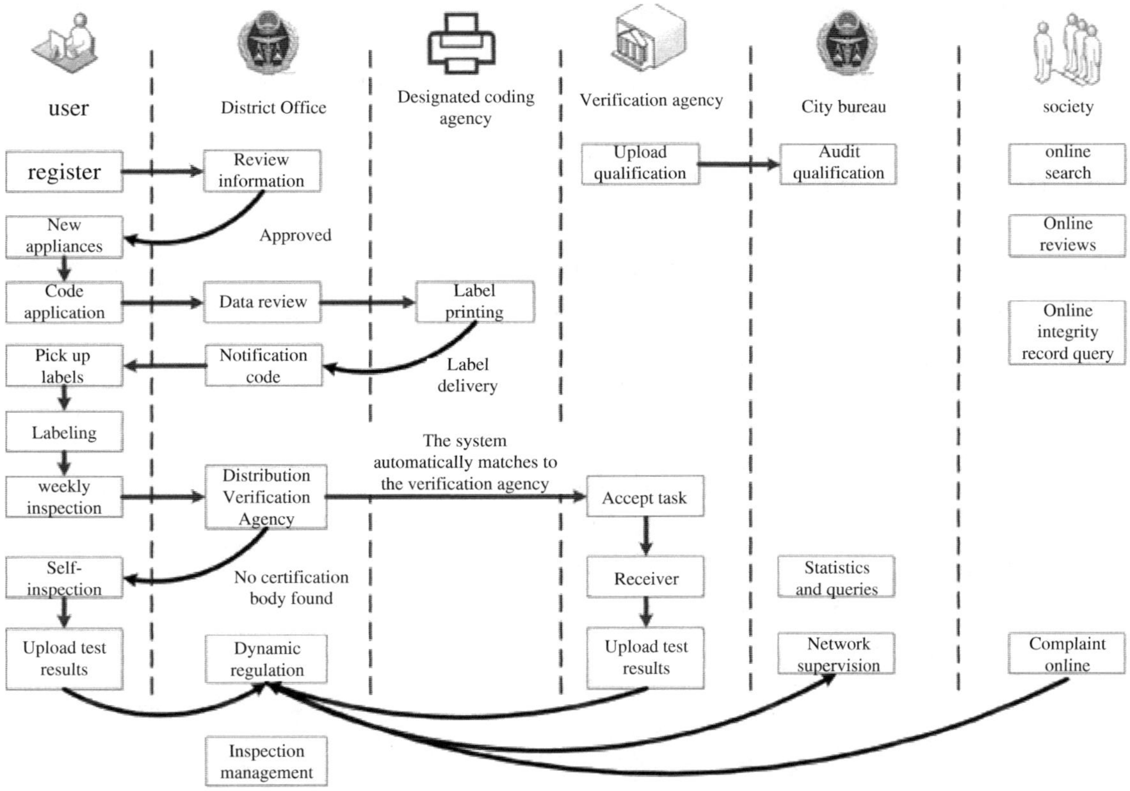


**Figure 2.** Strong check code flow chart

*4.2. system design*

First, the use of modern Internet of Things technology solves the problem of the ambiguity of the appliance information, and implements the tracking and management of the entire life cycle of the appliance; secondly, through the establishment of a mandatory verification of the basic database and data associations, a unified, standardized and automatic data Matching and intelligent error correction. At the same time, the management of strong inspection was extended to the technical level, and problems such as over-range inspection were eliminated; again, through the design of the code assignment system, functions such as online application, online approval, online inspection, and progress inquiry were realized. Fourth, through regular exchanges with certification bodies. Realized automatic data update. Finally, through the application of the handheld terminal, online inspection and appliance reception were realized.

## 5. Code Assignment Management System Development

The strong inspection code system includes five modules: "user management module", "metering basic information module", "code assignment module", "report inspection module", and "analysis and statistics module".

*5.1. User Management Module*

The user management includes enterprise users, inspection agencies, various supervisory departments and Xi'an Quality Management Supervision Bureau. This article mainly introduces the management of enterprise users. The management of enterprise users is verified by various regulatory authorities on the unified social credit code certificates or organizational certificates of enterprises in their jurisdictions, and it is clear that the measuring instruments they use are within the jurisdiction of their jurisdiction.

### *5.2. Measurement basic information module*

According to the "Medical Catalogue of Compulsory Verification of Work Measuring Apparatus", establish scientific names for each type of instrument, and at the same time establish the association between the scientific name and the name of the measurement parameter, the unit of measurement, and the accuracy level, so as to realize automatic loading of the measurement characteristic parameters of the device.

### *5.3. Code Assignment Module*

After the new measuring instrument is approved, the system will automatically assign a unique code to each measuring instrument, that is, the identity code of the measuring instrument, as shown in Figure 3.

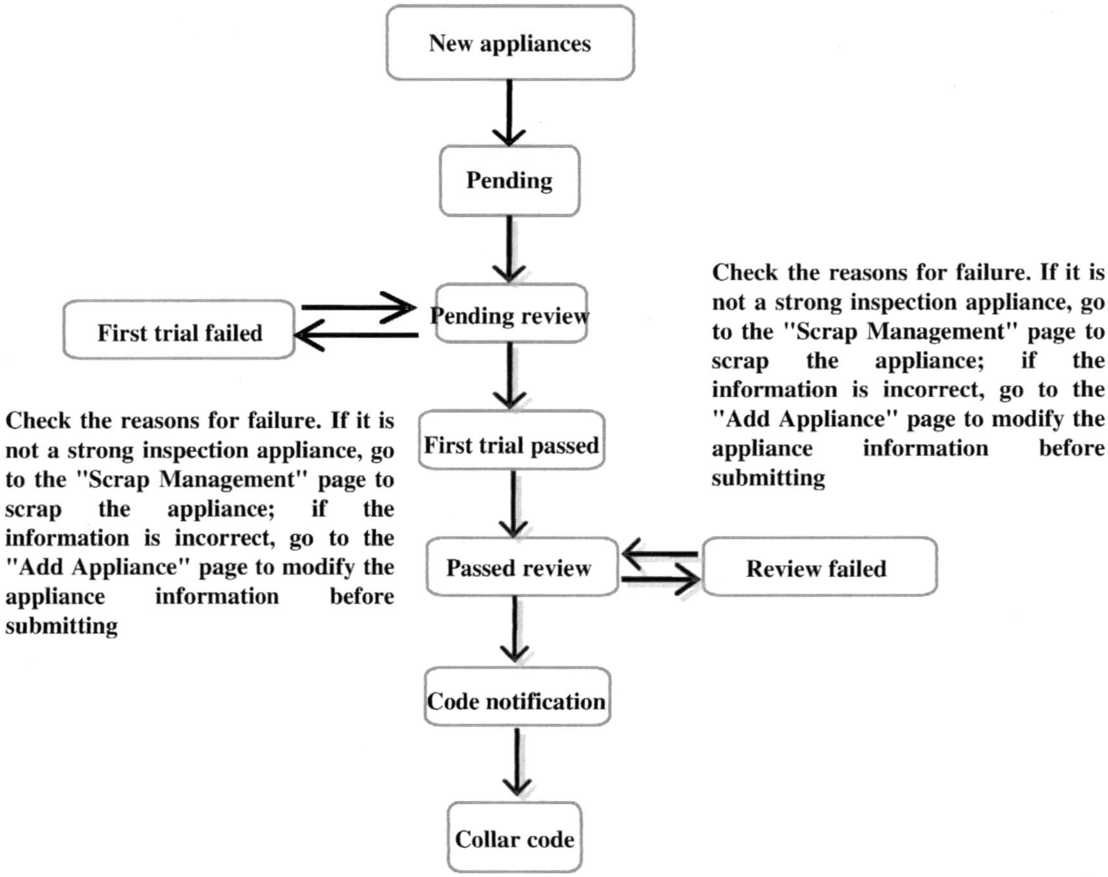


**Figure 3.** Code flow chart

### *5.4. Inspection module*

Within one month before the expiry date of the measuring instrument, the user can apply for an inspection online. After the application for the inspection is submitted, the system will initially match the qualification capabilities of the measuring instrument with the verification institution and find a suitable verification institution, and then submit it to the administrative agency Regional sub-bureaus are reviewed by the sub-bureaus and assigned to the verification agency. When the verification mechanism in the system can match the measuring instrument, the inspection process in the system is adopted, as shown in Figure 4.

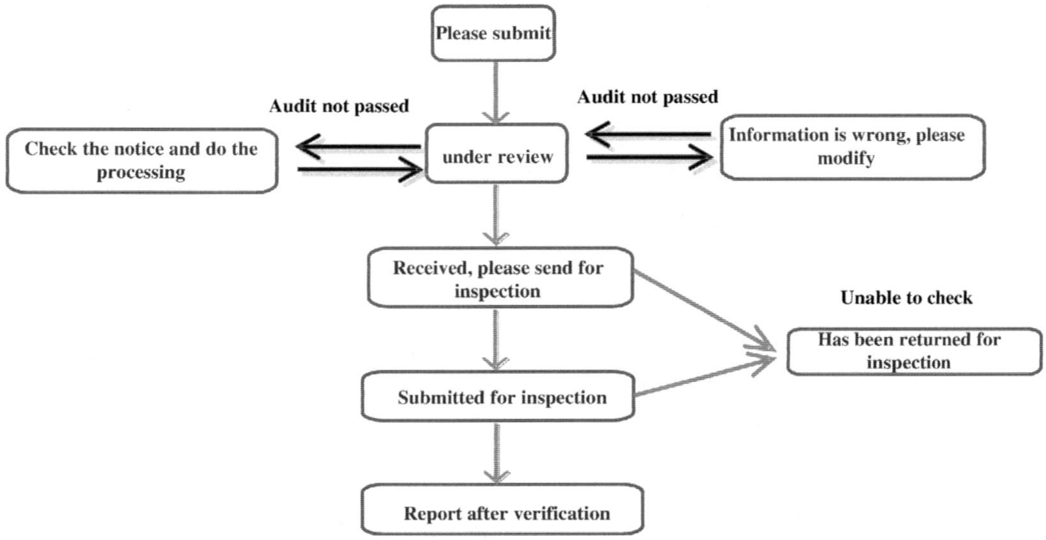


**Figure 4.** In-system inspection flow chart

If the system cannot find a verification organization that meets the verification capability, it applies for verification to a verification organization outside the system through the process of reporting outside the system, as shown in Figure 5.

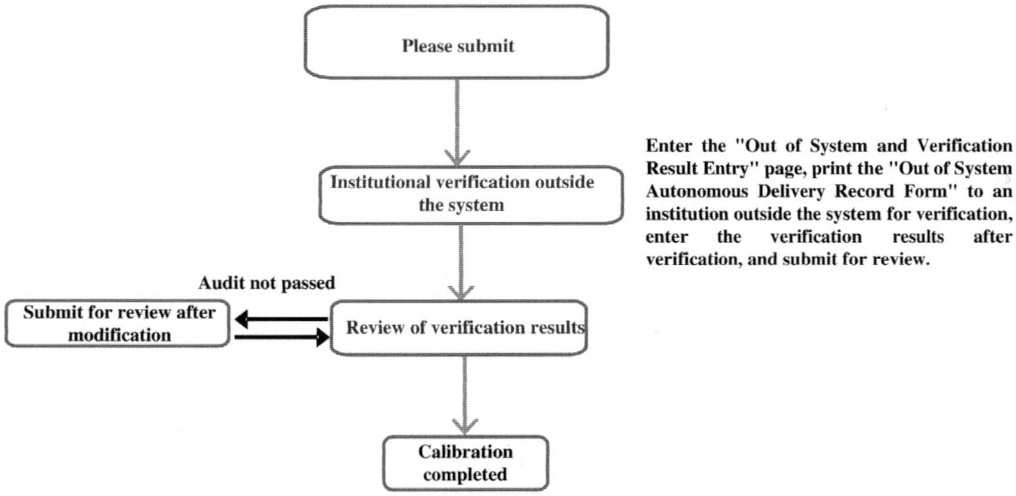


**Figure 5.** External system inspection flow chart

### *5.5. Analysis and Statistics Module*

Supervision departments can perform statistical analysis on the compulsory verification measurement instruments in the city through various statistical methods. Such as statistics by appliance use, appliance classification, regulatory area, administrative area, and inspection agency.

## 6. Running result

Since Xi'an's code assignment work was officially launched in November 2015, the city has completed code assignments in the areas of gas refueling, medical physicochemicals, stores, supermarkets, and trade markets, and a large number of concealed, missed, unreported, and rejected inspection equipment have been Detect and achieve lifetime control. In Xi'an, there has been a blowout increase in the number of cases, inspections, and enforcement of strong inspection meters. In addition, the system takes into account the needs and interests of all parties. It is mainly reflected in the following six aspects:

### *6.1. For users*
a). Implement the user's responsibility for filing the subject;

b). After the compulsory inspection is stopped, ensure that the appliances that really need to be inspected complete the inspection on time and quality;

c). Maximize and simplify the process to facilitate the masses;

### *6.2. For regulators*
a). Free from cumbersome censuses, filing, and data updates;

b). Real-time supervision, precise law enforcement, due diligence, and reduction of supervision risks for strong inspection equipment in the jurisdiction;

### *6.3. For measurement administration*
a). Accurate statistics of forced inspections in the jurisdiction, and provide the government with evidence of due diligence;

b). Real-time supervision and dynamic assessment of regulatory departments and inspection agencies to identify problems in a timely manner;

### *6.4. For metrology institutes*
a). Implementation of graded inspections to resolve bottlenecks caused by centralized inspections after free inspections are free of charge.

b). Find any missing items in the strong inspection project in time;

### *6.5. For the public*
a). Draw a barcode or log in this system to query the entire process information of the strong inspection appliance from sales to scrapping;

b). Violations can be reported online.

### *6.6. For other government departments*
a). Provide various types of strong inspection data to government departments in a timely and accurate manner;

b). Form a closed management loop and reduce regulatory risks.

## 7. Problem
a). Because the user needs to fill in the measurement characteristics of the appliance during the code assignment process. Although the system has established a large number of manufacturers, model specifications and measurement characteristics. However, due to the lack of collection summary and artificial intelligence, there are still many measurement characteristics that users need to fill in based on data or experience, and the system did not provide reference information filled in by the same model and manufacturer in the past.

b). In the process of data access of the certification body, the existing database of the certification body needs to be modified. Individual institutions are unwilling to share data due to their own considerations, and their cooperation is low.

c). FRID labels have problems with extremely small dimensions, surface contamination, and all-metal surfaces.

d). The handheld terminal has not been explosion-proof certified and cannot be used in places with explosion-proof requirements, such as gas stations and gas stations. Only after the relevant information is copied out, inspections can be conducted outside the site.

IOP Conf. Series: Materials Science and Engineering **914** (2020) 012019 doi:10.1088/1757-899X/914/1/012019

## 8. Conclusion

After three years of operation and improvement, the management system for compulsory verification measurement instruments has strengthened the management of compulsory verification measurement instruments in Xi'an, which greatly facilitates enterprises to apply for inspection. However, the above-mentioned problems still exist. To make the system more effective Convenient and intelligent. A more detailed measurement basic database should be established, and the system can provide the value range of the measurement characteristics of the appliance according to the scientific name and specifications selected by the user; at the same time, the design of the system should be considered from the perspective of the enterprise Instead of leaning on administrative management, optimize the process, making the process of coding and inspection of measuring instruments easier and more efficient.

## References

[1] Notification of the General Administration of Quality Supervision, Inspection and Quarantine on the Establishment of Compulsory Verification of Measuring Instruments. [Z] .2006
[2] Measures of Xi'an for measurement supervision and management. [Z] .2014-06-10
[3] JJF 1051-1996 Technical specifications for naming and classification codes of work measuring appliances [S]

IOP Conf. Series: Materials Science and Engineering **914** (2020) 012020 doi:10.1088/1757-899X/914/1/012020

# A New Insulate Method of Ship-Swaying Interference

**Min Tao, Hao Zhang, Yu Wang**

China Satellite Maritime Tracking and Control Department, Jiangyin 214431, China

**Abstract**. When Ship-borne radar tracking target need to solve the problem of ship shaking disturbance at first. This article analyzed the deficiency of the current methods for the isolation ship-swaying, proposed a new insulate method based on gyro swinging loop. Which used inertial space velocity of the gyro sensor signals instead of speed measuring machine speed to complete the closed-loop system, and used a kalman filter to reduce the toll of gyro output noise. The theoretical analysis and simulation results show that the method can greatly improve the degree of isolating ship rocking.

## 1. Introduction

The ship's roll isolation reflects the ship's radar servo system's ability to offset the ship's roll and stabilize the antenna beam's ground pointing. It is also an important design indicator of the ship's radar servo system. At present, the shipboard radar servo system has designed measures such as self-tracking, ship roll forward, and gyro stabilizing loop to achieve the isolation of the ship roll. The anti-shipping capability of the self-tracking loop is the capability of the servo system and has no effect on the tracking performance of the servo system. Therefore, it is the most stable and reliable anti-shipping solution. Due to the influence of factors such as system nonlinearity and inaccurate mathematical model, the degree of isolation of ship rolls that can be achieved is low. When the sea conditions are poor, the system's requirements for ship roll isolation cannot be met, and the target cannot be tracked normally. In order to improve the shortcomings of the gyro stabilizing ring scheme, this paper proposes a new method of isolating the ship rocking to meet the needs of the system.

## 2. Basic structure of shipborne radar servo drive

The shipboard antenna drive unit is divided into azimuth and pitch driving branches. The pitch branch adopts DC motor double-chain drive, and the azimuth branch adopts DC motor dual / quad-chain drive. The motor is controlled by electric anti-backlash method. Three-loop control is adopted in the design. At the same time, in order to achieve electrical clearance, it is equipped with a bias torque ring and a differential speed suppression ring to ensure the uniform force of multiple motors. Take the pitch branch as an example to briefly introduce the basic structure of the servo drive. The pitch branch is a dual motor drive system. In order to ensure high measurement accuracy and high dynamic performance, the typical three loop is used as the conventional servo system design. The current loop is the innermost loop, which is realized by modifying the current control loop of the motor controller; speed loop It is an intermediate loop and a loop designed independently. It synthesizes speed control information, bias torque information, differential suppression information, etc. to the current loop; the position loop is the outermost loop, which is a closed loop. Design control algorithms to achieve precise antenna control. Due to the characteristics of the working environment of shipboard measurement and control equipment, a gyro ring to suppress ship sway was added. Two gyros were fixed at specific positions of the antenna to sense the rotation speed of the antenna relative to the reference, and the induced electrical signal was processed and fed into the speed loop So as to achieve the purpose of overcoming ship sway. The block diagram of the system is shown in Figure 1.

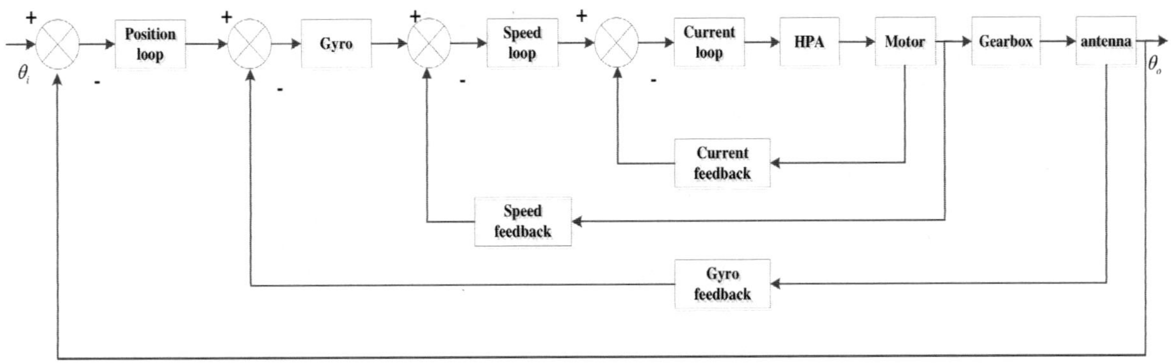


**Figure 1.** Block Diagram of Shipborne Radar Servo Control

### 3. Gyro feedback method to isolate the ship's rocking principle

The gyro feedback method is one of the most commonly used ship rock stabilization schemes, and can achieve a certain degree of ship rock isolation. The dual-axis rate gyroscope is installed on the reference plane of the pitch fork arm of the antenna base, and can sense the velocity information of the inertial space of the azimuth axis and the pitch axis. The velocity information is used for feedback closed-loop control to realize the design of the gyro stability loop. The principle is shown in Figure 2.

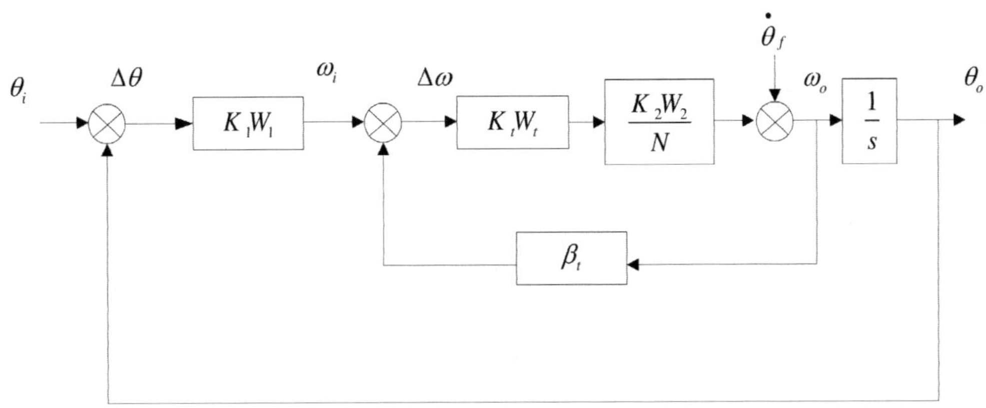

**Figure 2.** Functional block diagram of gyro feedback method

Supposing the disturbance isolation degree of the gyro stabilizing loop scheme is $L = L_{zt} + L_t$, there are:

$$L = 20\lg\frac{\theta_f}{\Delta\theta} = 20\lg(1+K_{zt}W_{zt}) + 20\lg(1+K_tW_t) = L_{zt} + L_t \tag{1}$$

After adding the gyro stabilization loop, although the overall ship roll isolation is improved, the frequency band of the original tracking loop is narrowed, the characteristics of the controlled object are deteriorated, and the tracking performance is reduced. Therefore, in practical applications, there is a "no

IOP Conf. Series: Materials Science and Engineering **914** (2020) 012020 doi:10.1088/1757-899X/914/1/012020

gyro-tracking" working mode, that is, when the ship's roll is not very serious, the gyro ring is disconnected and the ship's roll is isolated by the self-tracking ring to ensure tracking performance.

## 4. The principle of gyro speed ring isolation ship rocking

The basic idea of the gyro speed loop is to replace the speed signal of the tachometer with the inertial space velocity signal induced by the gyro to complete the speed loop closed-loop work. The inner and outer loops of the servo system are: current loop, gyro speed loop and position loop. In this way, the ship's turbulence is inside the gyro speed loop, which becomes its internal disturbance. According to the principle of automatic control, the loop has a strong suppression effect on the internal low-frequency disturbance during closed-loop control. The bandwidth of the gyro speed loop is designed at 20 Hz, and the disturbance frequency of the ship's roll is about 0.08 Hz. Therefore, the gyro speed loop has a strong suppression effect on the ship's roll. At the same time, the turbulence of the ship is also inside the position ring, which is also suppressed by the position ring. The block diagram of its loop is shown in Figure 3.

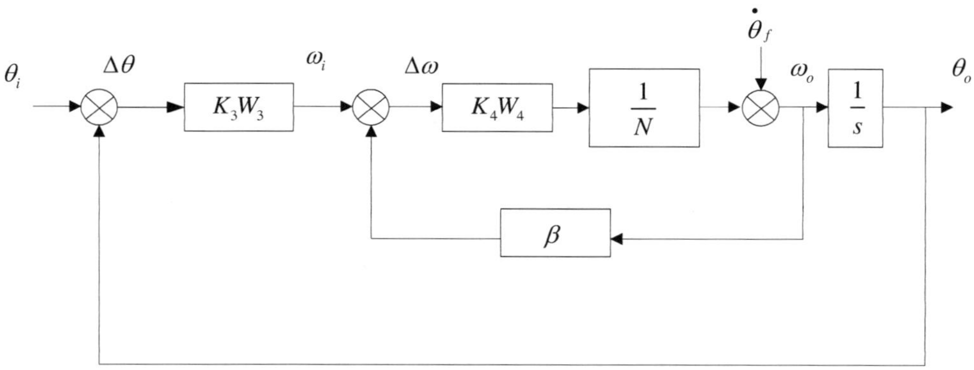


**Figure 3.** Functional block diagram of gyro speed loop

At this time, the error transfer function of the ship sway disturbance is:

$$E(s)=\frac{\Delta\theta}{\theta_f}=\frac{\Delta\theta}{\dot{\theta}_f\frac{1}{s}}=\frac{N}{(Ns+K_4W_4\beta S-K_3W_3K_4W_4)\frac{1}{s}}=\frac{1}{1+\frac{K_4W_4\beta}{N}-\frac{K_3W_3K_4W_4}{Ns}} \tag{2}$$

In the parameter design, in order to ensure the consistency with the original system position loop and speed loop parameters, you can make $\beta=\beta_v=v\beta_t$, where v is the conversion coefficient. The speed loop feedback coefficient is the same. In order to ensure that the forward channel transfer function of the gyro speed loop is consistent with the forward channel transfer function of the original speed loop, $K_4W_4=NK_vW_v$, where $K_vW_v$ is the original speed loop forward channel transfer function. Therefore, the open-loop and closed-loop transfer functions of the gyro speed loop are:

$$W(s)=K_4W_4\frac{1}{N}\beta=NK_vW_v\frac{1}{N}\beta_v=K_vW_v\beta_v \tag{3}$$

$$\phi(s)=\frac{K_4W_4\frac{1}{N}}{1+K_4W_4\frac{1}{N}\beta}=\frac{K_vW_v}{1+K_vW_v\beta_v} \tag{4}$$

After the parameter design, the open loop and closed loop transfer functions of the gyro speed loop are consistent with the original speed loop. $K_3W_3$ is the position loop regulator of this scheme. Designing $K_3W_3 = \frac{1}{N}K_1W_1$ can ensure that the open loop transfer function of the system position loop is also unchanged, so it will not affect The tracking performance of the system.

According to the mathematical model of the shipborne radar, the ship's roll isolation can be calculated as:

$$L_1 = 20\lg(1+\frac{1764(0.048s+1)}{s^2(0.0016+1)} - \frac{29615.5(0.008s+1)(0.5s+1)}{s^2}\times\frac{(0.048s+1)}{s^2(0.0016+1)}) \tag{5}$$

It can be seen that the gyro speed loop scheme obtained through theoretical analysis has a very high degree of rocking isolation, which fully meets the index requirements. However, in the actual use of the gyro, the influence of the vibration of the hull, the thermal noise generated by the internal signal processing of the gyro and the transmission process will cause unnecessary noise interference in the output data. These noises are amplified in the open-loop control May reduce the system tracking accuracy and stability. Therefore, the measurement data of the gyro must be filtered before it can be added to the system loop.

## 5. Kalman filter algorithm

### *5.1. Fundamental*

Kalman filtering is a linear minimum variance estimation method. The basic idea is to establish a state space model that uses signals and noise, use the previous value of the previous time and the observation value of the current time to update the estimate of the state variable, and calculate the estimated value of the current time. This filtering algorithm is suitable for real-time processing and computer operations.

Let the state equation and measurement equation of the stochastic linear discrete system be:

$$\begin{aligned} X_k &= \phi_{k|k-1}X_{k-1} + \Gamma_{k|k-1}W_k \\ Y_k &= H_kX_k + V_k \end{aligned} \tag{6}$$

Suppose $W_k$ and $V_k$ are uncorrelated zero-mean Gaussian white noise, ie:

$$\left.\begin{aligned} &E[W_k]=0 \\ &E[V_k]=0 \\ &E[W_k \bullet V_k^T]=0 \\ &E[W_k \bullet W_k^T]=Q_k, Q_k \geq 0 \\ &E[V_k \bullet V_k^T]=R_k, R_k > 0 \end{aligned}\right\} \tag{7}$$

The iteration process of the Kalman filter algorithm using state equations and measurement equations is as follows:

$$\begin{aligned}
\hat{X}_{k|k-1} &= \phi_{k|k-1}\hat{X}_{k-1|k-1} \\
P_{k|k-1} &= \phi_{k|k-1}P_{k-1}\phi^{T}_{k|k-1} + Q_k \\
K_k &= P_{k|k-1}H_k^{\,T}(H_kP_{k|k-1}H_k^{\,T} + R_k)^{-1} \\
P_k &= (1 - K_kH_k)P_{k|k-1} \\
\hat{X}_{k|k} &= \hat{X}_{k|k-1} + K_k(Y_k - H_k\hat{X}_{k|k-1})
\end{aligned} \tag{8}$$

### 5.2. *filter*
Because the output rate gyroscope is the rate of change in the angle of the antenna, the observation errors of the filter measurement equations are coupled to each other, resulting in the coupling of the filter measurement equations. Although the state equations of the filter are assumed to be independent of each other, in order to ensure the filtering effect, the filter still needs to be designed as a six-state filter, resulting in a very complicated filter calculation process. Since we use a rectangular coordinate system, based on relevant data and engineering experience, it can be approximated that the measurement errors are not related to each other, so the filters can be designed as two three-state filters, which will greatly simplify the filter calculation process It has no obvious effect on the filtering performance. Therefore, this paper uses this linearized decoupling filtering algorithm.

### 5.3. *Filter effect*
Taking the output voltage of the pitch gyro as an example, Kalman filtering is used to process the gyro output data. The filtering result is shown in Figure 4. The blue line in the figure is the output value of the gyro before filtering, the red line is the output value of the gyro after filtering, and the right figure is a partially enlarged view of the left figure. It can be seen from the figure that after Kalman filtering can effectively reduce the gyro noise.

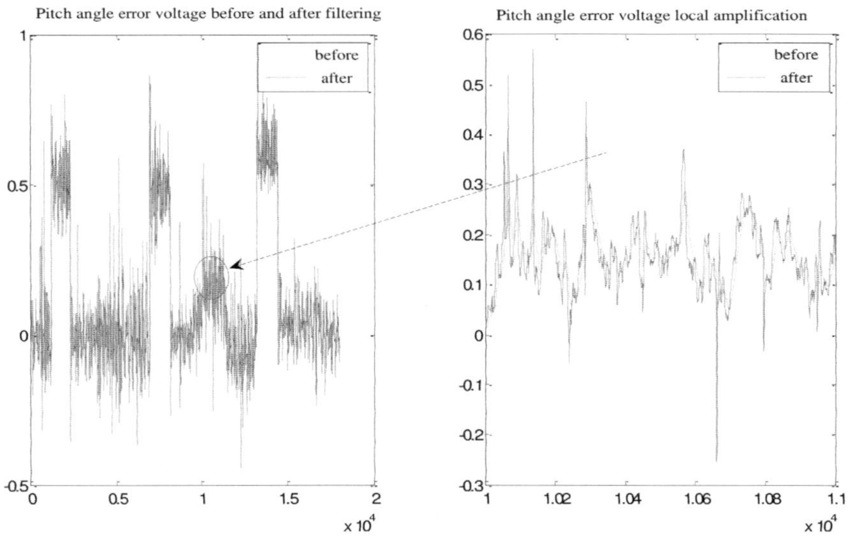


**Figure 4.** Output curve before and after pitch gyro filtering

## 6. Simulation analysis
The MATLAB / Simulink simulation environment is used to realize the simulation of the gyro speed loop scheme. In the gyro speed loop, the output of the rate gyro is the measurement of the angular

velocity of the azimuth and pitch axis in the inertial coordinate system. The Kalman filter algorithm is implemented in MATLAB with m language, and the function is called in MATLAB function module in Simulink. The speed output of the Kalman filter is added as feedback information to the speed loop simulation block diagram shown in Figure 5. The simulation data is output to the MATLAB workspace for processing, and the total ship roll isolation at this time is 67dB.

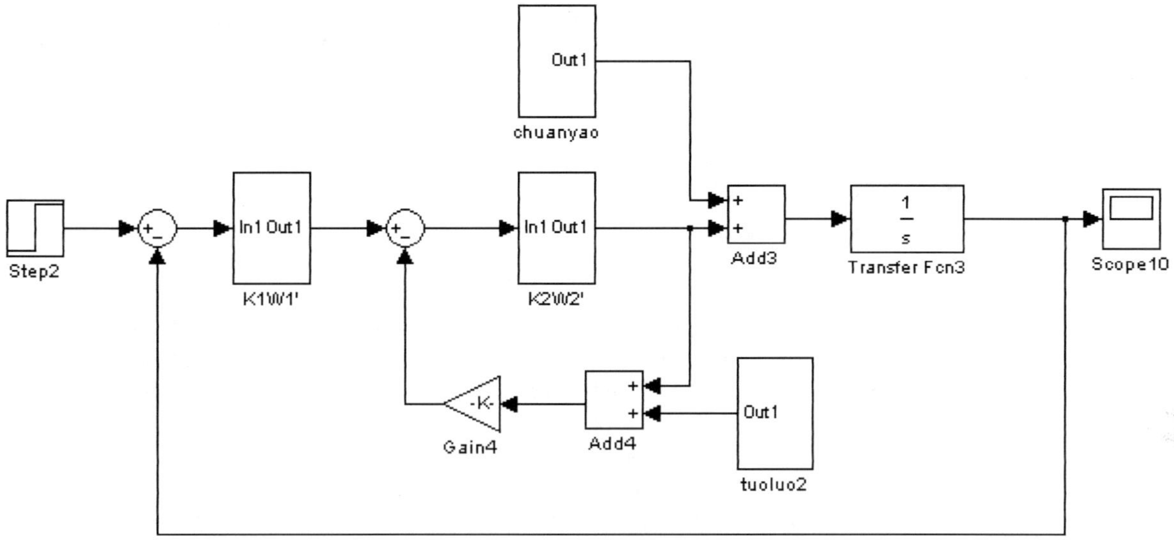


**Figure 5.** Gyro speed loop simulation block diagram

## 7. Conclusion

This paper analyzes the limitations of several current ship-shake isolation schemes, and proposes a new method of ship-shake compensation that uses the inertial space velocity signal induced by the gyro to replace the speed signal of the tachometer to complete the construction of the closed-loop gyro speed loop. Processing the noise in the output data of the gyro, and analyzing this method through simulation can greatly improve the system isolation without changing the tracking performance of the system.

## References

[1] Qu Yuanxin, Zhou Jinbiao, Xie Yong. Research on the automatic test method of ship's rocking isolation of shipborne measurement and control equipment [J]. Journal of Aircraft Measurement and Control. 2006, 25 (4): 62-65.

[2] Liu Sifang, Luo Youming. Research on Vc ++-based Ship Rocking Isolation Automatic Test [J]. Microcomputer Information. 2008, 24 (4): 36-37, 53.

[3] Wang Heng, Li Yonggang, Chen Liang, et al. Research and application of feedforward data processing method for surveying ship's yaw [J]. Journal of Aircraft Measurement and Control. 2011, 30 (2): 64-69.

IOP Conf. Series: Materials Science and Engineering **914** (2020) 012021 doi:10.1088/1757-899X/914/1/012021

# Design and implementation of Vibration Detecting System of Ship-borne Radar

**Min Tao, Hao Zhang, Yu Wang**

China Satellite Maritime Tracking and Control Department, Jiangyin, Jiangsu 214431

**Abstract**. Although redundancy design has been adopted by measure angle computer (MAC), but the switches between the host and the backup were by mankind which can't catch up with requirement of system real time. This paper analyses the task of primary-backup switching and confirm the time of auto switching are system fault and communication interruption. It proposes a new strategy by using the characteristic of dual -network of MAC by which add fault status words in net send data bags and changes the net send time series.This strategy doesn't need change hardware and only last one frame data on switching process which great enhancing the reliability of MAC.

## 1. Introduction

During the survey ship's mission at sea, the shipborne measurement and control radar has been working in complex and severe sea conditions such as surge, heavy load, strong impact, fatigue, and corrosion for a long time, which puts forward higher requirements for improving the reliability of the radar system. The sky servo feed sub-system is the key equipment of the shipborne measurement and control radar. Its main task is to always effectively track the target, complete the effective reception and transmission of the radio frequency signal, and send the angle measurement data to the ship central machine in real time through the measurement and control system Ethernet. In the process of equipment development, the use of modular design and the selection of high-quality, high-stability, low-failure components are the guarantee of system reliability. The computer is the core component of the Tianwei feed sub-system. It is responsible for system control, status monitoring, signal acquisition, information transmission, and data processing. High reliability is one of the important requirements for the computer. Redundancy design is an important means to improve the reliability of the system. Some components of the system are repeatedly configured. When the system fails, the redundantly configured components intervene and assume the work of the failed components, thereby reducing the system's failure time. Redundancy includes hot backup and cold backup of the system. The cold backup of the system generally has problems such as long switching time and delayed task timing, which does not meet the real-time requirements of the Tianshi feedback system. Although the angle measuring computer in the current day servo feed sub-system adopts a redundant design, the switch between the main and standby machines is still carried out through the table control buttons. Therefore, the judgment of the fault state will affect the timing of the system's master-slave switchover. In view of this situation, this paper uses the existing equipment status of the radar system to transform the angle measuring computer to realize the automatic switch of the angle measuring computer and improve the reliability of the system.

## 2. Working principle of angle measuring computer

The angle measuring computer is an important part of the sky servo feed sub-system, it is the center of data acquisition, data transmission and data calculation in the sky servo feed sub-system, and one of the key equipments in the tracking and measurement system. The main functions include:

①. Complete the operation control and status monitoring of the Tian servo feed sub-system;

②. Receive center computer, system monitoring station / radar data processor, inertial navigation, low-light TV and other equipment data to realize real-time data reporting;

③. Complete the management and control of the antenna working mode, and realize the remote control of the sub-system;

④. Complete antenna polarization control, program control of program-controlled attenuator and power combiner;

⑤. Complete temperature detection, vibration detection and central body humidity detection of each monitoring point of the Tian servo feed sub-system;

⑥. Complete real-time display and recording of measurement data, post-event data processing and text printing in corresponding format; complete sub-system assisted automated testing.

The angle measuring computer switches the main and standby machines through the tabletop key switch. The output signal of the switch is divided into three channels: one to the dual-machine switching unit, which controls the relay on the dual-machine switching board to realize the external transmission of the host serial port information; the other two After the PCI-1712 boards in the two computers are collected, the status is judged in the software code. The software first judges the A / B machine. If it is the A machine, the collected switch state is assigned to the Isonline function; if it is the B machine, the collected switch state is inverted and assigned to the Isonline function to realize the host network data External transmission. It can be seen that although the current angle measuring computer adopts a redundant design, the dual-machine switching is still manually implemented, and the timing of the switching depends on the degree of recognition of the system fault status by the post staff. Once the host machine fails and cannot switch the standby machine in time, it will cause data interruption and loss of control instructions, which affects the continuity of task data and poses a hidden danger to the safe operation of the system.

## 3. Angle measuring computer interface signals and classification

The angle measuring computer is the data center of the sky servo feedback system. There are many types of data and different interface forms. It is necessary to classify the interface signals and perform different forms of processing for different signal forms or sources when switching between two machines. For convenience of description, the two angle measuring computers are called A and B respectively. When the system is working, two computers receive data at the same time, according to the online status of A or B, the host sends data externally, but the standby does not send data.

## 4. Key technology of dual-machine switching

When the system is running, the master and backup machines switch because the host has failed or the host communication is interrupted and the standby machine works normally. In order to avoid unnecessary switching or the main and standby machines compete for priority, repeated switching causes the system to be unavailable. Real-time and accurate monitoring of the equipment status of the master and backup machines, and at the same time report their respective status to the other party. Therefore, the key technologies for dual-machine switching include fault monitoring and status reporting.

### *4.1. Fault monitoring*

Simply monitoring the fault state of the computer has no practical significance for the operation of the sky servo system. Since the angle measuring computer is the center of data interaction between the

IOP Conf. Series: Materials Science and Engineering **914** (2020) 012021 doi:10.1088/1757-899X/914/1/012021

equipment of the sub-system and the data transmission channels are not the same, the angle measuring computer can be divided The fault monitoring result of the system is used as the fault state of the computer to determine the timing of the two-machine switching. The main function of sub-system fault monitoring is to monitor the working status of each replaceable unit. When a fault occurs, it can quickly and accurately locate the fault location and the smallest replaceable unit. At the same time, it sends an alarm through sounds and indicators to remind the user Handle in time. The fault monitoring function is implemented by means of unit self-inspection, mutual inspection, and reporting. All faults and status information are collectively collected to the angle measuring computer for centralized processing and display. Failure monitoring technologies include: ①. The built-in test (BIT) function is set in the circuit unit of the B code shaft angle encoder and the polarization controller, and the status bit is set in the digital communication with the angle measuring computer. Through these methods, the operating status of each replaceable unit can be monitored. ②. Mutual inspection logic is set between the units to ensure the accuracy of fault monitoring. ③. Check the data continuity by adding the packet sequence number in the unit communication protocol. Therefore, real-time and accurate monitoring of various transmission information of the angle measuring computer can be realized.

### *4.2. Status report*

Although the angle measuring computer can realize fault monitoring, only after the monitoring results are reported to the other party can the other party's fault status be known, so as to decide whether to switch the master and backup machines. At present, "heartbeat" technology is generally used to monitor the working status of the other party. The A / B machine regularly sends the fault status, "heartbeat", through a dedicated channel, and then the two parties judge whether the other party is working properly based on the received heartbeat or whether enough heartbeat signals are received within a certain time. Since the angle measuring computer needs to receive and send data to the internal network and the external network at the same time, and the reception and transmission of data are completed in the same time period, the fault status can be sent to the same network at the same time Status), by analyzing the data received in the same cycle, the fault status of the other party can be known, and then the online status of the A / B machine can be automatically switched.

## 5. Implementation method of dual-machine switching

### *5.1. Network timing*

Intranet data exchange is carried out at a frequency of 20 Hz. The timer sends interrupt signals such as 1 Hz, 20 Hz, 20 Hz transmission, and 20 Hz reception to the computers on the network. The 20 Hz transmission signal is different for each network node, and each node is at the corresponding 20 Hz Send a data message when the transmission interruption time comes. The 20Hz reception interrupt signal indicates that all network node data has been sent. The 20Hz interrupt and 1Hz interrupt are used for time synchronization and generating the number of beats required in the network data. At present, the angle measurement computer first judges the online status of the main and standby machines, and then sends the number of the network. Therefore, the number of angle measurement computers sent to the intranet is arranged at the 21ms of the 50ms repetition period. In order to ensure that the two computers send numbers simultaneously without causing data loss, the timing of the two computers sending numbers must be adjusted to different times, which can be achieved by connecting the reserved timing signal of the timing board to one of the angle measuring computers.

### *5.2. Data transmission*

In the internal network, each sub-system device sends data packets through broadcast or multicast through SOCKET, and performs data exchange according to a fixed time sequence. The exchange frequency is 20 Hz. In order to report the fault monitoring results to the intranet, it is necessary to modify the format of the data frame sent by the network and add the fault status word to the original data frame. The length of the data frame sent by each device on the internal network is 256 bytes. By querying the internal information transmission convention, the format of the data frame is shown in Table 1. Because the maximum length of the data content of each device is only 192 bytes, there is enough space in the padding part to add the fault status word.

**Table 1.** Intranet data frame format

| Unified message format | Message length | Subsystem number | Data length | Data content | Filled part |
|---|---|---|---|---|---|
| 6bytes | 2bytes | 1byte | 2bytes | N bytes | 245-N bytes |

The angle measuring computer already has the functions of fault monitoring and system inspection, so it is only necessary to convert the monitoring results into fault status words. The fault status word is defined as follows (here only represents the received data), filled to the 253-256 bytes of the data frame. Since the total length of the data frame is not changed, it will not cause any impact on the transmission and reception of network data.

Bit0: voltage signal failure, 0 means no failure, 1 means failure;
Bit1: Pulse signal failure, 0 means no failure, 1 means failure;
Bit2: Serial information failure, 0 means no failure, 1 means failure;
Bit3: Network data failure, 0 means no failure, 1 means failure.

### *5.3. Data reception*

Each device in the internal network adopts an event-driven mechanism to receive data, and stores all received data in the receive buffer. When the receive interrupt signal sent by the timer arrives, it means that all devices have been sent. The data is processed. In order to detect the data transmission status of each device, you can define the first 2 bytes of the receive buffer to indicate the status of the received data packet. The corresponding relationship is as follows. In each bit, 0 indicates that the data is invalid and 1 indicates that the data is valid. Among them P3 represents the data transmission situation of A machine, P4 represents the data transmission situation of B machine.

**Table 2.** Packet status

| Bit | 15~11 | 10 | 9 | 8 | 7 | 6 | 5 | 4 | 3 | 2 | 1 | 0 |
|---|---|---|---|---|---|---|---|---|---|---|---|---|
| Node | Stain | P4 | P23 | P22 | P21 | P15 | P12 | P9 | P7 | P3 | P2 | P1 |

### *5.4. Two-machine switch*

After the system is running, the angle measurement computer scans the status of the main and backup machines, data packet status and fault status in each cycle. According to the logical relationship between the three, the fault status and communication status of the main and backup machines can be judged. It can realize automatic switching between the main and standby machines. When working in real time, the two computers simultaneously receive data from the external network and the internal network, and perform corresponding real-time data processing at the same time. After the processing is completed, the data is sent to other computers on the internal network at the specified intranet transmission time nodes, but in Only one computer sends data to the external network in the same cycle. After the host is started, in the first 50ms cycle, only the "Enable Host" command is sent to the internal network, and the second cycle begins to send data to the internal network and the external network at the same time. At the same time, the network status is monitored at any time. Interruption, that is, send the "switch standby command" to the intranet in the same cycle, the next cycle will no longer send data to the external network, the backup machine monitors the network status at any time, after receiving the host intranet data is abnormal or not received in the host When the network and the external network data, replace the host to send data to the external network. The switching process is shown in Figure 1.

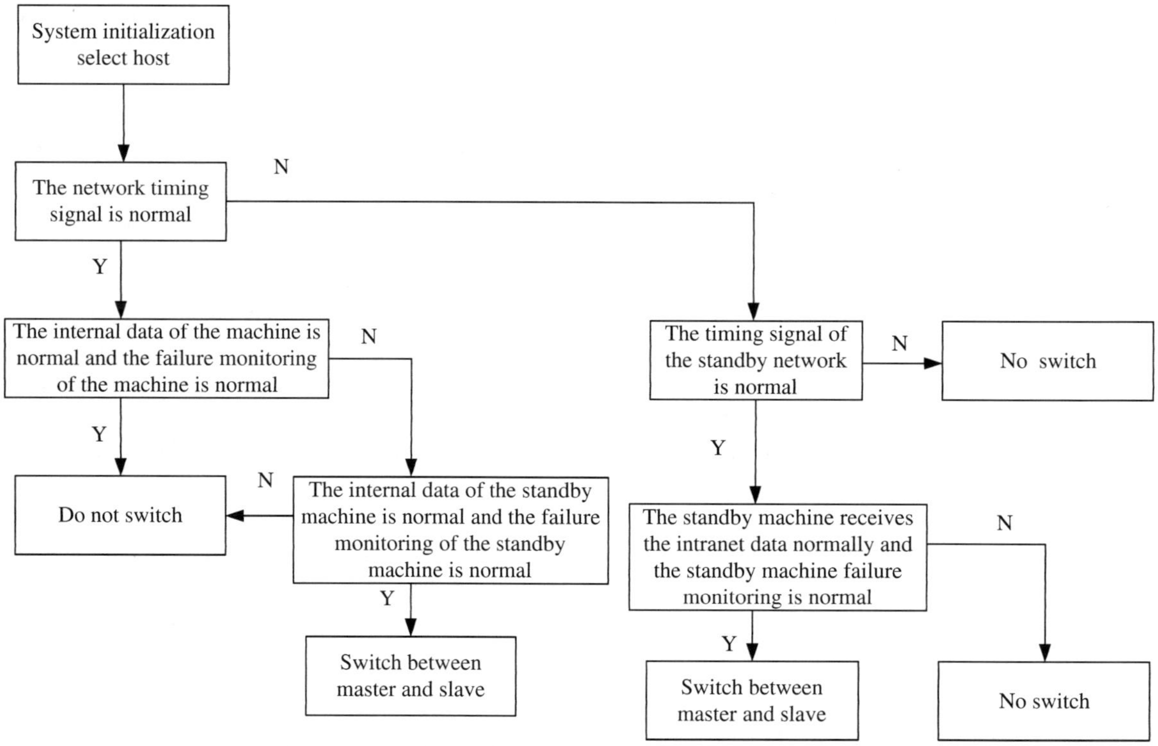


**Figure 1.** Flow chart of dual-machine switching

### 5.5. *Result analysis*

In order to verify the data loss during the two-machine switchover process, we analyzed the data received by the external network center machine. During the test, the network cable of the external network of the A machine was manually pulled out. The data received by the central machine is shown in Table 3 below. It can be seen from the table that due to the period of network interruption, machine A is used to send a command to switch the standby machine to stop sending data to the external network. Therefore, the central machine receives data as a reframe, and the next cycle starts to send data to the external network The data received by the central unit is back to normal.

**Table 3.** The center receives data

| Time | Azimuth | Pitch angle | distance | Azimuth error | Pitch error | AGC | Data Sources |
|---|---|---|---|---|---|---|---|
| 10:25:16.300 | 13.256° | 28.032° | 963.032 Km | 0.02V | 0.01V | 19 dB | machine A |
| 10:25:16.350 | 13.324° | 28.104° | 962.859 Km | 0.01V | -0.01V | 19 dB | machine A |
| 10:25:16.400 | 13.409° | 28.179° | 962.648 Km | -0.02V | 0.02V | 21 dB | machine A |
| 10:25:16.400 | 13.409° | 28.179° | 962.648 Km | -0.02V | 0.02V | 21 dB | machine A |
| 10:25:16.500 | 13.586° | 28.267° | 962.107 Km | 0.03V | 0.02V | 21 dB | machine B |
| 10:25:16.550 | 13.647° | 28.346° | 961.892 Km | 0.01V | 0.01V | 22 dB | machine B |

## 6. Conclusion

In view of the shortcomings of the current angle measurement computer dual-machine cannot be automatically switched, which is difficult to meet the real-time requirements of the system, the task of dual-machine switching and the fault monitoring content of the main and backup machines are introduced. The conditions for automatic switching are determined as system failure and communication interruption, and According to the current equipment status, using the characteristics of the dual-network communication of the angle measurement computer, the dual-machine automatic switching is realized. When the dual-machine switching operation is performed, only the 50ms period of the switching is used to process the switching command without data transmission, which is extremely Improve the reliability of the system. This method does not require the establishment of a dedicated heartbeat channel, no hardware changes, and only requires partial changes to the software and timing relationships.

## References

[1] Li Kang, Chen Xuejun, Shi Xiang, et al. Application of time-frequency analysis in fault diagnosis of shipborne antenna and ship power system [J]. Telemetry and Remote Control, 2011, 32 (4): 33-39.

[2] Yang Ling, Huang Hai, Hou Ying. Design of dual-system backup of embedded system based on distributed storage [J]. Computer and Modernization. 2007, (2): 16-21.

[3] Wang Jiangjiang, Li Zhiqiang, Zhao Liang. Research on active and standby switching of dual-machine hot backup system [J] Railway Communication Signal.

[4] Lu Jinyi. Angle measurement subsystem of aerospace measurement and control system [M]. Xi'an: 39 Institute of China Electronics Technology Group Corporation. 2006: 102-136.

[5] Yang Yucheng, Wu Wansheng, Le Bin. Research on the key technology of system master-slave switch [J]. Avionics Technology, 2013, 44 (1): 20-23.

[6] Di Chang, Xia Zhang, Qiong Liu, Ge Gao, and Yue Wu. "Location based robust audio watermarking algorithm for social TV system." In *Pacific-Rim Conference on Multimedia*, pp. 726-738. Springer, Berlin, Heidelberg, 2012.

[7] Di Chang, Xia Zhang, and Yue Wu. "A Multi-Source Steganography for Stereo Audio." *Journal of Wuhan University (Natural Science Edition)*, 2013(3): 277-284.

[8] XIa Zhang, Di Chang, et al. "An Audio Steganography Algorithm Based on Air-Channel Transmitting." *Journal of Wuhan University (Natural Science Edition)* 57, no. 6 (2011): 499-505.

[9] Xia Zhang, Di Chang, et al. "Tree-like Dimensionality Reduction for Cancer-informatics." In *IOP Conference Series: Materials Science and Engineering*, vol. 490, no. 4, pp. 042028. IOP Publishing, 2019.

[10] Wang Fei, Cao Guijun. Protection scheme for heartbeat failure of dual-machine hot backup system [J]. Railway Computer Application, 2013, 22 (3): 35-37.

IOP Conf. Series: Materials Science and Engineering **914** (2020) 012022 doi:10.1088/1757-899X/914/1/012022

# Real-Time Object Detection based on Divergent Activation

**Hui Liang[1, a], Yueyuan Wang[1, b], Yuan Ma[2, c] and Kewen Liu[2, *]**

[1]Xi'an Satellite Control Center, Xian, China

[2]School of Information Engineering, Wuhan University of Technology, Wuhan, China

*Corresponding author e-mail: liukewen@whut.edu.cn, [a]lianghwui @163.com, [b]576622378@qq.com, [c]yuan.ma@whut.edu.cn

**Abstract**. Object detection is a research hotspot in the field of computer vision and has significant research value in the fields e.g. human-computer interaction, video surveillance, automatic driving, face recognition, medical imaging and etc. This paper proposes a novel real-time object detection algorithm based on YOLOv2 and divergent activation which aims to extract semantically complementary and discriminative features. The introduced divergent activation mechanism can be divided into differential divergent activation and hierarchical divergent activation that obtains the spatial complementary features by fusing different spatial distribution information of objects at the same level and activates a complete range of objects by fusing complementary semantics of objects from different levels, respectively. Experimental results on PASCAL VOC2007 and PASCAL VOC2012 show that our proposed algorithm reached 81.0% on mean average precision(mAP), indicating the effectiveness of the proposed algorithm.

## 1. Introduction

Object detection refers to locating and identifying the region of interest efficiently and accurately in images. With the booming evolution of computer science, it has become a research hotspot in the field of computer vision and has significant research value in the fields e.g. human-computer interaction, video surveillance, automatic driving, face recognition, medical imaging and etc. [1]

In recent years, the convolutional neural network (CNN) has achieved success in the field of object detection, and CNN-based algorithms have become mainstream ideas for designing object detection algorithms. Compared with the traditional algorithms, CNN-based algorithms have better feature representation ability and higher accuracy. It can learn favourable high-level feature representations, which can be effectively used for object location and recognition, with sufficient training data.[2]

Based on our observation, the feature extraction part of the YOLOv2[3] algorithm network tends to extract only local features, ignoring the integrity of the object, resulting in a decrease in locating accuracy, which in turn affects detection accuracy. To solve the above problem, this paper proposes a novel real-time object detection algorithm based on YOLOv2 and divergent activation which aims to extract semantically complementary and discriminative features.

## 2. Related Work

Due to the speed limitation of algorithms based on region proposals, this paper mainly focuses on algorithms based on regression.

In 2016, Redmon et al. proposed YOLOv1 [4], which divides the input image into S*S cells. Each cell is responsible for regressing B boxes (including the location and classification) whose center points fall within the cell. The B boxes share the same category probability. YOLOv1 can predict at once whether all boxes contain the objects' confidence and category probability with a single glimpse, and output detection results, with the sacrifice of a little accuracy, whose detection speed deployed on Titan X can reach 45 FPS.

Although YOLO greatly improves the detection speed, there still exist many problems that should be solved. Combining the advantages of YOLO and Faster R-CNN [5] algorithm, Liu et al. adopted multi-scale feature extraction for location regression and object classification to adapt to multi-scale detection tasks, and proposed SSD [6]. Liu et al. proposed that low-level large-scale feature maps contain more low-level texture details, suit-able for learning small-scale targets, and high-level small-scale features contain more semantic details, suitable for learning large-scale targets. Furthermore, Liu et al. replaced a convolutional layer instead of a fully connected layer at the output of the YOLO to regress the detection results and set several prior boxes with different scales and aspect ratios to adapt to different shapes.

To solve the aforementioned problems of YOLOv1, Redmon et al. proposed an improved YOLOv2 [3] who has about six innovations as following. Firstly, batch normalization layers were added after each convolutional layer to accelerate network convergence and reduce over-fitting. Secondly, before training on detection datasets, high-resolution input images are used to train the classification network to help the network better adapt. Thirdly, prior boxes proposed in Faster R-CNN and SSD were introduced to YOLOv2, and K-means was used to select the number and aspect ratio of prior boxes, which improves the recall rate. Fourthly, the offset of prior boxes is constrained to [0, 1] to ameliorate the difficulty of convergence. Fifthly, a pass-through layer was proposed to rearrange high-resolution feature maps and aggregate them with low-resolution feature maps in channel dimension, which effectively improves the accuracy of small object detection. Moreover, a multi-scale training strategy is adopted, and the size of input images continuously change during training, so that the algorithm can adapt to different inputs, making the algorithm more robust.

To further counterpoise accuracy and speed, Redmon et al. proposed an improved YOLOv3 [7] algorithm based on YOLOv2. Compared with YOLOv2, YOLOv3 replaced the Softmax classifier with an independent Logistic classifier and used cross-entropy loss to guide training, so that YOLOv3 has multi-label classification capabilities. At the same time, YOLOv3 expands the pass-through layer first pro-posed in YOLOv2. Multi-scale fusion is also adopted to further enhance the performance of small objects. Furthermore, the basic network is updated, and a large number of 3×3 and 1×1 convolutional layers are added to deepen the depth and improve the accuracy.

## 3. Proposed Method

In this section, the object detection model based on divergent activation was introduced in details.

### *3.1. Divergent Activation*

Xue et al. proposed a divergent activation mechanism [8] in 2019 to solve the problem that CAM (Class Activation Mapping) [9] mainly focuses on local discriminative regions to improve the performance of object classification but ignores the integrity, thereby affects locating accuracy. The introduced divergent activation mechanism can be divided into differential divergent activation (DDA) and hierarchical divergent activation (HDA). HDA mainly focuses on semantic complementary features from the perspective of the hierarchy level, and DDA mainly focuses on mutually exclusive visual features from the perspective of space.

Hierarchical divergent activation (HDA).The purpose of HDA is to activate similar features that are suppressed and locate the full object extent. The basic idea of HDA is to merge two subclasses into the same parent class, and train a classifier for the parent class. The resulting characteristics of the parent class include the similar characteristics of the two subclasses. And continue to merge into a new parent class, will further activate more features. The loss function of HDA is defined as:

IOP Conf. Series: Materials Science and Engineering **914** (2020) 012022 doi:10.1088/1757-899X/914/1/012022

$$Loss_{HDA} = -\sum_{h} \frac{1}{C^h} \sum_{c} y_c^h \log\left(p_c^h\right) \tag{1}$$

Where $L_h$ is the loss of the $h^{th}$ class hierarchy. $C^h$ is the number of the classes in $h^{th}$ class hierarchy. $y_c^h$ is the label of the $c^{th}$ class where $c \in C^h$.

Discrepant divergent activation (DDA). The purpose of DDA is to activate spatial complementary information between objects in a single hierarchical category. In order to achieve the above purpose, a single category feature map is expanded into K feature maps, and the loss function is used to make the K feature maps as different as possible, The loss function of DDA is defined as:

$$Loss_{DDA} = \sum_{1 \le k \le k' \le K} C\left(A_c^k, A_c^{k'}\right) \tag{2}$$

Where $A_c^k$ denotes the $k^{th}$ activation map for the $c^{th}$ class. $C\left(A_c^k, A_c^{k'}\right)$ denotes the cosine similarity between $A_c^k$ and $A_c^{k'}$.

### 3.2. *Overall Network Structure*

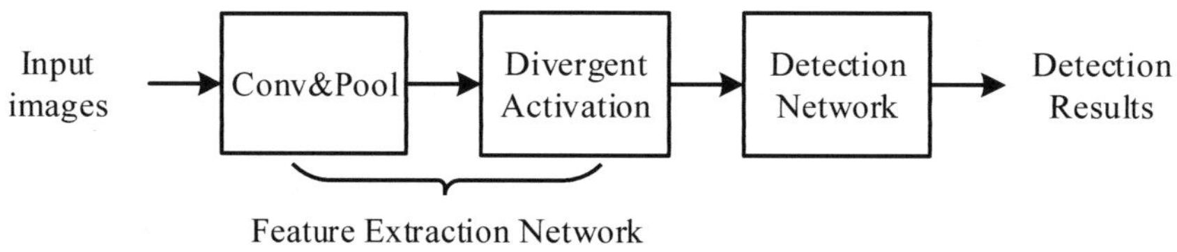


**Figure 1.** Overall Network Structure.

The overall network structure of the proposed algorithm based on YOLOv2 and divergent activation is shown in Figure 1, which consists of two parts, namely feature extraction network and detection network. Feature extraction network is used to learn the feature representations of input images, to provide a basis for object classification and locating. Detection network predicts category, location and corresponding confidence of each object, according to input features.

### 3.3. *Structure of Feature Extraction Network*

As shown in Figure 2, we introduce a divergent activation on the basis of the feature extraction network based on YOLOv2. The introduced divergent activation aims to extract semantically complementary and discriminative features to solve that the feature extraction part of the YOLOv2 algorithm network tends to extract only local features, ignoring the integrity of the object, resulting in a decrease in locating accuracy, which in turn affects detection accuracy. The network structure of the feature extraction network is shown in Figure 2.

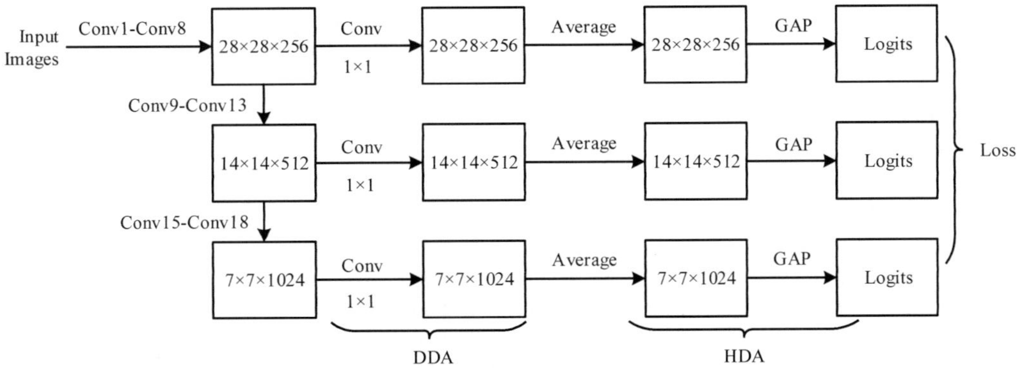


**Figure 2.** Network Structure of Feature Extraction Network.

We first extract multi-scale features after the 8th, 13th and 18th convolutional layers of Darknet-19 and fuse by 1×1 convolution to obtain K feature maps belonging to the same category, and then send to HDA and DDA modules. In the DDA module, concatenation operation is used to combine feature maps, and then DDA loss is used to minimize the similarity, thereby achieving spatial complementarity between each feature map. In the HDA module, feature maps of each level are generated by averaging K feature maps of each category. Global average pooling is used to generate un-normalized probabilities supervised by HDA loss, so as to achieve semantically complementary feature maps of each level.

### *3.4. Structure of Detection Network*

The network structure of the detection network is shown in Figure 3. The input of detection network is the fusion by pixel average of semantically complementary and discriminative HDA features whose size is 7×7×1024. The detection network is consistent with that of YOLOv2, and is composed of four convolutional layers. The parameters of the first three convolutional layers are set to 3×3×2014, the parameters of the last convolutional layer are set to 1×1 to obtain the final output feature map. The output feature maps can be decoded into classification and location confidence as detection results.

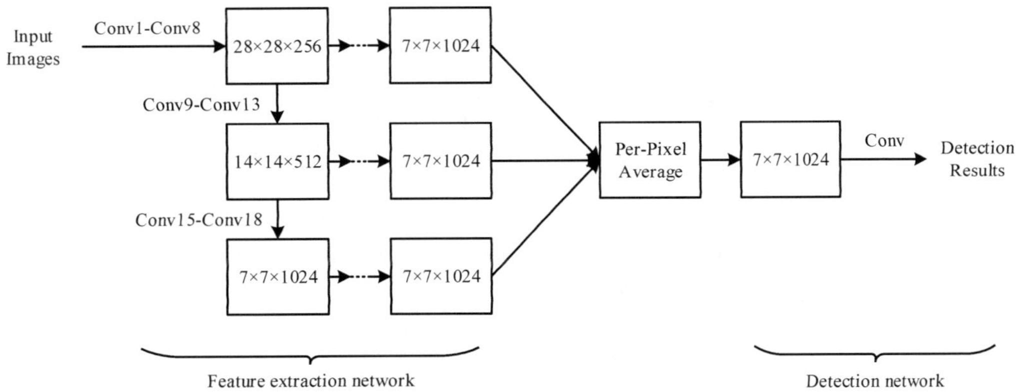


**Figure 3.** Network Structure of Detection Network.

### *3.5. Loss Functions*

The loss function of the proposed algorithm consists three weighted parts is defined as:

$$L_{all} = Loss + Loss_{HDA} + \lambda Loss_{DDA} \tag{3}$$

Where $Loss_{HDA}$ represents loss of HDA modules, $Loss_{DDA}$ represents loss of DDA modules, $Loss$ represents locating loss and classification loss to guide training, and $\lambda$ represents a balance factor.

## 4. Experiments

In this section, we mainly describe our experiment implement details and some comparisons of experimental results.

### 4.1. Implementation details

Training datasets used in this paper are ImageNet [10], PASCAL VOC2007 and PASCAL VOC2012 [11]. For evaluation, we mainly use mean Average Precision (mAP) and frame per second (FPS) to measure the performance. It took several hours to train with one TiTan X.

Training phase is divided into two stages. In stage one, we added cascading convolutional, max-pooling and Softmax layer to train the feature extraction network on ImageNet to equip favorable feature extraction ability. In stage two, we added a detection network to fine-tune on PASCAL VOC2007 and PASCAL VOC2012 to obtain detection results.

For parameter settings, $\lambda$ in loss function is set to 0.6. For optimization, it is optimized by ADAM optimizer with $\beta_1 = 0.9, \beta_2 = 0.999$. Learning rate in stage one is set to 1e-4 and will decay by half when there is no more promotion for 10000 iterations. Learning rate in stage two is set to 5e-5 and will decay by half when there is no more promotion for 5000 iterations.

### 4.2. Ablation Studies

The ablation studies on PASCAL VOC are used to evaluate the effects of proposed DA modules.

Study of K in DDA. As shown in Table 1, we studied the influence of K in DDA on detection precision. Our proposed algorithm achieved the best results when K is set to 8. If K is too small, it is hard to achieve spatial complement without enough spatial difference. If K is too large, the model complexity will increase a lot, and there will be a risk of over fitting, decreasing mAP.

**Table1.** Study of K in DDA.

| K | Training set/Testing set | mAP (%) |
|---|---|---|
| $2^0$ | VOC07+12/VOC07 | 78.8 |
| $2^1$ | VOC07+12/VOC07 | 79.2 |
| $2^2$ | VOC07+12/VOC07 | 79.5 |
| $2^3$ | VOC07+12/VOC07 | <u>79.7</u> |
| $2^4$ | VOC07+12/VOC07 | 79.4 |
| $2^5$ | VOC07+12/VOC07 | 79.2 |

Study of $\lambda$ in DDA. As shown in Table 2, we studied the influence of $\lambda$ in equation 1 on detection precision. Our proposed algorithm achieved the best results when $\lambda$ is set to 0.01. We set K=8 and $\lambda$=0.01 as default settings, in the following experiments.

**Table 2**. Study of $\lambda$ in DDA.

| $\lambda$ | Training set/Testing set | mAP(%) |
|---|---|---|
| 0.005 | VOC07+12/VOC07 | 79.6 |
| 0.010 | VOC07+12/VOC07 | <u>81.0</u> |
| 0.025 | VOC07+12/VOC07 | 80.9 |
| 0.050 | VOC07+12/VOC07 | 80.1 |
| 0.100 | VOC07+12/VOC07 | 80.7 |
| 0.200 | VOC07+12/VOC07 | 79.9 |
| 0.500 | VOC07+12/VOC07 | 79.6 |
| 1.000 | VOC07+12/VOC07 | 79.3 |

Study of HDA. As shown in Table 3, it can be noticed that the derivative algorithm DarkNet-19+HDA increased by 1.5%, and DarkNet-19+DDA increased by 1.1% on mAP, compared to YOLOv2. The above experimental results verified the effectiveness of HDA and DDA modules for object detection. However, since the main purpose of HDA is to activate a complete range of objects, improving locating accuracy, it is more efficient than DDA. The algorithm DarkNet19+HDA+DDA increased by 2.4% mAP compared with YOLOv2 with the sacrifice of a little running time. However, it's acceptable and can still meet real-time needs.

**Table 3.** Study of HAD.

| Configuration | Training set/Testing set | mAP (%) | FPS |
|---|---|---|---|
| DarkNet-19 | VOC07+12/VOC07 | 78.6 | 40 |
| DarkNet-19+HDA | VOC07+12/VOC07 | 80.1 | 34 |
| DarkNet-19+DDA | VOC07+12/VOC07 | 79.7 | 36 |
| DarkNet19+HDA+DDA | VOC07+12/VOC07 | 81.0 | 32 |

### *4.3. Comparison with state-of-the-art algorithms*

**Table 4.** Comparison with state-of-the-art algorithms.

| Algorithm | Training set/Testing set | mAP (%) | FPS |
|---|---|---|---|
| Faster R-CNN[3] | VOC07+12/VOC07 | 76.4 | 2.4 |
| MR-CNN[12] | VOC07+12/VOC07 | 78.2 | 0.03 |
| R-FCN[13] | VOC07+12/VOC07 | 80.5 | 9 |
| SSD513[14] | VOC07+12/VOC07 | 80.6 | 6.8 |
| DSSD513[15] | VOC07+12/VOC07 | **81.5** | **5.5** |
| RefineDet512[16] | VOC07+12/VOC07 | 80 | 24.1 |
| YOLOv2[5] | VOC07+12/VOC07 | 78.6 | 40 |
| **Ours** | VOC07+12/VOC07 | <u>81.0</u> | <u>32</u> |

Comparison with state-of-the-art algorithms is shown in Table 4 where bold represents the best and underlined represents the second best. Clearly from Table 4 that our proposed algorithm can achieve superior results among all the algorithms except DSSD513. However, its FPS is merely 5.5, it's also worth noting that our proposed algorithm only lacks 0.5% on mAP.

## 5. Summary

In this Paper, we propose an improved YOLOv2 target detection algorithm based on divergent activation. The algorithm introduces a divergent activation mechanism based onYOLOv2 which aims to extract semantically complementary and discriminative features. Experiment results based on

IOP Conf. Series: Materials Science and Engineering **914** (2020) 012022 doi:10.1088/1757-899X/914/1/012022

PASCAL VOC2007 and PASCAL VOC2012 datasets indicate our proposed algorithm can effectively improve the mAP on the premise of ensuring real-time performance.

**References**


[1] Szegedy C, Toshev A, Erhan D. Deep neural networks for object detection [C] //Advances in neural information processing systems. 2013: 2553-2561.

[2] Yu Jinyong, Ding Pengcheng, Wang Chao. Overview:Application of Convolution Neural Network in Object Detection [J]. Computer Science, 2018, 45 (S2): 17-26.

[3] Redmon J, Farhadi A. YOLO9000: better, faster, stronger [C] //Proceedings of the IEEE conference on computer vision and pattern recognition. 2017: 7263-7271.

[4] Redmon J, Divvala S, Girshick R, et al. You only look once: Unified, real-time object detection [C] //Proceedings of the IEEE conference on computer vision and pattern recognition. 2016: 779-788.

[5] Ren S, He K, Girshick R, et al. Faster r-cnn: Towards real-time object detection with region proposal networks [C] //Advances in neural information processing systems. 2015: 91-99.

[6] Liu W, Anguelov D, Erhan D, et al. SSD: Single shot multibox detector [C]// European conference on computer vision. Springer, Cham, 2016: 21-37.

[7] Redmon J, Farhadi A. Yolov3: An incremental improvement [J]. arXiv preprint arXiv:1804.02767, 2018.

[8] Xue H, Liu C, Wan F, et al. DANet: Divergent Activation for Weakly Supervised Object Localization [C] //Proceedings of the IEEE International Conference on Computer Vision. 2019: 6589-6598.

[9] Zhou B, Khosla A, Lapedriza A, et al. Learning deep features for discriminative localization [C] //Proceedings of the IEEE conference on computer vision and pattern recognition. 2016: 2921-2929.

[10] Russakovsky O, Deng J, Su H, et al. Imagenet large scale visual recognition challenge [J]. International journal of computer vision, 2015, 115 (3): 211-252.

[11] Everingham M, Van Gool L, Williams C K I, et al. The pascal visual object classes (voc) challenge [J]. International journal of computer vision, 2010, 88 (2): 303-338.

[12] Gidaris S, Komodakis N. Object detection via a multi-region and semantic segmentation-aware cnn model [C] //Proceedings of the IEEE international conference on computer vision. 2015: 1134-1142.

[13] Dai J, Li Y, He K, et al. R-fcn: Object detection via region-based fully convolutional networks [C] //Advances in neural information processing systems. 2016: 379-387.

[14] Liu W, Anguelov D, Erhan D, et al. Ssd: Single shot multibox detector [C] //European conference on computer vision. Springer, Cham, 2016: 21-37.

[15] Fu C Y, Liu W, Ranga A, et al. Dssd: Deconvolutional single shot detector [J]. arXiv preprint arXiv:1701.06659, 2017.

[16] Zhang S, Wen L, Bian X, et al. Single-shot refinement neural network for object detection [C] //Proceedings of the IEEE conference on computer vision and pattern recognition. 2018: 4203-4212.

# Economic Adaptation Grade Evaluation of Integrated Transport System Based on Matter-Element Extension with ANP

**Jinhui Zhang***

School of Traffic and Transportation, Beijing Jiaotong University, Beijing 100044, China

*Corresponding author e-mail: zhang_sdxs@126.com

**Abstract**. Based on the concept of coordinated development between transportation and economy, a set of evaluation index system is constructed from total scale, transportation structure and service level, which is to measure economic adaptability of transport system. Then Matter-Element Extension with ANP algorithm is put forward to calculate the economic adaptation grade of integrated transport system with empirical analysis of Yunnan province, which has verified that this model is applicable.

## 1. Introduction

The development of transport is closely related to the level of the national economy in this region. A consistent consistency interval of judgment in the decision support system of the Analytic Hierarchy Process (AHP) was proposed [1]. And an indicator system between public transport and economy is established which verified that the two were positively related [2]. Yulong Pei used an improved AHP algorithm to calculate the index weights [3]. Da Zheng applied AHP and entropy weight method to calculate the weight [4]. Yuqi Yang constructed an index system and quantitatively analyzed it based on the information entropy theory [5]. Yuxia Chen proposed 20 indicators and gray correlation model for empirical analysis [6].

Therefore, the research mainly involves the design of evaluation index system, the calculation of weight and evaluation method. It is known from existing research that the proportion of quantitative indicators is insufficient and it also does not consider the interaction between the indicators at the same layer, so this paper designs a more reasonable indicator system and chooses matter-element extension theory with analytic network process (ANP) to determine economic adaptation level of transport system.

## 2. Matter-Element Extension with ANP Algorithm

In extension theory, Matter-Element can organically combine name, feature and value of objects to transform qualitative concept into quantitative values, which is to calculate the weight parameters of indexes, and then obtain the correlation degree and finally determine the evaluation level. The index weight adopts ANP algorithm, which is a target decision method composed of two clusters, namely the control layer and the network layer, which can reflect the mutual relationship between the elements in the index set. Specific steps of the algorithm are as follows:

**Step 1:** Taking $P_s(s=1,2,\cdots,m)$ as the criterion, and taking $e_{jk}(k=1,2,\cdots,n_j)$ in $\boldsymbol{C}_j$ as sub-criteria, the influence of $e_{ik}$ in $\boldsymbol{C}_i$ on $e_{jk}(k=1,2,\cdots,n_j)$ is calculated.

**Step 2:** Eigenvectors are calculated and normalized to obtain the weight vectors, then the hypermatrix $\boldsymbol{W}$ is calculated.

$$\boldsymbol{W}=\begin{bmatrix} \boldsymbol{W}_{11} & \boldsymbol{W}_{12} & \cdots & \boldsymbol{W}_{1N} \\ \boldsymbol{W}_{21} & \boldsymbol{W}_{22} & \cdots & \boldsymbol{W}_{2N} \\ \vdots & \vdots & \ddots & \vdots \\ \boldsymbol{W}_{N1} & \boldsymbol{W}_{N2} & \cdots & \boldsymbol{W}_{NN} \end{bmatrix} \tag{1}$$

**Step 3:** Based on $P_s(s=1,2,\cdots,m)$ and the importance of $\boldsymbol{C}_j(j=1,2,\cdots,n)$ ,judge matrix and eigenvector are calculated, and normalized weighted matrix $\boldsymbol{A}$ is obtained. Hypermatrix $\boldsymbol{W}$ is weighted calculated, then $\hat{\boldsymbol{W}}_{ij}$ and limit supermatrix $\boldsymbol{W}^{\infty}$ are obtained.

$$\hat{\boldsymbol{W}}_{ij}=a_{ij}\boldsymbol{W}_{ij}(i=1,2,\cdots,N,j=1,2,\cdots,N) \tag{2}$$

$$\boldsymbol{W}^{\infty}=\lim_{t\to\infty}\hat{\boldsymbol{W}}^{t} \tag{3}$$

**Step 4: To** determine rank field $U$ ,evaluation index set $C$ ,matter-element, and calculate the classical and extension al matter elements.

$$R_j=\left(U_j,C,V_j\right)=\begin{bmatrix} U_j & c_1 & \left\langle a_{j1},b_{j1}\right\rangle \\ & c_2 & \left\langle a_{j2},b_{j2}\right\rangle \\ & \vdots & \vdots \\ & c_n & \left\langle a_{jm},b_{jm}\right\rangle \end{bmatrix} \tag{4}$$

**Step 5:** With the weight of the secondary index and the correlation degree of adaptive grade, the correlation degree of the primary index can be calculated.

$$\boldsymbol{K}\left(c_i\right)=\left[w_{i1},w_{i2},\cdots,w_{ip}\right]\cdot\begin{bmatrix} k_1\left(c_{i1}\right) & k_2\left(c_{i1}\right) & \cdots & k_m\left(c_{i1}\right) \\ k_1\left(c_{i2}\right) & k_2\left(c_{i2}\right) & \cdots & k_m\left(c_{i2}\right) \\ \vdots & \vdots & \cdots & \vdots \\ k_1\left(c_{ip}\right) & k_2\left(c_{ip}\right) & \cdots & k_m\left(c_{ip}\right) \end{bmatrix} \tag{5}$$

**Step 6:** According to the weight of the first-level index and its relevance to the adaptation level, the relevance of the element to be evaluated was calculated.

$$\boldsymbol{k}(N)=\left[w_1,w_2,\cdots,w_n\right]\cdot\begin{bmatrix} k_1(c_1) & k_2(c_1) & \cdots & k_m(c_1) \\ k_1(c_2) & k_2(c_2) & \cdots & k_m(c_2) \\ \vdots & \vdots & \cdots & \vdots \\ k_1(c_n) & k_2(c_n) & \cdots & k_m(c_n) \end{bmatrix} \tag{6}$$

$j$ is called the adaptability level of $N$ , if $k_{j0}(N)=\max\limits_{j=\{1,2,\cdots m\}} k_j(N)$ .

## 3. Construction of Index System

This paper designs an index system from three dimensions: total scale, transport structure, and service level. The total scale includes the coordination degree of transport facilities and investment scale and the integrated transport capacity. The combination of industrial structure and internal structure reflects the level of transport structure. The service level includes the level of travel service, management and construction, and safety emergency level. The index elements in each index cluster have dependencies and feedbacks. For example, the intensity of passenger transport is affected by the level of transport network. The specific evaluation index system is shown in Figure 1.

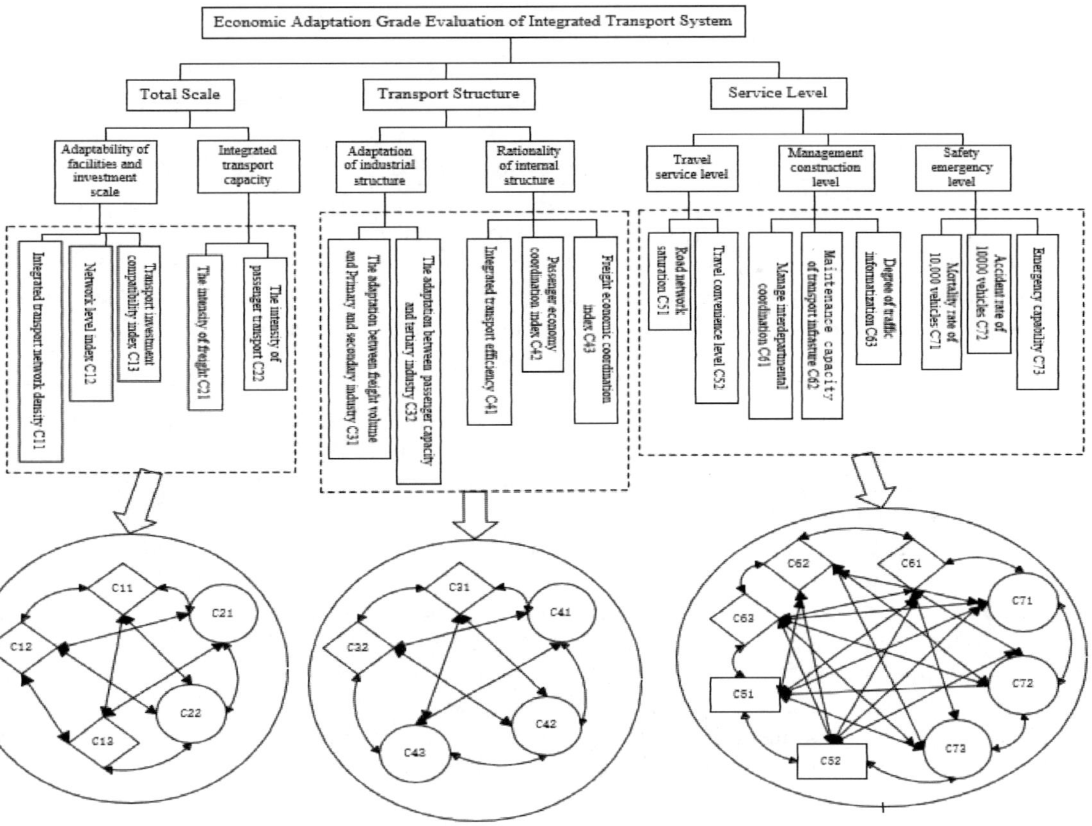


**Figure 1.** The hierarchical evaluation index system

## 4. Results

Based on the above research, the evaluation hierarchy is divided to five grades respectively I, II, III, IV, V. Level I represents the highest level, and level V indicates that transport system and economic system does not match.

**Table 1.** Rate assessment result

| | Level I | Level II | Level III | Level IV | Level V | Level |
|---|---|---|---|---|---|---|
| Total Scale | -0.6292 | -0.4513 | 0.14 | 0.1401 | -0.0195 | IV |
| Transport Structure | -0.5242 | -0.3996 | -0.22 | -0.0099 | -0.0437 | IV |
| Service Level | -0.4875 | -0.3422 | -0.19 | 0.0117 | -0.2032 | IV |
| Adaptation Grade | -0.56335 | -0.4102 | -0.18 | -0.06375 | -0.0679 | IV |

According to the indicator system of Figure 1, and based on the statistical data of Yunnan Province from 2010 to 2019, all indicators are quantitatively calculated and qualitatively analyzed. The economic adaptation grade of Yunnan Province is calculated as level IV. The specific data is shown in Table 1. Comparing the results of this study with the current situation of Yunnan Province, the evaluation results are more reasonable, which shows that this research method is reliable and effective.

## 5. Conclusion

Based on the above research, the following conclusions are drawn:

(1) The calculation of index weight by replacing AHP with ANP changes the independent hierarchy structure of AHP, which can fully consider the mutual influence of each index factor cluster, whose results are relatively more accurate.

(2) Matter-Element extension theory can combine qualitative and quantitative methods and overcome the difficulties of uncertainty and multiple attributes between transport system and economic system.

In conclusion, economic adaptation grade evaluation based on Matter-Element Extension with ANP can avoid the limitation of single method. After evaluating the economic adaptation level of the integrated transport system in Yunnan province, the results are satisfactory. But the index system constructed in this paper is not perfect, and further quantitative study is needed to obtain a more effective judge matrix. At the same time, the universality of the system needs further discussion, and whether there is a better comprehensive weight calculation method remains to be studied.

## References

[1] Pavlous S Kanaroglou, William P Anderson. Economic Impacts of Highway Infrastructure Improvements Part1, Conceptual Framework [J]. Transport Geography, 1998, 6 (3): 15-20.

[2] Kiyohito Utsunomiya. Social capital and local public transportation in Japan [J]. Research in Transportation Economics. 2016(02): 74-81.

[3] Pei Yulong, et al. Novel AHP algorithm to evaluate adaptability of transport system to urban sustainable development [J]. Journal of Jilin University,2008, 38(1): 42-47.

[4] Zheng Da. Research on the economic adaptability evaluation of regional comprehensive transport system [D]. Beijing: Beijing Jiaotong University, 2014.

[5] Yang Yuqi. Information entropy based integrated transport development and regional industrial structure adaptability [J]. Shandong Science, 2015, 28(4): 46-51.

[6] Chen Yunxia. Coupling coordinated development and optimization path of tourism economy and transportation [D]. Xian: Changan University, 2019.

IOP Conf. Series: Materials Science and Engineering **914** (2020) 012024 doi:10.1088/1757-899X/914/1/012024

# 8-Bit NCL Asynchronous Multiplier based on Radix-4 Booth Algorithm

**Weixiang Deng[1, a], Wenxiang Cheng[1, b], Jie Cheng[2, c], Leibin Ni[2, d] and Anping He[1, *]**

[1]School of information Science and Engineering, Lanzhou University, Lanzhou, China
[2]Data Center Technology Laboratory, Huawei Technologies Co., Ltd., Shenzhen, China

*Corresponding author e-mail: heap@lzu.edu.cn, [a]594766943@qq.com, [b]363999029@qq.com, [c]chengjie8@huawei.com, [d]nileibin@huawei.com,

**Abstract**. Multiplier is one of the key modules in signal processing circuit and processor, and its robustness is especially important in the complex environment. Due to the lack of robustness of single-rail logic, a Null Convention Logic (NCL) Multiplier based on the Radix-4 Booth 8-bit fast parallel structure is proposed in this paper. The Radix-4 Booth algorithm reduces the number of partial products to lower the computing time, and the "sign generate" algorithm simplifies sign extension bit computation. In order to demonstrate the effectiveness of the proposed hardware implementation scheme, we implement an improved 8-bit NCL multiplier on vivado platform, the speed of which is 14% higher than that of the traditional Wallace NCL 8-bit multiplier.

## 1. Introduction

After clock frequency turned to GHz's order of magnitude, the design method of the synchronous circuit will face more and more challenges in the deep sub-micron field, such as, clock skew, clock jitter, and power consumption, etc [1]. To mitigate these problems, the asynchronous clock-less circuit methodology was designed and becoming prevalent in recent years. The clock-less feature of the asynchronous circuit shows the following merits: low power dissipation, quasi-delay-insensitivity, low noise, high reliability, etc [2]. Therefore, the asynchronous circuit is propitious to Systems-on-a-Chip (SoCs) design which is insensitive to delay while sensitive to power dissipation and performance. Among asynchronous design approaches, Null Convention Logic (NCL) [3] is the most robust method with a low-power advantage, greater optimization potential and better integration ability, and is gaining more and more attention. Therefore, this paper discusses the design of asynchronous multiplier based on NCL with the purpose of high reliability.

The multiplier is a vital component in a signal processing circuit and its performance will greatly affect the chip's performance. Asynchronous circuit has been used in multiplier design, and the current research focuses on the speed and area of multiplier.To increase the multiplier's speed, many scholars have proposed plenty of high-efficiency algorithms and the corresponding circuit structure. Two of the most famous algorithm and structures are the Booth algorithm [4] and the Wallace tree structure [5]. By separating the multiplier into several multiplier segments, the classical Booth algorithm transforms the multiplication problem into the sum of the partial product of multiplicand and multiplier segments.

Specifically, in the Booth algorithm, based on the data character of binary system multiplier section, each multiplicand's multiplication can be mapped to the equivalent shifting and subtraction operation. In [6], the authors improved the Booth encoding, employing a more optimized shift strategy, to simplify the multiple sum (difference) of the traditional Booth algorithm to a single time, which shows higher performance for the multiplier with smaller bit width. Ohkubo et al uses MUX to implement Radix-4 Booth coding circuit, and uses NAND gate structure instead of and-or structure in the process of partial product generation[7], which reduces the delay of coding circuit. Moreover, in [8], the author improved the encoding circuit by reducing the total operation time of the encoding circuit under the same scenario, and has more advantages in fast generating of partial product.

And for the signed multiplication, one needs to extend the sign of the partial product, and the "sign generate" algorithm [9] can be used to reduce the number of compressors used in the multiplier. Furthermore, by making use of Wallace tree structure, the submission progression can be compressed into only two operands, and the final result can be obtained by adding the two operands through a ripple carry adder (RCA).

This paper is organised as follows. The theory of NCL is provided in Section 2 and it is necessary to appreciate the structure of NCL circuit. Section 3 introduces our optimized design of Radix-4 Booth based on NCL. Section 4 presents the implementation of NCL multiplier design. In Section 5, we discuss our results and compare the proposed NCL multiplier and NCL multiplier based on Wallace tree in terms of delay. And Section 6 concludes the paper.

## 2. Null Convention Logic

NCL is a quasi-delay-insensitive asynchronous circuit design method which embeds the status into the data itself in order to eliminate timing dependence [10]. Through this method the circuit will operate correctly no matter when the input of the NCL circuit inputs is available; therefore, it only needs very few timing analysis for the NCL circuit to ensure correct operation. For example, dual-rail signal A consists of two wires, $A_0$ and $A_1$, where $A_0=1$, $A_1=0$ represent logic 0, denoted as DATA0; $A_0=0$, $A_1=1$ represent logic 1, denoted as DATA 1; $A_0=0$, $A_1=0$ represent empty status, denoted as NULL which meaning that the value of A is not yet available; $A_0=1$, $A_1=1$ represent illegal status which will not appear while the circuit is regular working. In addition, the valid data is any combination of DATA0 and DATA1. During the circuit working, when the circuit output is null, the input is open to allow the valid data to enter the circuit from the input; on the other hand, when the output is valid data, the input is open to allow the Null signal to enter the circuit from the input. After that, the circuit is ready for the next valid data (Null signal).

The NCL circuit consists of 27 fundamental gates, as shown in Table 1, which constitute all function sets consisting of four or less variables, thus providing better optimization ability. As shown by Fig.1 (a), THmn gate is the primary type of threshold gate, in which, n is the input number of THmn gate, m is the threshold gate value which means that the output will be asserted when at least m inputs are asserted and deasserted when all inputs are deasserted; otherwise, it will keep its previous state.

Another kind of threshold gate is called weighted threshold gate in which one input with multiple wires may connect to this gate and threshold value of four is illustrated in Fig.1(b) described the TH44w2 gate. Note that, the first input consists of two connective wires, and hence its corresponding weight is two and the Boolean logic function is Z=ABC+ABD+ACD. Usually, in $THmnw_1w_2...w_k$ gate, the first k inputs are weighted as $w_1$, $w_2$, …, $w_k$, and the default is 1.

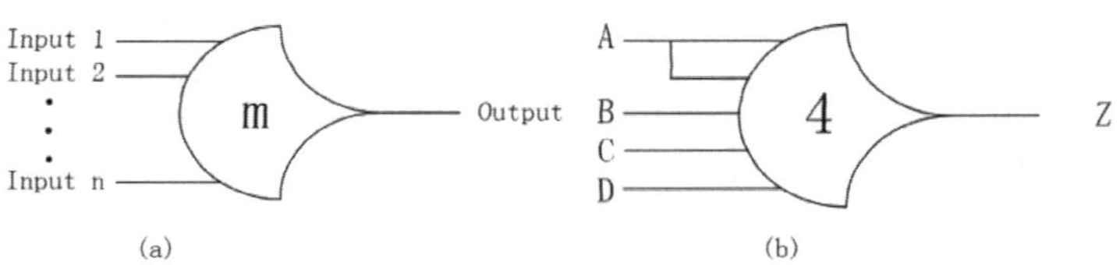


**Figure 1.** (a): THmn threshold gate[10]; (b): TH44w2: Z=ABC+ABD+ACD.

**Table 1.** 27 fundamental NCL gates [10].

| NCL Gate | Boolean Function | NCL Gate | Boolean Function |
|---|---|---|---|
| TH12 | A+B | TH34w3 | A+BCD |
| TH22 | AB | TH44w3 | AB+AC+AD |
| TH13 | A+B+C | TH24w22 | A+B+CD |
| TH23 | AB+AC+BC | TH34w22 | AB+AC+AD+BC+BD |
| TH33 | ABC | TH44w22 | AB+ACD+BCD |
| TH23w2 | A+BC | TH54w22 | ABC+ABD |
| TH33w2 | AB+AC | TH34w32 | A+BC+BD |
| TH14 | A+B+C+D | TH54w32 | AB+ACD |
| TH24 | AB+AC+AD+BC+BD+CD | TH44w322 | AB+AC++AD+BC |
| TH34 | ABC+ABD+ACD+BCD | TH54w322 | AB+AC+BCD |
| TH44 | ABCD | THxor0 | AB+CD |
| TH24w2 | A+BC+BD+CD | THand0 | AB+BC+AD |
| TH34w2 | AB+AC+AD+BCD | TH24comp | AC+BC+AD+BD |
| TH44w2 | ABC+ABD+ACD | | |

## 3. The Proposed NCL Based Improved Booth Algorithm

In this section, for the convenience of the following description we first introduce the classic Radix-4 Booth algorithm [11] which was devised by O.L.MacSorley. When we define $A=a_7a_6...a_0$ and $B=b_7b_6...b_0$ which represent multiplier and multiplicand, respectively, our aim is to compute the product $PP=A\cdot B$.The calculation formula can be written as follows:

$$\begin{aligned} PP &= A \times B = A \times (b_7 b_6 ... b_0) \\ &= A \times (-b_7 \times 2^7 + \sum_{i=0}^{6} b_i 2^i) \\ &= A \times [\sum_{i=0}^{3} (b_{2i-1} + b_{2i} - 2b_{2i+1}) 4^i] \\ &= \sum_{i=0}^{3} A(b_{2i-1} + b_{2i} - 2b_{2i+1}) 4^i \end{aligned} \tag{1}$$

Each items A $(b_{2i-1}+b_{2i}-2b_{2i+1})4^i$ so-called partial product, and the PP is the sum of four partial products. Moreover, the encoding scheme is designed as in Table 2, and the coefficient of the partial product is $b_{2i-1}+b_{2i}-2b_{2i+1}$.

The Radix-4 Booth coding circuit is given in [9], but there are more gate delay and circuit area when we mapping these circuit to NCL circuit directly. Therefore we redesign NCL Booth coding circuit based on Radix-4 coding rules of Booth algorithm which was first proposed by O.L.MacSorley. The Fig.2 (a) shows the result and in this figure X, Neg represent the coefficient and sign of the partial product respectively and $a_ja_{j-1}$ is the two consecutive bits of the multiplicand. Furthermore, $PP_j=0$ corresponds to X=0; $PP_j=a_j$ corresponds to X=1; $PP_j=a_{j-1}$ corresponds to X=2; and $PP_j$ is negative when $M_{2i+1}=1$, Neg=1. The Karnaugh map(K-map) for partial product generation component is shown in Fig.2(b).

IOP Conf. Series: Materials Science and Engineering **914** (2020) 012024 doi:10.1088/1757-899X/914/1/012024

**Table 2.** Radix-4 Booth algorithm encoding rule.

| $M_{2i+1}M_{2i}M_{2i-1}$ | PP |
|---|---|
| 000 | 0 |
| 001 | +A |
| 010 | +A |
| 011 | +2A |
| 100 | -2A |
| 101 | -A |
| 110 | -A |
| 111 | 0 |

| $M_{2i+1}$ | $M_{2i}$ | $M_{2i-1}$ | X | Neg | $PP_j$ |
|---|---|---|---|---|---|
| 0 | 0 | 0 | 0 | 0 | 0 |
| 0 | 0 | 1 | 1 | 0 | $a_j$ |
| 0 | 1 | 0 | 1 | 0 | $a_j$ |
| 0 | 1 | 1 | 2 | 0 | $a_{j-1}$ |
| 1 | 0 | 0 | 2 | 1 | $-a_{j-1}$ |
| 1 | 0 | 1 | 1 | 1 | $-a_j$ |
| 1 | 1 | 0 | 1 | 1 | $-a_j$ |
| 1 | 1 | 1 | 0 | 1 | 0 |

(a)

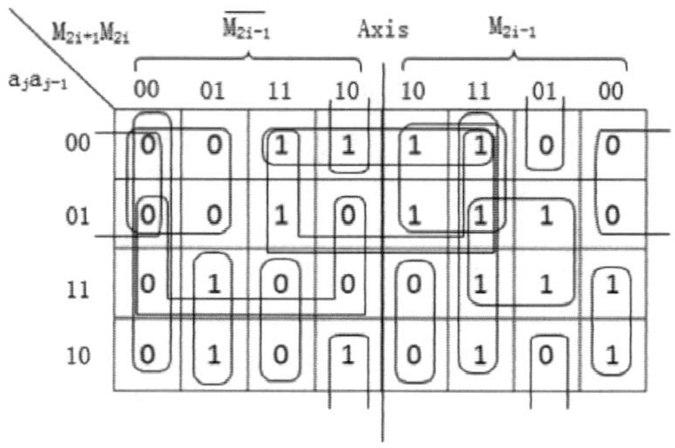


(b)

**Figure 2.** (a): Improved Booth encoding rule; (b): K-map for improved Booth PP generation component.

And the simplified sum-of-products expressions of each output can be obtained as follows:

$$PP_j^1 = A_j^0 M_{2i+1}^1 M_{2i}^1 + A_j^0 M_{2i+1}^1 M_{2i-1}^1 + M_{2i+1}^1 M_{2i}^1 M_{2i-1}^1 + A_{j-1}^1 M_{2i}^1 M_{2i-1}^1 + A_j^1 M_{2i+1}^0 M_{2i}^1 M_{2i-1}^0 + A_{j-1}^0 M_{2i+1}^1 M_{2i}^0 M_{2i-1}^0 + A_j^1 M_{2i+1}^0 M_{2i}^0 M_{2i-1}^1 \quad (2)$$

$$PP_j^0 = A_j^0 M_{2i+1}^0 M_{2i-1}^0 + A_{j-1}^1 M_{2i}^0 M_{2i-1}^0 + M_{2i+1}^0 M_{2i}^0 M_{2i-1}^0 + A_j^0 M_{2i+1}^0 M_{2i}^0 + A_j^1 M_{2i+1}^1 M_{2i}^1 M_{2i-1}^0 + A_{j-1}^0 M_{2i+1}^0 M_{2i}^1 M_{2i-1}^1 + A_j^1 M_{2i+1}^1 M_{2i}^0 M_{2i-1}^1 \quad (3)$$

More importantly, in order to ensure that the PP generation component is input complete[10], each product term of an output must contain any of the rails of $M_{2i}$. Therefore, to make the circuit input-complete with respect to $M_{2i}$, $M_{2i}$ must be added to all product terms in which it is missing in either $PP_j^0$ or $PP_j^1$, but not both. So we add $M_{2i}$ to $PP_j^0$ by ANDing the product term with logic 1 which yields the following equations:

$$PP_j^0 = A_{j-1}^1 M_{2i}^0 M_{2i-1}^0 + M_{2i+1}^0 M_{2i}^0 M_{2i-1}^0 + A_j^0 M_{2i+1}^0 M_{2i}^0 + A_j^0 M_{2i+1}^0 M_{2i}^1 M_{2i-1}^0 + A_j^1 M_{2i+1}^1 M_{2i}^1 M_{2i-1}^0 + A_{j-1}^0 M_{2i+1}^0 M_{2i}^1 M_{2i-1}^1 + A_j^1 M_{2i+1}^1 M_{2i}^0 M_{2i-1}^1 \quad (4)$$

## 4. Design implementation

Based on the output logic equation, $PP_j^1$ and $PP_j^0$ can be mapped to seven TH44 gates, three TH54w22 gates, three TH13 gates, one TH33 gate and one TH14 gate. After mapping, the proposed NCL circuit still only needs three gate delays and fifteen NCL logic gates, which reduces 3 NCL gates and 2 gate delay when compared with direct mapping, and the resulting optimized NCL improved Booth PP generation circuit is shown in Fig.3.

Since adding 1 only occurs when Neg = 1, we can directly incorporate four sign bits $s_0, s_1, s_2, s_3$ into the empty low bit calculation of partial product in the compression process. Because all the partial product need to extend the sign to the highest bit of the array, which leads to the increase of the number of the compressors, the raising of the area and power consumption of the multiplier. Therefore the "sign generate" algorithm is used in the design.

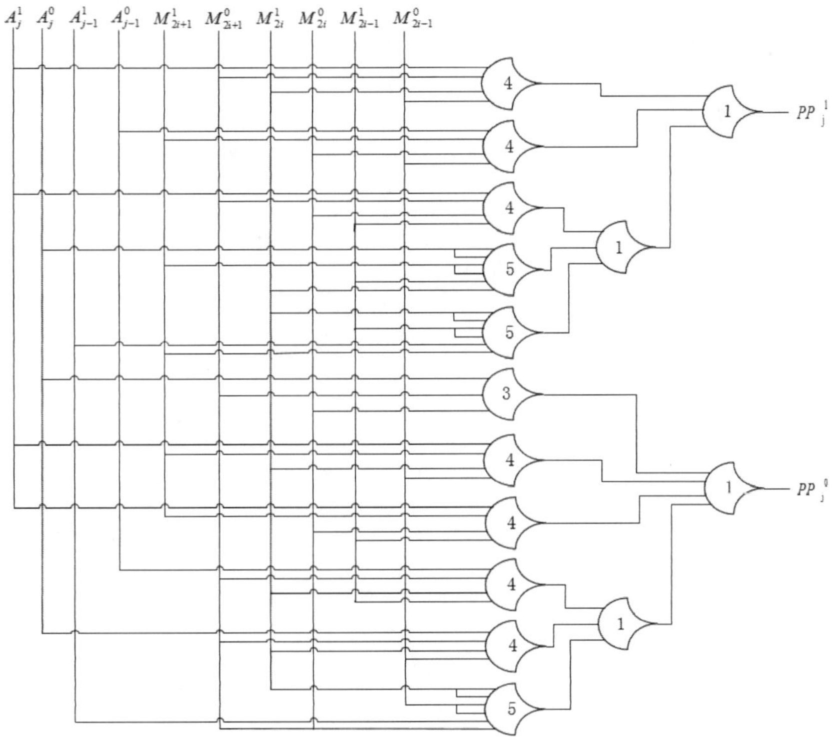

**Figure 3.** NCL improved Booth PP generation circuit.

According to the algorithm, all sign extension bits are added to get the sum, "Sgn", and the calculation formula can be written as follows:

$$Sgn = (s_0 \sum_{i=9}^{15} 2^i) \cdot 2^0 + (s_1 \sum_{i=9}^{13} 2^i) \cdot 2^2 + (s_2 \sum_{i=9}^{11} 2^i) \cdot 2^4 + (s_3 \sum_{i=9}^{9} 2^i) \cdot 2^6 \tag{5}$$

We use $\sum_{i=j}^{k} 2^i = 2^{k+1} - 2^j$, $s_i = 1 - \overline{s_i}$ in the derivation above. Therefore we can simplify it as follows:

$$\begin{aligned} Sgn &= [s_0(2^{16} - 2^9)] \cdot 2^0 + [s_1(2^{14} - 2^9)] \cdot 2^2 + [s_2(2^{12} - 2^9)] \cdot 2^4 + [s_3(2^{10} - 2^9)] \cdot 2^6 \\ &= -(s_0 \cdot 2^9 + s_1 \cdot 2^{11} + s_2 \cdot 2^{13} + s_3 \cdot 2^{15}) + 2^{16} \cdot (s_0 + s_1 + s_2 + s_3) \end{aligned} \tag{6}$$

Since the second term of formula(5) only affects the seventeenth bit and the above, it can be omitted, therefore, we can get:

$$\begin{aligned} Sgn &= -(s_0 \cdot 2^9 + s_1 \cdot 2^{11} + s_2 \cdot 2^{13} + s_3 \cdot 2^{15}) \\ &= \overline{s_0} \cdot 2^9 + \overline{s_1} \cdot 2^{11} + \overline{s_2} \cdot 2^{13} + \overline{s_3} \cdot 2^{15} - (2^9 + 2^{11} + 2^{13} + 2^{15}) \\ &= \overline{s_0} \cdot 2^9 + \overline{s_1} \cdot 2^{11} + \overline{s_2} \cdot 2^{13} + \overline{s_3} \cdot 2^{15} - (2^9 + 2^{10} + \ldots + 2^{15} + 2^9) + (2^9 + 2^{10} + 2^{12} + 2^{14}) \\ &= \overline{s_0} \cdot 2^9 + \overline{s_1} \cdot 2^{11} + \overline{s_2} \cdot 2^{13} + \overline{s_3} \cdot 2^{15} + (2^9 + 2^{10} + 2^{12} + 2^{14}) \end{aligned} \tag{7}$$

After coding, a total of 4 partial products (n/2=4) are generated, and we can get the final product by adding these partial products. To accelerate the operation speed and reduce the area of NCL multiplier, we adopt the Wallace tree structure composed of 4:2 compressors [12] which is made up of two full adders.

After generation the partial products, we use Wallace tree to sum the final product PP, as shown in Figure 8, which is composed of NCL full adders and half adders [13]. Where W, X, Y, Z are the four partial products, s0, s1, s2, s3 are the four signs of the partial products. After compression, the four partial products are summed by the RCA to produce the 16 bit product. It should be noted that a RCA was used instead of a carry look-ahead adder (CLA) because the performance of the asynchronous circuit depends on average-case delay but the worst-case delay in the synchronous circuit. As for the N-bit adder, the average delay of CLA and RCA is O (logN) [14]. Additionally, NCL RCA circuit has more advantages in area than CLA under the same conditions, Therefore, RCA is the preferred choice [15].

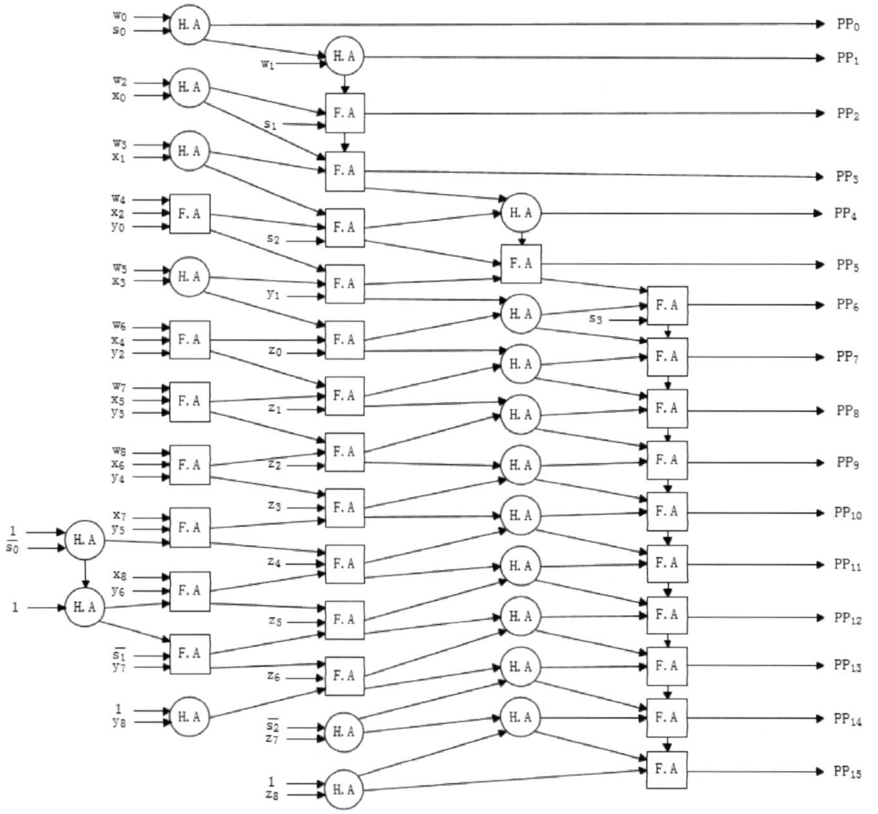


**Figure 4.** Wallace tree and RCA.

## 5. The Simulation and Performance Analysis of NCL Multiplier

In this paper we design and implement an 8-bit NCL multiplier on the Vivado platform and the hardware description language is Verilog-1995. Furthermore, we compare the delay of the designed NCL multiplier with that of the traditional 8-bit NCL multiplier based on Wallace. As shown in Table 2, the worst delay time of the two multipliers is given. It should be noted that when the partial product is compressed, the two final operands after compression are generated serially from the lowest to the highest bit, so the final sum operation and compression operation will overlap in the operation time to a certain extent, hence, we combine the delay of the two operations. According to the Table 2, it can be concluded that although the delay of the NCL multiplier designed in this paper is about three times that of the NCL multiplier based on Wallace while generating the partial product, the compression and adder delay of the former are much less than that of the latter. The comparison results show that the delay of the multiplier designed in this paper is reduced by 14%. Moreover, it's parallel computing when the partial product is generated, and there is no connection between delay and bit-width of multiplier. Therefore, when the number of multiplier bits is larger, the advantages of the NCL multiplier designed in this paper will be more significant.

**Table 3.** The delay of multiplier.

| NCL Multiplier | Delay of PP generation | Delay of compression and RCA | Total delay |
|---|---|---|---|
| The proposed NCL multiplier | 7.6ns | 15.2ns | 22.8ns |
| The NCL multiplier based on Wallace | 2.5ns | 24ns | 26.5ns |

## 6. Conclusion

In this paper we designed and implemented an 8-bit NCL multiplier based on Radix-4 Booth algorithm, and this NCL improved Booth circuit reduced coding time and area through the optimization of K-map. The Radix-4 Booth algorithm will reduce a half the number of the partial product, and the sign-generate algorithm will simplify the sign extension of the partial product. The final product is obtained by using the 4:2 compressor and the RCA. Compared with the 8-bit NCL multiplier based on Wallace tree, the calculation speed is reduced by 14%, and the larger the number of multiplier bits, the more significant the advantage will be.

However, compared with the multipliers currently used, 8-bit NCL multiplier has limited processing capacity. Therefore, the later work in this paper has carried out the research and development of 64-bit NCL multiplier based on this designed Booth algorithm which improves the calculation efficiency by clearly dividing the module function and increasing the circuit throughput by establishing a scientific pipeline under the condition of ensuring the minimum area and delay.

**Acknowledgments**

This work is supported by Huawei Technologies Co., Ltd, partly supported by Guangxi Science Technology Project (AB17129012) and the Fundamental Research Funds for the Central Universities (Grant No. lzujbky-2017-193) and by National Natural Science Foundation of China (No. 61772006) and the special fund for Bagui Scholars of Guangxi.

**References**

[1] Bandapati S K , Smith S C . Design and characterization of NULL convention arithmetic logic units[J]. Microelectronic Engineering, 2007, 84(2):280-287.

[2] S.C. Smith S C , Demara R F , Yuan J S , et al. Optimization of NULL convention self-timed circuits[J]. Integration the VLSI Journal, 2004, 37(3):135-165.

[3] Fant K M , Brandt S A . NULL Convention LogicTM: a complete and consistent logicfor asynchronous digital circuit synthesis[C]// International Conference on Application Specific Systems. IEEE, 2002.

[4] BOOTH A.D.A signed binary multiplication technique[J].Quarterly Journal of Mechanics and Applied Mathematics,1951,4(2):236-240.
[5] Wallace C S . A Suggestion for a Fast Multiplier[J]. IEEE Trans. Electron. Comp. 1964, 13.
[6] HE An-ping, LIU Xiao-qing, CHEN Hong. Study of 8-bit Booth Asynchronous Multiplier based on two-phase handshake protocol with bounded bundle data[J]. Acta Electronica Sinica,2018,46(04):961-968.(in Chinese)
[7] Ohkubo N , Suzuki M , Shinbo T , et al. A 4.4 ns CMOS 54×54-b multiplier using pass-transistor multiplexer[J]. 1995, 30(3):251-257.
[8] Cui Xiaoping.32 ×32 bit fast multiplier based on modified booth algorithm[J]. Electronic Measurement Technology, 2007(01):82-85.(in Chinese)
[9] Millar B , Madrid P E , Jr E E S . A fast hybrid multiplier combining Booth and Wallace/Dadda algorithms[C]// Symposium on Circuits & Systems. IEEE, 1992.
[10] Smith S C , Di J . Designing Asynchronous Circuits using NULL Convention Logic (NCL)[J]. Synthesis Lectures on Digital Circuits and Systems, 2009, 4(1):1-96.
[11] MACSORLEY O.L. High-speed arithmetic in binary computers[J].Proc.IRE,1961,49:67-91.
[12] Weinberger A . 4:2 carry-save adder module[C],1981.
[13] Smith S C , Demara R F , Yuan J S , et al. Optimization of NULL convention self-timed circuits[J]. Integration the VLSI Journal, 2004, 37(3):135-165.
[14] Joshi M V , Gosavi S , Jegadeesan V , et al. NCL Implementation of Dual-Rail 2S Complement 8Ã 8 Booth2 Multiplier using Static and Semi-Static Primitives[C]// Region 5 Technical Conference, 2007 IEEE. IEEE, 2007.
[15] S.C. Smith. Development of a large word-width high-speed asynchronous multiply and accumulate unit[J]. Integration, the VLSI Journal,2004,39(1).

IOP Conf. Series: Materials Science and Engineering **914** (2020) 012025 doi:10.1088/1757-899X/914/1/012025

# Design of Emergency Evacuation Scheme for Louvre

**Zhuoqun Zhang, Chen Wang, Yinzhe Wang**

Information Science and Technology College, Dalian Maritime University, Dalian, Liaoning, 116000

**Abstract.** In this paper, we used four steps in analyzing the optimal model of evacuation within the Louvre. For step 1, we refer to the traffic concept in the Dynamic Network Flow Model to define the export throughput. For step 2, considering that the original floor plan is segmented with pixel points as a unit with high complexity and little meaning, we simplify the floor plan when the model is guaranteed to be true. For step 3, we first mark the exit node, make its dual graph, and draw the point chain containing the exit node, which can divide the original image into several pieces according to the area. Breadth-First Search from each egress node can mark themselves with color, ownership and other information. For Step 4, in order to get the optimal segmentation which satisfies the definition of reasonable segmentation edge set, we use Genetic Algorithm to solve the optimization problem. A sequence consisting of the sequence numbers in the set of Edge-Cut is encoded as a gene in the Genetic Algorithm and iterated by the Theory of Chromosome Cross-Variation to obtain an optimal solution. Finally, we have designed a scientific and complete emergency evacuation plan.

## 1. Introduction

Located on the North Bank of the Seine River in the center of Paris, the Louvre is the first of the four museums in the world [1]. As one of the largest and most popular art museums in the world, the Louvre has been attracting more and more visitors in recent years. Between January and December 2017, the number of visitors to the Louvre rose sharply to 8.1 million, up 10 per cent from 2016 [2]. After the tourism boom of 2016, the Louvre has more visitors and is under pressure from more tourists. Although in recent years, the interior of the Louvre has been carefully designed and renovated [3]. The exhibition space and exhibits have been rearranged. The exhibition area has been increased, the display effect has been greatly enhanced. But there are still unavoidable safety risks in the venue.

At the same time, the increase in the number of terrorist attacks in France and the instability of the domestic environment have also increased the security risks of the Louvre to a certain extent. For example, on 3 February 2017, a man armed with a knife attempted to break into the Louvre Museum and attacked soldiers stationed there.

## 2. Model Development

### 2.1. *Simplification of Model*

When we create the model, we consider that the segmentation of the planar graph has some difficulties. At the same time, in the actual evacuation plan people generally use the form of zoning planning. As a result, the floor plans of each floor of the Louvre were extracted, simplified and partitioned. In this

IOP Conf. Series: Materials Science and Engineering **914** (2020) 012025 doi:10.1088/1757-899X/914/1/012025

process, we removed the unnecessary factors in the floor plan, and divided the exhibition hall into a certain small area. At the same time, we ignore the area of the corridor, only consider the connectivity of the corridor to the graph, and regard the corridor as the edge connecting each pavilion. Through the above steps we further abstract out each area, viewing the pavilion as a node, the channel as a side. The pavilions and passages on each floor converge to form an undirected graph. In this way, it is easier to analyze the model in the next step if the model is true.

In this model, we set the corresponding properties for each node and edge. The information carried by each node includes: area, crowd density, predecessor and successor. Each side contains information such as endpoints, traffic, and length. For ease of recording and understanding, we distill the contents of each node into Figure1.

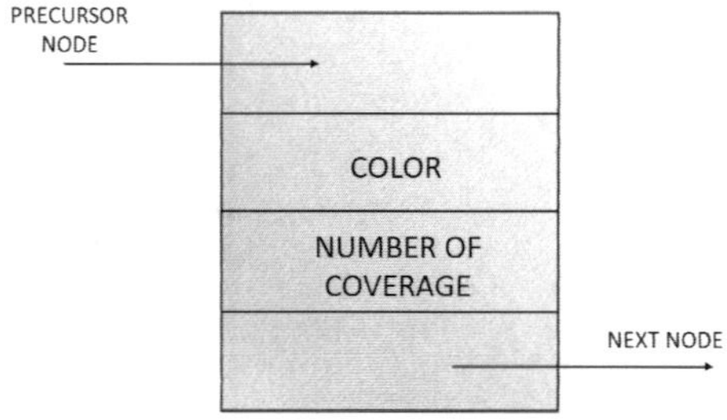


**Figure1.** Details of nodes

After refining the attributes shown above for each node, we summarize the abstract edges such as elevator, staircase, corridor for different channels. From this we created the connected graph as shown in the following figure according to the different position and internal structure of each floor. The five floors of the Louvre are laid out in plan, and the elevator connects each floor. The graph includes a total of 148 partitions, that is, 148 nodes, 166 edges. Based on the simplified floor plan, the undirected graph retains the information of the building structure and the crowd density. Thus, the model can be simplified without distortion and can be easily analyzed by graph theory.

### *2.2. Analysis of Cut Edge and Dual Graph*

*2.2.1. Cut Edge.* An undirected graph constructed from the above simplified model, instead of an extended exit flow plane, the exit nodes would mark the depth or breadth-first search of the graph as well as the passing nodes. We select some edges into the cut edge set E, then there is such a set. After removing the edges in the set, the original graph is perfectly divided into several connected point sets, and each point set has only one exit point. We call such a set E a reasonable set of edges.

A reasonable set of divided edges is shown in Figure2.

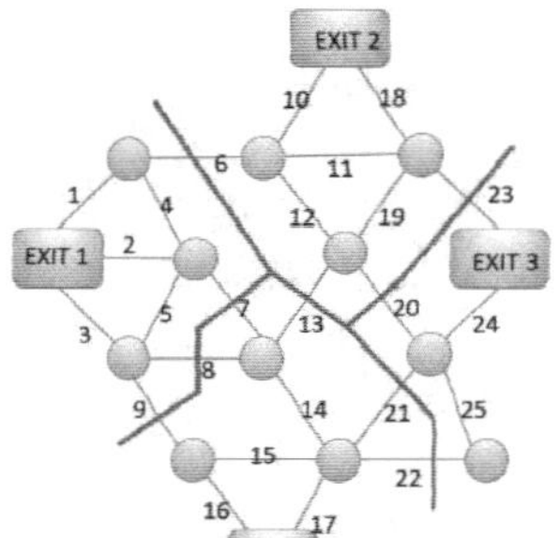


**Figure2.** Divided edges

*2.2.2. Dual Graph.* To facilitate our analysis of the internal relations of a reasonably segmented edge set, we apply a regular deformation to the graph above:

(1) The edges of the original graph are regarded as the set of points of the new graph.

(2) The Grid (topological plane), which is formed by the edges and points of the original graph, is regarded as the edge set of the new graph.

The new graph is called the Dual Graph of the original graph. Take the Figure 7 above as an example. Its Dual Graph is Figure3 as follows:

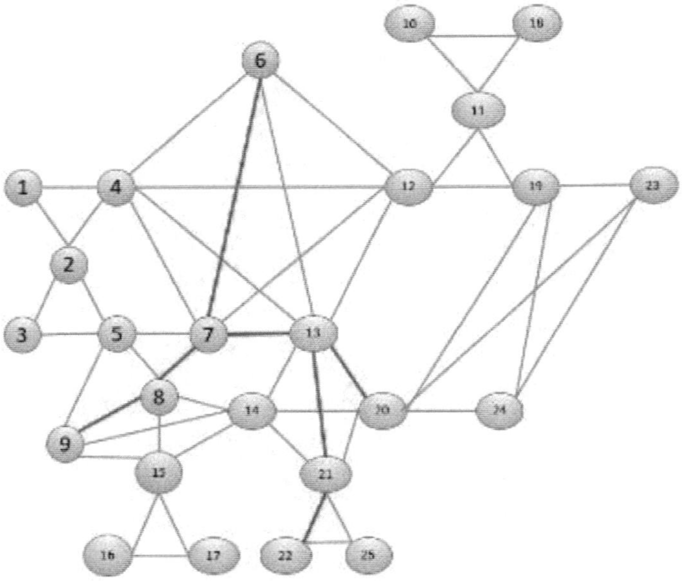


**Figure3.** Dual Graph

The above-mentioned deformation can be expressed more strictly as a kind of graph operation. And the operation is reversible:

$$G' = inv(G) \quad G = inv(G') \tag{1}$$

It can be seen that in the original graph, there is no obvious connection between the reasonable segmentation edge sets. But in the dual graph, all the corresponding points in the reasonable segmentation edge sets exist in the form of point chains in the dual graph. It is convenient for us to analyze the edge set from the angle of graph theory.

From the point of view of dual graph, for any kind of segmentation, we can use a dual graph point set 01 sequence, that is, the edge set of the original graph 01 sequence. You can specify rules for several graphs to assign values to nodes in the original or dual graph

*2.3. Genetic Algorithm*

*2.3.1. Brief view.* From the cut edge theory of the graph above, we need to find a reasonable cut edge set is not enough, it is not a necessarily segmentation. That is, the segmentation of the corresponding evacuation plan to get the overall evacuation time is not necessarily small. So we have to look for reasonable (satisfying the definition of reasonable edge) and better segmentation. We need an algorithm to solve this optimization problem-genetic Algorithm. The Best Block model is the best evacuation time, so the best evacuation time formula is:

$$\min(\max(f(G(Ex_i),\rho))) \tag{2}$$

Among them, the connected graph $G(Ex_i)$ is a unicom diagram containing the exit nodes obtained by reasonably splitting the edge sets. $f(G(Ex_i),\rho)$ means he time required for all personnel to be evacuated by the exit in the area of $G(Ex_i)$ covered by the population density $\rho$.

Since the Louvre's plan is divided into 148 key nodes, and 166 edges guarantee its connectivity and building structure, an undirected graph is generated. We combine cut edges as gene sequences. A better solution can be obtained theoretically by iteratively reasonable and superior chromosomes.

*2.3.2. Encode and Initialization.* The serial number of the side of the cut edge is taken as the gene sequence $A=[\ldots,e_i,e_{i+1}\ldots]$. Through the graph theory BFS, the cut edge number in the figure is found to correspond to the encoding process. The decoding process means that some of the edges in the annotation map are not passable. At the same time, several export nodes are expanded at the same time. They are dyed in the original picture. Each round randomly selects an area under the jurisdiction of the egress node for expansion. Until there is no unstained node in the original picture. Finally, the segmentation map is encoded to obtain a set of gene sequences.

*2.3.3. Adaptive Function.* For each sequence $A_i$, the corresponding undirected graph segmentation method and the connected block point $Ex_i$ containing each exit $G(Ex_i)$ can be obtained. As mentioned earlier, the pros and cons of a split can be evaluated by the overall time taken for the evacuation plan corresponding to that split. And the shorter the segment, the better solution we will have. The fitness function can be selected:

$$\begin{cases} F(A_i)=e^{\frac{1}{t_{total}}} \\ t_{total}=\max(f(G(Ex_i),\rho)) \end{cases} \tag{3}$$

So how to calculate $f(G(Ex_i),\rho))$ becomes the main calculation problem of fitness function. At present, the main algorithms to simulate crowd evacuation and traffic are dynamic network flow model and cellular automata model. Both of them can get better simulation of the traffic model within a certain scope of their application.

*2.3.4. Select the cellular automaton according to the Louvre.* Although there is no exact length data, the floor plan of the Louvre website gives the details of the structure and distribution of the exhibition area. Because the cellular automata can simulate the influence of the building structure on the crowd flow. It can modify the evolution rules of the cellular automata according to the characteristics of different crowd, the simulated crowd is more robust. So we choose to use cellular automata to simulate evacuation, and calculate the time of evacuation.

*2.3.5. Type-directed gene-directed cells.* For different cells, there are mainly different types (normal people, disabled people) and different blocks. As a result, their transfer equation is also different. Consider the cellular automata as a discrete model in both space and time. We need to divide the original plan of the Louvre into a uniform two-dimensional grid. But considering the multi-storey structure of the Louvre, in the stairs and elevator access is three-dimensional structure. We only consider extending the three-dimensional grid in the stairs and elevators. (though in proportion to the actual size of the plan).

Each grid is the equivalent of a cell. Each grid point corresponds to a position hazard, according to the Position Hazard of the grid point in its domain, the next time the mobile state is determined. Each time step is 0.5 seconds, and the cells throughout the building are renewed every time.

The danger level for each grid point is equal to the serial number of its breadth-first search in the exit grid set. The breadth-first search formed by the blocks of each graph assigns a danger value to the cells in the region, and each cell moves first to the low risk squares around it, and does not move if its value is lowest. As the formulary follows:

$$\begin{cases} d(x_i, y_i, z_i) = \begin{cases} bfs \\ +\infty & (x_i, y_i, z_i) \notin G(Ex_i) \end{cases} \\ D = \min\{d(x-1, y, z), d(x+1, y, z), d(x, y-1, z)......d(x_i^t, y_i^t, z_i^t)......\} \\ (x^{t+\Delta t}, y^{t+\Delta t}, z^{t+\Delta t}) = (x_{D_{\min}}^t, y_{D_{\min}}^t, z_{D_{\min}}^t) \end{cases} \tag{4}$$

Among them, $d(x_i^t, y_i^t, z_i^t)$ on behalf of the danger level of $(x_i^t, y_i^t, z_i^t)$. When the cell does not belong to the Connected Block, the danger value is set to positive infinity. When a cell or wall is occupied in the grid, the hazard value is set to positive infinity. At the same time each cell has a different type, called $type(x, y, z)$ , corresponding to its surrounding 8 grid, the current cell in the surrounding 8 grid feasible path can go, otherwise it can not go. This is type-driven. For example, $type(x, y, z)$ on behalf of the type of the formulary $access\ [type(x, y, z)][x-1][y][z] = 1$. Then next time the cell goes to location $(x-1, y, z)$.

In combination with gene and type orientation, the rules of cellular transfer are as follows:
Set time and the formulary follows:

$$(x_i^t, y_i^t, z_i^t) = \begin{cases} (x+1, y, z), (x-1, y, z), (x, y-1, z), (x, y+1, z), (x, y, z-1), (x, y, z+1) \\ (x+1, y+1, z), (x+1, y-1, z), (x-1, y+1, z), (x-1, y-1, z) \\ (x, y+1, z+1), (x, y+1, z-1), (x, y-1, z+1), (x, y-1, z-1) \\ (x+1, y, z+1)...... \\ ...... \end{cases} \tag{5}$$

There are 27 subsequent states. For the convenience of calculation and selection, the following state of a certain type of cell is assumed:

When it's impassable:

$$access[type(x, y, z)][x_i][y_i][z_i] = +\infty \tag{6}$$

When it's passable

$$access[type(x, y, z)][x_i][y_i][z_i] = 1 \tag{7}$$

The final choice is as follows:

$$\begin{cases} D = \min\{access[type(x^t, y^t, z^t)][x_i^t][y_i^t][z_i^t] \bullet d(x_i^t, y_i^t, z_i^t)\} \\ (x^{t+\Delta t}, y^{t+\Delta t}, z^{t+\Delta t}) = (x_{D_{\min}}^t, y_{D_{\min}}^t, z_{D_{\min}}^t) \end{cases} \tag{8}$$

*2.3.6. Selection function.* The fitness of each chromosome group is obtained from the fitness function, and the probability of each chromosome being selected is:

$$p(A_i) = \frac{F(A_i)}{\sum_i^N F(A_i)} \tag{9}$$

For Selection, we use the Roulette Wheel Algorithm; there are two roulette wheel algorithm to select two adaptive individuals, cross mutation.

*2.3.7. Crossover.* For the encoding of gene sequence in this model, it is the serial number of the cut edge mapped to a split graph. The edge-cutting sequence only reflects the non-connectivity of this edge. To get the corresponding segmentation graph, you need to consider the structure of the graph itself. If the form of a simple exchange of genes, then the resulting offspring decoding the division is not necessarily a reasonable division, the image can be said that the offspring have a genetic disease, can not survive. In order to maintain the number of population has to do continuously roulette selection algorithm. This wastes a lot of computation and time, and can lead to more intractable problems. That is, the offspring of good individuals may die from genetic diseases, so that good genes can not be retained, which is certainly not optimal.

Therefore, the individual crossover should take into account the structure of the original graph. Set $G_{decode}(A)$ as the segmentation map obtained by decoding the gene sequence. If two gene $A_x$ and $A_y$ sequences are selected by the Roulette Algorithm, $G_{decode}(A_i)$ and $G_{decode}(A_j)$ the segmentation graph they generated.

The specific cross-cutting processes are as follows:

(1) The segmentation graph after decoding two sequences is obtained.

(2) For the split graph of the genes of the offspring, we take the split graph generated by the two parents, and alternate the positions of each exit node $G(Ex_i)$, and merge $A_{x,t+1}$ into the decoding graph of the offspring:

$$G_{decode(A_{x,t+1})} = \bigcup_i^N G_{decode(A_x)}(Ex_i) \tag{10}$$

(3) The details of the merge operation $G_{decode(A_{x,t+1})}$ are that the initial count of the nodes in is 0, and the number of times they are overwritten in $G_{decode(A_x)}(Ex_i)$ during the merge is recorded. Suppose a gene sequence is decoded to produce a simple segmentation model -- a square with only four corners. The black lines represent splitters, and the colored squares represent different gene sequences.

(4) The best-case scenario $G_{decode(A_{x,t+1})}$ is a reasonable split graph in which all nodes are covered $G_{decode(A_x)}(Ex_i)$ exactly once. Then you can code $G_{decode(A_{x,t+1})}$ directly to a reasonable and healthy offspring.

(5) It is unreasonable that some nodes are covered many times, that is, the graph $G_{decode(A_{x,t+1})}$ is not completely separated, there are two or more exit nodes in the region connected together. As shown in Figure4.

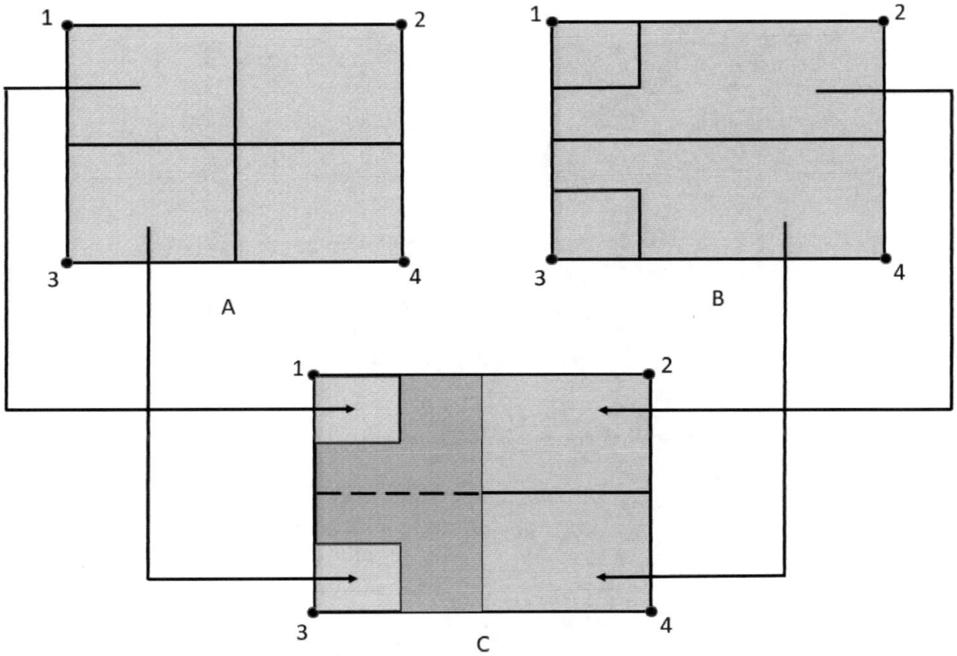


**Figure 4.** Area nodes

Where A and B represent two parent genes obtained by decoding the segmentation, a simple cross rule for the offspring of C1, C3 export segmentation way choose to inherit the parent. The C2, C4 export segmentation way to inherit the parent B. As can be seen from the gray part of figure C above.

So, we're going to do a contraction adjustment on each one. Put each exit node into the queue and do the breadth-first search. First search the reasonable range of each exit node $G_{decode(A_{x,t+1})}$. That is, the red and blue part of C above. Then the Middle Node count greater than 1 node edge expansion. Until there are no such violations.

Finally, the dual map of the original map, coding, we can get a reasonable gene sequence after adjustment. The process and results are shown in Figure5 and Figure6.

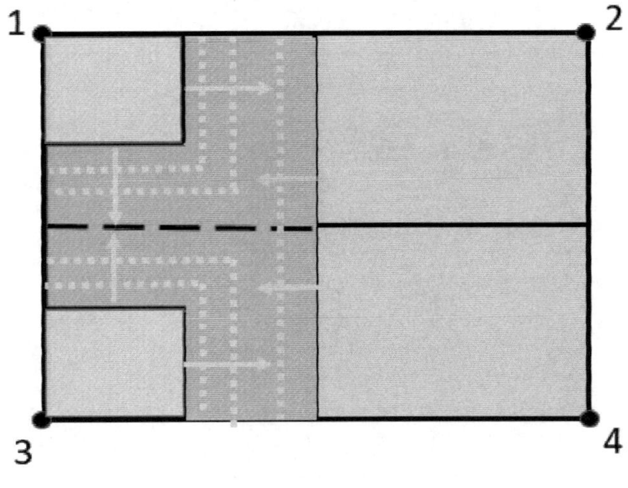


**Figure 5.** Coding proccss

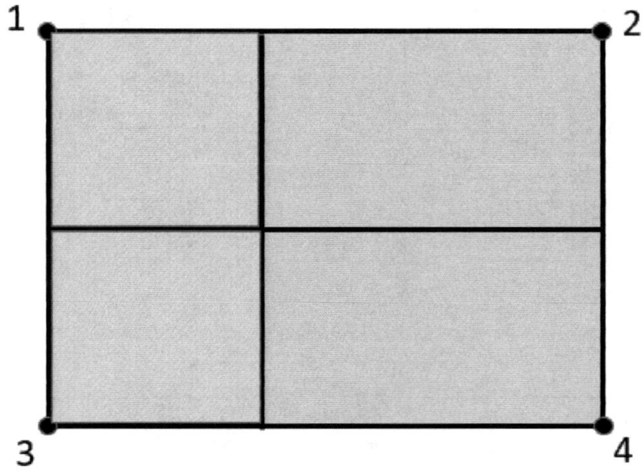


**Figure 6.** Coding results

Another unreasonable is that some of the nodes $G_{decode(A_{x,t+1})}$ are not covered by $G_{decode(A_x)}(Ex_i)$, that is, the original graph is divided into more than N (the number of exit nodes). Where C2 is crossed, exit 1,3 splitting inherits parent B, and exit 2,4 splitting inherits parent A. See the blank portion of figure C above, the extra portion of the C2 decoding map that has been split.

The process is similar to the point above and each $G(Ex_i)$ needs to be widened. We take all the exit nodes as the starting point, complete the reasonable part of the breadth-first search, and then extend the edge of the node $G_{decode(A_{x,t+1})}$ with the count less than 1. Until there are no such violations.

Finally, the dual map of the original map, coding, we can get a reasonable gene sequence after adjustment. The process and results are shown in Figure8 and Figure9.

*2.3.8. Termination Conditions.* Set the maximum number of iterations N=2000. If during its iteration, the difference between the optimal value of the offspring and the optimal value of the parent is less than $eps$, the evolutionary process can also be terminated.

### *2.4. Evacuation Sequence*

**Table 1.** Evacuation Sequence

| EXIT | Area number |
|---|---|
| **Glass pyramid** | 100~105、130~137、160~186、236~305、308~316、500~526 |
| **Carrousel Gallery** | 200~218、219~230、535~564、800~803 、811~835、836~864 |
| **Lions Gate** | 400~403、405~423、424~433、700~711、712~717、728~734、900~913 |
| **Art door** | 317~338、339~348、601~622、633~640、641~663、917~931、936~952 |

## 3. Conclusion

Based on the study of the Louvre emergency evacuation strategy, we combine the above established models and the various analytical methods provided. We make the following specific recommendations for the evacuation route of the Louvre:

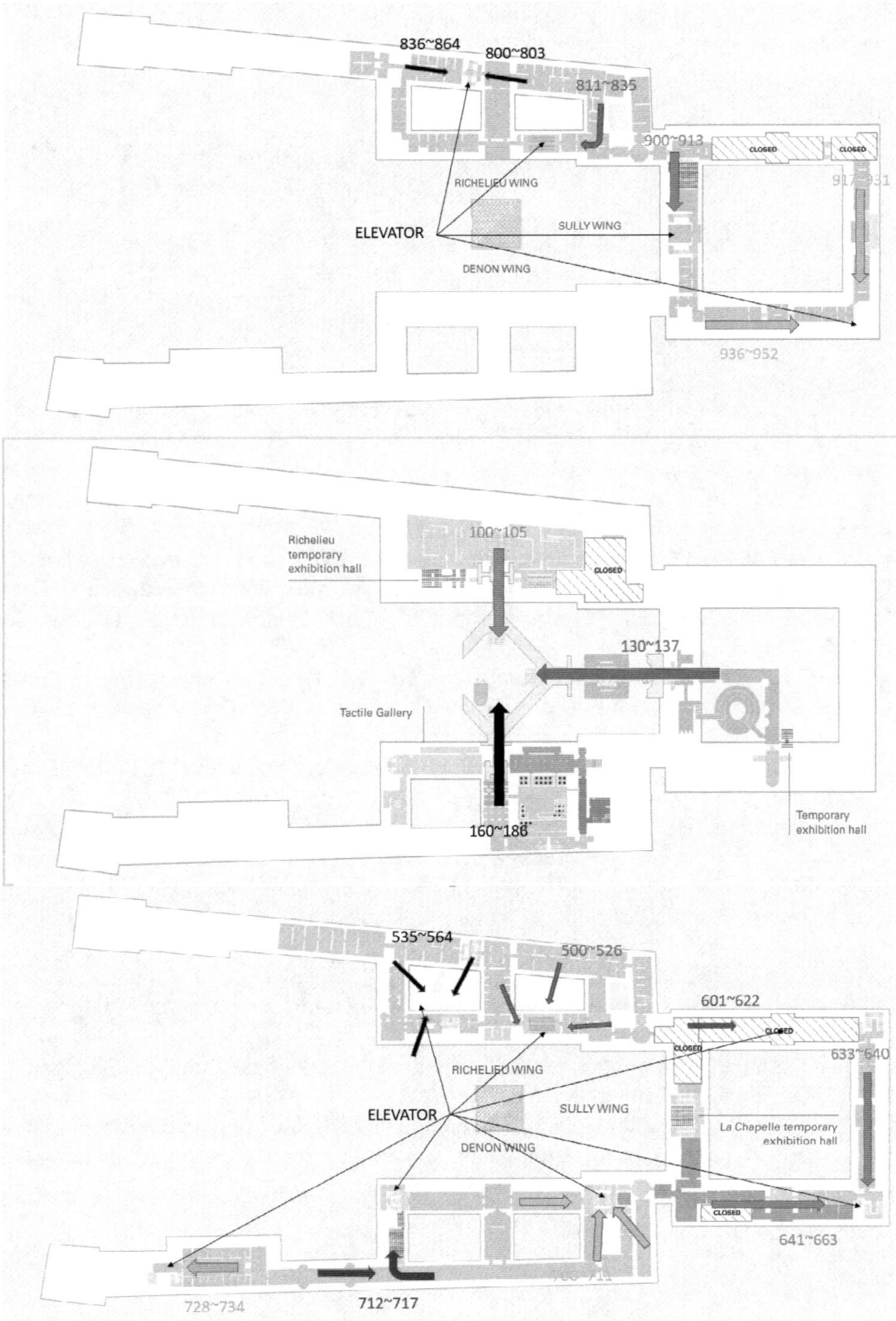

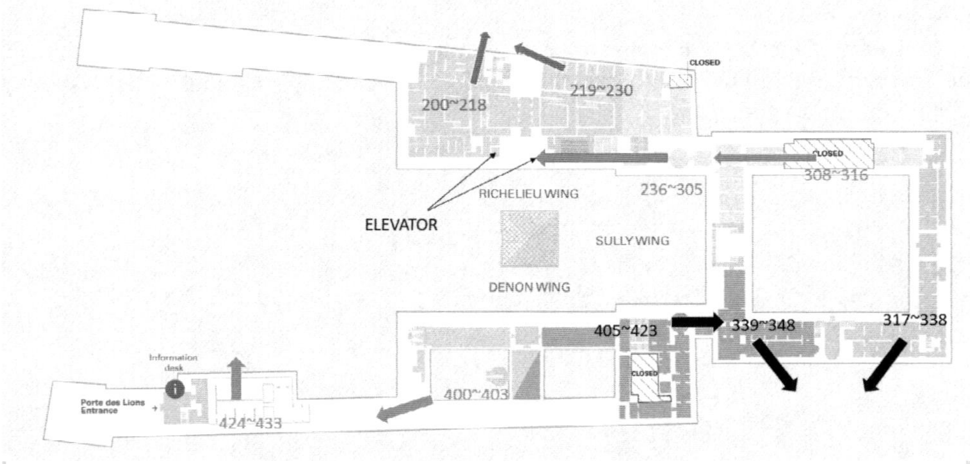


**Figure 7.** Evacuation route suggestions

## References

[1] Haoran Ma, Liang Pang. Study on urban emergency evacuation model based on Genetic Algorithm [P]. Remote Sensing, Environment and Transportation Engineering (RSETE), 2011 International Conference on, 2011.

[2] Zhang, Lei,Wang, Jinhuan,Shi, Qiongyu,Zang, Ting. A new dynamic cellular automaton model for pedestrian evacuation process in stadium [P]. 2013.

[3] Yan Xu, Hai-jun Huang, Li-jun Tian. Simulation of Exit Choosing in Pedestrian Evacuation Using a Cellular Automaton Model Based on Surrounding Pedestrian Density [P]. Computational Sciences and Optimization (CSO), 2011 Fourth International Joint Conference on, 2011.

[4] Zhang Xinlong, Chen Xiuwan, Li Huaiyu, Li Fei.A Person Evacuation Model for Improved Cellular Automata [J]. Journal of Wuhan University (Information Science Edition), 2017, 42(09): 1330-1336.

[5] Zheng Mao, Zheng Wei. Research on Ship Evacuation Scheme Based on Graph Theory Algorithm [J]. Computer Simulation, 2018, 35(10): 393-397.

[6] HUANG Wangxing, HE Youyan. Indoor fire evacuation model based on improved cellular automata [J]. Fire Science and Technology, 2016, 35(11): 1562-1565.

[7] Zhang Xinlong, Chen Xiuwan, Li Huaiyu, Li Fei. A Personnel Evacuation Model for Improved Cellular Automata [J]. Journal of Wuhan University (Information Science Edition), 2017, 42(09): 1330-1336.

[8] WANG Xudong, ZHONG Ming, LIU Shaobo. Study on evacuation efficiency of multi-story subway stations based on microscopic simulation [J]. Chinese Journal of Safety Science, 2017, 27(09): 51-56.

[9] Wang Yuna. Research on evacuation path of subway station based on dynamic system optimal model [D]. Chang'an University, 2017.

[10] XIONG Lichun, YANG Libing, CHEN Jianhong, ZHENG Haili, SHI Dongping. Study on Personnel Evacuation Optimization Model Based on Graph Theory and Queuing Theory [J]. Journal of Safety and Environment, 2014, 14(03): 166-171.

IOP Conf. Series: Materials Science and Engineering **914** (2020) 012026 doi:10.1088/1757-899X/914/1/012026

# Study on the Establishment and Optimization of Sandcastle Infrastructure Model

**Zhixuan Han[1], Yuqiu Gao[2], Shuyu Ren[1], Depin Chen[3]**

[1]Nanjing University of Information Science & Technology Mathematics and statistical institute, Nanjing, Jiangsu, 210044
[2]Nanjing University of Information Science & Technology Changwang School of Honors, Nanjing, Jiangsu, 210044
[3]Nanjing University of Information Science & Technology School of Automation, Nanjing, Jiangsu, 210044

**Abstract**. Sand castle was created by the tourists on the beach for fun, then how to set up a stable sand castle under the condition of waves and tidal waves to get the honor of "sandcastles builder" has been their main concern. In this paper, considering the formation principle of sandcastle and the effect of the uncertain factors and the effect on the stability of the sand castle, the sandcastle infrastructure model is established and the 3D geometric model of the sand pile is optimized step by step. We start with sand heap discretization, the impact process of waves and tides is regarded as the external force acting on the sand, and the stress analysis shows that there are three kinds of stress conditions: no external force, one external force and two external forces. Then from the point to surface, the stress energy of the whole layer of sand is analyzed according to the two aspects of energy transfer dissipation and internal force failure dissipation. The optimal bearing capacity plane is approximately pentagon by annealing algorithm. Finally, from the surface to the solid, taking the self-weight of the sand pile, the effective external force action and the seawater erosion action as the constraint conditions, the optimal tilt angle of the plane stack is calculated to be 36.87° by measuring the risk degree of the overlaying slope Angle. Therefore, the oblique pentagonal body with an inclination angle of 36.87° is relatively stable.

## 1. Introduction

Building sandcastles is a pleasure created by beachgoers on the casual beach, who use tools, toys and imagination to assemble simple piles of sand into complex architectural models which are known as sandcastles. The sandcastle is usually formed by cutting and molding on a pile of single, indescribable wet sand dunes, gradually forming an identifiable three-dimensional geometry as part of its infrastructure, and then sketching different architectural shapes on this basis. But inevitably, with the waves' joint and the rising tides will erode the sandcastle. Influenced by the geometry of the sandcastle's foundation, even sandcastles of the same size and the same distance from the beach respond differently to waves and tides.

IOP Conf. Series: Materials Science and Engineering **914** (2020) 012026 doi:10.1088/1757-899X/914/1/012026

## 2. Stress Analysis of Sand Grains

### *2.1. Analysis of External Force*

Due to the complicated mechanism of the tide ashore lapping and sand-piling, to further simplify the model, we regard the surf and tidal ashore lapping process as an external force $F_0$ acting on the sea surface of the sandcastle.

### *2.2. Analysis of Internal Forces*

The sand grains standing beside the beach all the year round are often eroded by the seawater, so we consider that in a short time, the seawater does not react to the sand grains by chemical erosion, and the water content in the sand grains gap is evenly distributed and remains unchanged.

In fact, the sand pile is composed of the sand particles, the water and air in its interstitial phase. This paper uses the concept of "circular approximation", Two adjacent sand grains are taken as an example for stress analysis, equivalent to consider these two solid material in bending ball water - the contact boundary condition of gas, close contact between grains of sand, and the main components of the sand is $SiO_2$, which is a kind of hydrophilic material. Therefore, according to the phase equilibrium theory can draw interface, under such condition, due to the gap between the difference of hydraulic pressure and atmospheric pressure makes a matrix suction between sand, as shown in Fig1(a), among them, $r_1$ says radius of curvature of concave to the water phase, which describes the ball tends to the trend of the water phase interface, $r_2$ says convex to the radius of curvature of the water phase. It shows the increasing trend of water phase pressure, and they control the change of water pressure from opposite aspects.

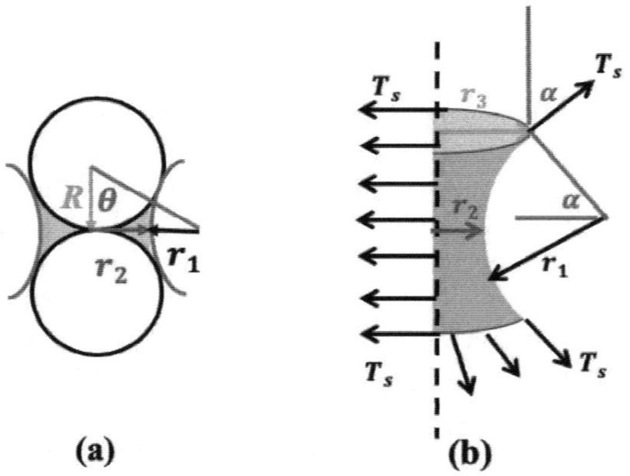


**Fig 1.** Stress between grains of sand

Consider the stress balance of the free body in the horizontal direction, as shown in Fig1(b), the free body is affected by three forces: the surface tension $(T_s)$ of the water at the interface, represented by $r_1$, is the energy on per unit length required when the water interface is damaged, which is a constant value under the condition of constant temperature, and generates a positive horizontal force. The surface tension $(T_s)$ at the interface represented by $r_2$ ,produces negative horizontal force. The two forces acting on the interface. The contact angle $\alpha$ is the angle between the tangent line on the gas-water interface and the straight line on the water-solid interface.

The projection of surface tension in the positive horizontal direction is:

$$F_1 = 4r_3T_s sin\alpha \tag{1}$$

The projection of surface tension in the positive horizontal direction is:

$$F_2 = -4r_1T_s sin\alpha \tag{2}$$

The difference between water pressure and air pressure, that is, the projection of matric suction $F_i$ is expressed as:

$$F_3 = 4r_1r_2F_i sin\alpha \tag{3}$$

From the equilibrium of three forces, the matric suction is:

$$F_i = T_s(\frac{1}{r_1} - \frac{1}{r_2}) \tag{4}$$

Let the particle radius be $R$, and according to the approximate expression relation proposed by $Dallavalle$. In this paper introduced the filling angle to describe the relation among $R$、$r_1$、$r_2$. As shown in Fig3, we can obtain the following relation formula:

$$r_1 = R(\frac{1}{cos\theta} - 1), r_2 = Rtan\theta - r_1 = R\left(tan\theta - \frac{1}{cos\theta} + 1\right), 0 \leq \theta \leq 85° \tag{5}$$

And then, the relationship between matric suction and filling angle can be obtained:

$$F_i = \frac{T_s}{R}\frac{cos\theta(sin\theta+2cos\theta-2)}{(1-cos\theta)(sin\theta+cos\theta-1)} \tag{6}$$

When considering the construction model, we assume that the water content between the sand grains in the structure is saturated and unchanged, which is equivalent to the fixed contact Angle of the sand grains. Therefore, we regard the force $F_i$ in the sand grains as a constant, and the destroyed energy of the matrix force is also fixed as a constant.

It is known that the radius of sand grains is $0.2063mm$, the surface tension of seawater at 20℃ is about $72.25mN/m$, and the filling angle of saturated water is $45°$. Therefore, the matric suction of the particles was calculated about $247.639mpa$.

## 3. Analysis of Sand Pile Energy

According to the above analysis of forces, it can be known that the sand is affected by external and internal forces at the same time. When the waves or tides come ashore, regardless of the remaining force of the waves and ebb, there is an external force on the upper surface of the sand pile for oblique upward transmission. Because of the same force and the energy transmission direction, we use the "slicing method" to analyze the energy dissipation of the model layer by layer and then stack it layer by layer to obtain the energy consumption efficiency of the entire model.

### *3.1. Analysis of Horizontal Energy Consumption of Sand pile*

We takes the $k$th middle layer section for consideration. As shown in fig2, when an external force acts on a certain sand, the direction of the external force is the $y$ axial direction, and the $y$ direction perpendicular to the axis is the $x$ axial direction, the transmission plane coordinate system is established. The force and energy will be transmitted along the upper half plane of the axis, which is represented on the diagram as particles transmitting force and energy to the three particles above the $x$. To further simplify the model, we only consider the force and energy transmission process in the direction of the external force.

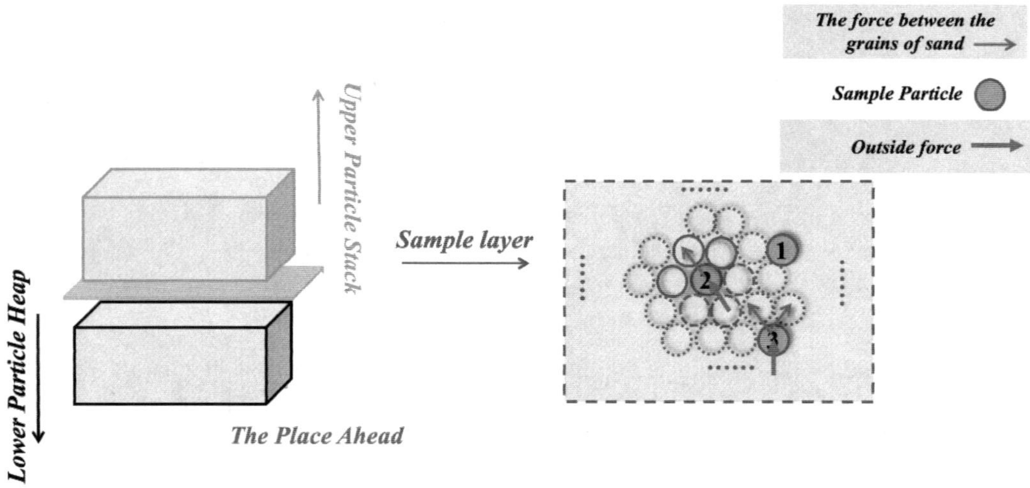


**Fig 2.** Schematic diagram of stress analysis of sand grains: Geometry as an example, non-final sandcastle foundation model

Therefore, according to the plane arrangement of sand particles, it can be shown that there are three kinds of forces on the plane that can transfer particles:

Steady-state particles (1st particle in the figure): Particles without force and energy transfer are only bonded to adjacent sand particles by the internal force. The set of all steady-state particles is recorded as $\boldsymbol{A}$;

Semi-steady-state particles (2nd particles in the figure): Particles in the direction of force and energy transmission can only carry force and energy to the adjacent sand particles along the transmission axis direction. The set of all semi-steady-state particles is recorded as $\boldsymbol{B}$;

Unsteady particles (3rd particles in the figure): When there is no nearby sand particle that can transmit force along the direction of the transmission axis, and it can transmit nearby particles, the particle will decompose the force and energy to the upper surface of the transmission axis coordinate system Two particles carry force and energy, and form a new coordinate system of the transmission axis. The set of all non-steady-state particles is recorded as $\boldsymbol{C}$**;**

At the same time, we consider that in the transmission process, each time there is an energy decay process, for the convenience of calculation, we set the constant coefficient$\delta$.In addition, based on the force analysis of the sand particles, it can be known that the matrix suction force of the bonding particles along the energy transmission direction is destroyed during the transmission process, and each piece of destruction energy $\varepsilon$ is set to be constant.

We consider each layer as a $m \times n$ 0-1 matrix$P_k(k = 1,2 \ldots \ldots, l)$. When there are sand particles in the $i$th position of the $jth$ row, $p_{i,j+1,k} = 1$ and vice versa, $p_{i,j+1,k} = 0$ .We suppose that the initial energy transferred is the initial energy which is harvested by the sand particles exposed to the impact of the seawater. Based on the above analysis, we obtain the energy harvest of the particles in the ith and jth positions of the kth layer:

$$e_{i,j,k} = \begin{cases} 0, & p_{i,j+1,k} \in \boldsymbol{A} \\ \delta e_{i-1,j1,k} + \varepsilon, & p_{i,j+1,k} \in \boldsymbol{B} \\ \delta\left(e_{i-1,j-1,k} + e_{i-1,j,k}\right) + 2\varepsilon, & p_{i,j+1,k} \in \boldsymbol{C} \end{cases} \tag{7}$$

Then the total capacity of the kth layer is $Energy_k = \sum_{i=1}^{m} \sum_{j=1}^{n} e_{i,j,k}$

### *3.2. Analysis of Vertical Energy Consumption*

The vertical stacking structure of the sand pile is related to the inclination angle of the sand beach. Meanwhile, we can determine the difference in the number of sand particles in each layer by the inclination angle $\beta$, that is, the size, and stack each layer to obtain the entire three-dimensional geometric model.

We use pressure to consider its strength:

$$P = \frac{F}{S_{effect}} \tag{8}$$

Among them, $S_{effect}$is the effective area where the external force acts, $S_{effect} = S_{real}\ sin\beta$, and $S_{real}$ is the real contact area between the geometry and the ocean tide.

At the same time, based on the $Wiesenfeld$'s sand pile model, it can be known that after the sand pile reaches the "critical" load-bearing state, the sand particles piled on the sand pile are in an integrated state, and the newly accumulated sand will disturb the sand pile due to gravity. The transmission in the whole sand pile makes the structure of the sand pile fragile, and eventually the sand pile collapses. Therefore, we need to judge the load-bearing capacity when considering the sand layer stacking process.

From the aspects of pressure and weight of sand discharge, this article has developed a measure of the degree of danger of stacking sand layers to help us determine the range of inclination angle.

$$\zeta = k_1 \frac{1}{S_{effect}} + k_2 E\left(\frac{\sum_{k=1}^{l} w_k}{S_k}\right) + +k_3 E(-\log\left(\frac{S_k}{S_1}\right)) \tag{9}$$

Assuming that $w_k$ is the weight of the $k$ layer. It can be seen from the figure5 that $\Delta x_i$ represents the most heterodyne value of the $k$ layer and $k+1$ layer. $eps$ is the offset term. $k_1$ is the amount of external force applied. $k_2$ is the weight value for the load bearing size of the sand pile. The slower the slope, the smaller the value

The second term is the average of the pressure generated by its own gravity on the next layer, and then even if the volume is certain, the slope is slower and the height is higher, but because the slope is very slow, the pressure on the next layer is very low, so it's still getting smaller and smaller.

The third term is the damage term of seawater to its horizontal layer, because the more the slope changes, the smaller the horizontal layer is, and the smaller the damage will be greater, so this number is getting larger and larger. $S_k$ is the area of the k layer, and $S_1$ is the area of the bottom layer.

## 4. Optimization Model

### *4.1. Determination of Horizontal Optimal Cross Section*

For the plane layer, the energy brought by external forces will destroy its stability, that is, the more total energy it obtains, the more easily its stacked structure will be destroyed. Therefore, in order to make its structure the most stable, the total energy of the plane layer is the least under the same external forces:

$$minEnergy_k = \sum_{i=1}^{m} \sum_{j=1}^{n} e_{i,j,k} \tag{10}$$

From the above equation, we can calculate the optimal plane layer under the condition of fixed sand grains as follows:

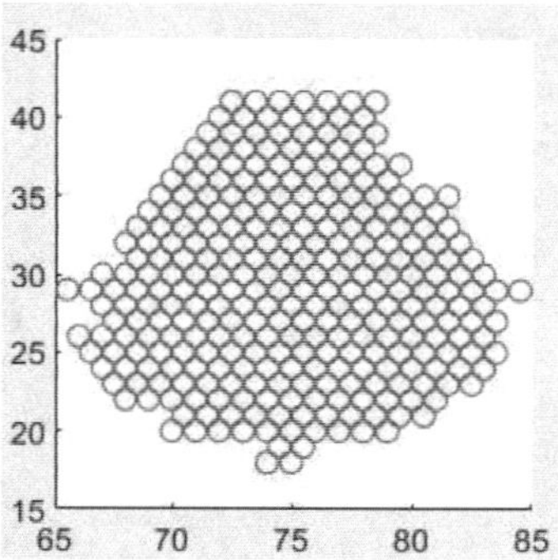


**Fig 3.** Optimal grain distribution plan

It can be seen that in the horizontal section, the plane which is approximately in the shape of a spindle has the optimal load-bearing capacity of applying pressure to the front.

*4.2. Determination of the Optimal Vertical Inclination Angle*

According to the risk coefficient measurement standard, in order to make the sand pile most stable, it is equivalent to making the sand pile least dangerous:

$$min\zeta = min[k_1 \frac{1}{S_{effect}} + k_2 E\left(\frac{\sum_{k=1}^{l} w_k}{S_i}\right) + k_3 E(-\log\left(\frac{S_i}{S_1}\right))] \quad (11)$$

Finally, we selected the minimum risk as the optimal inclination Angle of the vertical structure:

$$\beta = \arctan \frac{R}{\Delta x_i + eps} \quad (12)$$

According to the optimal plane layer and the optimal slope Angle, the sand grain ratio of each layer is calculated, and the optimal model of sand castle is finally determined.

By fig4, we found that the smaller the cross section in the horizontal direction, the higher the risk, and that for the slope in the vertical direction, in the case of the total volume and the bottom area, there is a slope that is the best value of the damage and the best. Through the intersection of two intersections, we get the optimal slope 0.75,, as a result, we get the angle $\beta = \arctan(0.75) = 36.87°$

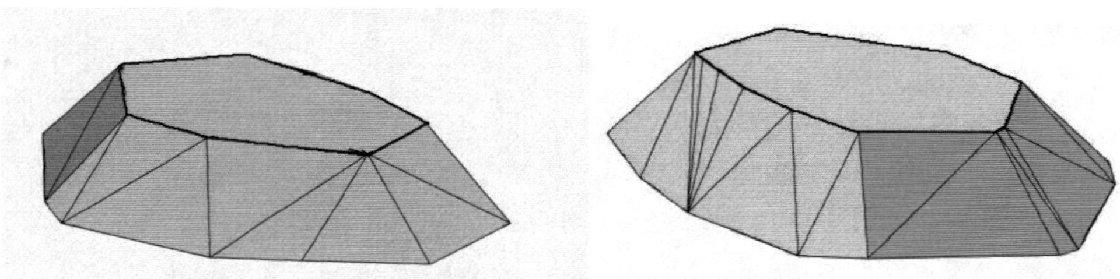

(a) Left view of geometry (b)Main view of geometry

**Fig 4.** Optimal 3D Geometry

## 5. Conclusion

According to the actual situation, the sand pile is made up of many sand particles. Therefore, we discretize the sand pile at first in this paper, and regarding it as a structure composed of many highly rigid and frictional spherical particles of the same shape and size. For the sand castle construction model, we fix the number of sand particles in the sand pile and make an explanation that the condition that the water content between the sand grains are saturated and unchanged. At this time, only the influence of geometry on the stability of the sand pile is taken into consideration.

**References**

[1] Li dayong, zhang yukun, liu weiwei. Application and understanding of the principle of effective stress in engineering problems [J]. Mechanics & practice, 2013,35 (02) 92-94.

[2] Zhou hui-guang, zhang yi, ren yu-peng, xu xing-bei, xu guo-hui. Experimental study on shear force of sediment muddy flow on fixed sand bed [J]. Journal of ocean university of China, 209,49 (S2): 92-98.

[3] https://www.cnblogs.com/gucheng/p/10104873.html

IOP Conf. Series: Materials Science and Engineering **914** (2020) 012027 doi:10.1088/1757-899X/914/1/012027

# Application of artificial intelligence in data processing

**Xingnan Jiang[1], Yueying Wang[2], Bowen Deng[3]**

[1]College of Finance and Economics, Jiangxi University, Nanchang, Jiangxi, 330000, China
[2]College of Information Engineering, Minzu University of China, Beijing, 102200, China
[3]School of Information and Communication Engineering, Harbin Engineering University, Harbin, Heilongjiang, 150000, China

**Abstract**. Aiming at the problems of large scale, variety and difficulty in data analysis, a fuzzy neural network algorithm based on artificial intelligence technology is proposed to analyze and process data. This algorithm combines the learning characteristics of the neural network and the fault-tolerant advantages of the fuzzy system, uses k-man's algorithm to cluster the sample input space, generates the membership matrix, and uses the neural network algorithm to train the data, and finally completes the target output. The analysis results can meet the existing engineering application requirements. The research content of this paper is of great significance to improve the utilization level of power engineering data and engineering management ability.

## 1. Introduction

With the continuous innovation and development of computer science and Internet technology, artificial intelligence is no longer a novelty. It has been applied in many fields and has a very important impact on people's production and life [1]. The development and improvement of artificial intelligence has effectively improved people's daily work efficiency and promoted the development and progress of social civilization. When applying artificial intelligence technology, traditional data processing technology is difficult to meet its development needs [2].

With more and more subdivided scenes of artificial intelligence technology, the fourth wave of industrial revolution brought by artificial intelligence has become a surging trend. Many traditional industries rely on ARTIFICIAL INTELLIGENCE to empower the industrial structure [3]. In the constant emergence, ARTIFICIAL INTELLIGENCE is also subtly changing all aspects of life, such as biometrics, video recognition, content review, and intelligent security [4].

In view of the problems of large amount of data, many types of power engineering, and difficult processing, this paper proposes a fuzzy neural network algorithm based on artificial intelligence technology to process and analyse power engineering data [5, 6]. The fuzzy neural network algorithm is a kind of artificial intelligence technology algorithm, which can analyze and process a large number of random historical data, and extract useful information and data laws from it, so as to complete the prediction of engineering cost and guide engineering project management [7-9]. The research content of this article is of great significance for improving the level of power engineering management in my country [10, 11].

IOP Conf. Series: Materials Science and Engineering **914** (2020) 012027 doi:10.1088/1757-899X/914/1/012027

## 2. Data mining algorithm

The data mining method based on artificial intelligence technology refers to the process of extracting useful information from a large amount of engineering data. Since its role is a large number of actual recorded data, it is faced with problems such as large data volume, large data noise, and difficulty in rule extraction. Generally, in order to better extract useful information, a complete data mining technology usually includes: selection sampling, data processing, data conversion, mining model, and data evaluation process.

In actual power engineering, in order to solve the data processing and information extraction work, the commonly used data mining techniques are: statistical analysis method, decision tree method, neural network method, fuzzy logic algorithm and genetic algorithm. Because the fuzzy neural network algorithm combines the learning advantages of the neural network and the logic fault tolerance of the fuzzy system, it has important applications in power engineering data processing and engineering cost prediction.

When applying neural networks to process and predict power engineering data, the first issues to be addressed are the acquisition of fuzzy system parameters and the identification of fuzzy rule parameters. Among them, the acquisition of fuzzy system parameters includes the determination of the number of fuzzy rules and the calculation of membership. Due to the complexity and diversity of the actual situation, when the fuzzy neural network algorithm is used, the data space needs to be divided into several fuzzy data sets by clustering method, and then the membership function can be obtained by training and the required output. Figure 1 shows a schematic diagram of a simple fuzzy neural network model.

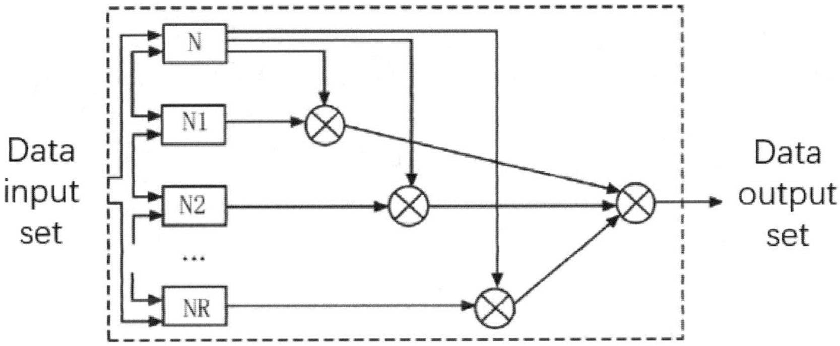


**Figure 1.** Schematic diagram of fuzzy neural network model

## 3. Data preprocessing

When analyzing power engineering data, a large amount of data involved in the project needs to be screened purposefully, and the largest possible is to find a data set representing the attributes of the engineering data on the basis of maintaining the original data information.

Power engineering data is usually stored in numerical form, so this paper uses Bayesian classifier as the evaluation function. Bayesian method is a conventional method for classifying large amounts of data based on mathematical statistical theory. Assuming that the data set is S, the sample attributes are represented by X1, X2, ..., Xn, and at the same time, the existing data types are represented by C1, C2, ..., Ck, then for a new sample data, it belongs to a certain category Cj The probability is:

$$P(X|C_j)=(n_{C_j}h)^{-1}\sum_{X:C(X)=C_j}K(\frac{X-\mu_j}{h}) \tag{1}$$

In the above formula, h = σ is the bandwidth, ncj is the number of samples, and K = g (X, 0, 1) is the Gaussian distribution function. The posterior calculation probability calculation formula of the improved Bayesian classification algorithm is:

$$P(X|C_j)=\frac{1}{n_{C_j}}\sum_i g(X,0,1) \tag{2}$$

$$g(X,0,1)=\frac{1}{\sqrt{2\pi}}e^{-\frac{X^2}{2}} \tag{3}$$

G is a Gaussian density function used to represent the probability of data distribution. The mean is 0 and the variance is 1. Taking a certain power engineering data as an example, 100 data nodes are evaluated, and the data attributes corresponding to the nodes are regularized. The input data and output data attributes of the power engineering are finally obtained as follows:

Input set: Voltage level, line number, transport distance, terrain coefficient, line length;

Output set: Transportation engineering, basic engineering, wiring engineering, attachment engineering.

## 4. Data model establishment

According to the analysis results of power engineering data in the previous section, due to the diversity of engineering data, the input and output of data are multiple, so the nature of data analysis is a nonlinear mapping problem, so when choosing an artificial intelligence algorithm The fuzzy neural network algorithm can be used as the core algorithm of engineering data processing, so as to establish the estimated model of power engineering cost.

The combination of fuzzy neural network algorithm and conventional neural network algorithm for data learning mechanism and fuzzy system inference ability, not only has the ability to obtain data rules, but also has a strong network fault tolerance ability, unique in dealing with complex nonlinear data Advantage.

## 5. Model simulation

The processing of power engineering data. The original data is 200 pieces of historical data after data preprocessing, the input data contains 5 attributes, and the output data contains 4 attributes. Then use the Kmeans method to classify the power engineering data. After the data classification is completed, the fuzzy neural network needs to be used to adjust the membership function. During the calculation, the sample is divided into 4 parts

All samples are divided into 3 parts, of which the number of samples in the training set, test set and verification set are 120, 40 and 4, respectively, and the operation is stopped after 500 iterations. At the same time, in order to verify the correlation of the results, the linear relationship between the output of the fuzzy neural network algorithm and the actual network output is obtained. The fuzzy neural network algorithm can be used to analyze and process power engineering data, and to a certain extent, the data rules obtained by the fuzzy neural network can be used to predict the power engineering cost level. The fuzzy neural network is used to extract data rules for 200 sets of historical data, and then the project cost prediction is made for a specific project.

## 6. Conclusion

This paper presents a fuzzy neural network algorithm based on artificial intelligence technology to analyze and process power engineering data. The K-man's algorithm is first used to perform cluster analysis on the sample input space and produce the corresponding membership matrix, and then the

IOP Conf. Series: Materials Science and Engineering **914** (2020) 012027 doi:10.1088/1757-899X/914/1/012027

neural network algorithm is used to train the data and perform regression analysis on the sample data. The artificial neural network-based fuzzy neural network algorithm proposed in this paper has important application value and guiding significance for the analysis and processing of power engineering data.

**References**

[1] Nakamoto, S. *Bitcoin: A peer-to-peer electronic cash system*; Manubot: 2019.
[2] Mantelero, A. J. C. L.; Review, S., AI and Big Data: A blueprint for a human rights, social and ethical impact assessment. 2018, *34* (4), 754 - 772.
[3] Singh, M. P.; Huhns, M. N. J. I. E., Automating workflows for service order processing: Integrating AI and database technologies. 1994, *9* (5), 19 - 23.
[4] Kowalski, R., AI and software engineering. In *Artificial Intelligence and Software Engineering*, Ablex Publishing: 1991; pp 339 - 352.
[5] Artikis, A.; Bamidis, P. D.; Billis, A.; Bratsas, C.; Frantzidis, C.; Karkaletsis, V.; Klados, M.; Konstantinidis, E.; Konstantopoulos, S.; Kosmopoulos, D. In *Supporting tele-health and AI-based clinical decision making with sensor data fusion and semantic interpretation: The USEFIL case study*, International workshop on artificial intelligence and NetMedicine, 2012; p 21.
[6] Yao, X. J. P. o. t. I., Evolving artificial neural networks. 1999, *87* (9), 1423 - 1447.
[7] Basheer, I. A.; Hajmeer, M. J. J. o. m. m., Artificial neural networks: fundamentals, computing, design, and application. 2000, *43* (1), 3 - 31.
[8] Zhang, G.; Patuwo, B. E.; Hu, M. Y. J. I. j. o. f., Forecasting with artificial neural networks:: The state of the art. 1998, *14* (1), 35 - 62.
[9] Bratko, I., *Prolog programming for artificial intelligence*. Pearson education: 2001.
[10] Luger, G. F., *Artificial intelligence: structures and strategies for complex problem solving*. Pearson education: 2005.
[11] Bond, A. H.; Gasser, L., *Readings in distributed artificial intelligence*. Morgan Kaufmann: 2014.

IOP Conf. Series: Materials Science and Engineering **914** (2020) 012028 doi:10.1088/1757-899X/914/1/012028

# Application of neural network based on self-organizing incremental learning in anti-interference

**Shaodong Hu[1], Xuechen Wu[2], Bo Xue[3]**

[1]School of Management Science, Chengdu University of Technology, Chengdu, Sichuan, 610059, China
[2]School of Computer Science and Technology, Chongqing University of Posts and Telecommunications, Chongqing, 400065, China
[3]School of Physics, University of Electronic Science and Technology of China, Chengdu, Sichuan, 610054, China

**Abstract**. In view of the shortcomings of traditional neural network in the actual information fusion process, a data anti-interference method based on self-organizing incremental learning neural network is proposed, which is used to cluster and represent the dynamic input data online without prior knowledge. In this paper, the neuron distribution, dynamic node adjustment, topology representation and denoising process of the network are described in detail. Finally, the multi-source heterogeneous data collected by the sensor can be self-adaptive dimensionality reduction and self-organization learning. At the same time, it has strong robustness to noise data, so that it can continuously learn new patterns in data flow. SOINN model is suitable for supervised learning, associative memory, pattern-based reasoning, manifold learning and other learning scenarios. It can also be extended to the application fields of unbalanced, highly complex and nonlinear big data prediction.

## 1. Introduction

The demand problems brought by social development make the design of intelligent human-computer interaction system more and more important. If robots can accurately recognize human actions and understand the information expressed by human actions, it will further improve the communication ability between machines and humans. Incremental learning refers to a learning system that can learn new knowledge from new samples and save most of the knowledge that has been learned before. It is similar to human's own learning mode, because in the process of growing up, people learn and receive new things every day, and will not forget what they have learned.

Self-organizing incremental learning neural network (SOINN) is a kind of neural network model based on competitive learning [1, 2], which can do incremental unsupervised learning for data without fixed topology structure. SOINN uses a group of neurons distributed in the feature space to approximate the density distribution of input data. SOINN's learning algorithm is fully incremental, which is not only reflected in the online learning process, but also the incremental learning results. It has the following advantages: (1) it can learn new knowledge from new data; (2) the previously processed data does not need to be repeated Processing; (3) only one training observation sample is seen and learned at a time; (4) most of the previously learned knowledge can be saved while learning new knowledge; (5) - the training observation sample is discarded once the learning is completed. Since the adaptive similarity threshold is used to identify the input patterns that have not been learned before, the network can dynamically generate new neurons to represent the input patterns without affecting the previous learning results.

Because of these characteristics, SOINN, as a general competitive neural network model, has been applied in various fields, including robot intelligence, computer vision, and anomaly detection and so on. It is also more suitable for all kinds of learning problems, including associative memory, density estimation, and manifold learning and so on. This paper combs and summarizes the basic idea of SOINN and its application in artificial intelligence.

## 2. The basic principle of SOINN algorithm

SOINN [3] is a two-layer competitive neural network (excluding the input layer), which clusters and represents the input data in a self-organized way. The working process is shown in Figure 1.

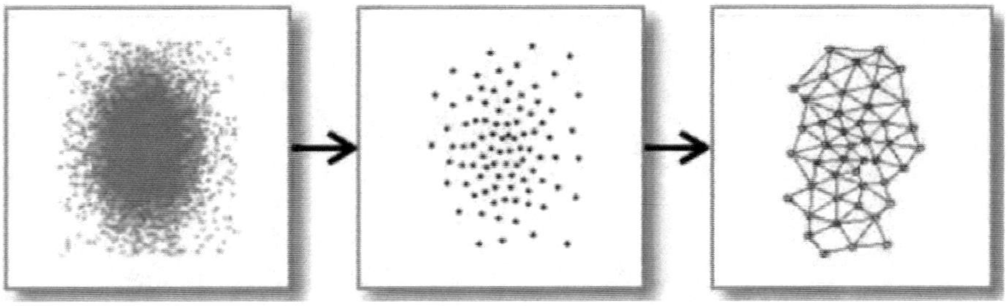

**Figure 1.** two level competitive learning of SOINN

(1) The first layer network receives the input of the original data and adaptively generates prototype neurons to represent the input data online. These nodes and their connections reflect the distribution and topology of the original data. It should be noted that neurons in SOINN, SOM, GNG networks are different from neurons in common feed forward neural networks. They can be regarded as vectors in a certain feature space, and their weights are coordinate representations of neurons in that space.

(2) The second layer estimates the distance between classes and the distance within classes of the original data according to the results of the first layer network, and then uses the neurons generated in the first layer as input to run the SOINN algorithm again to stabilize the learning results. When there are multiple clusters and noises in the input data, SOINN can still generate reliable neuron nodes to represent each cluster in the input data; at the same time, the topological structure of the sub graph reflects the nature of the original data distribution, and its working effect is shown in Figure 2 below.

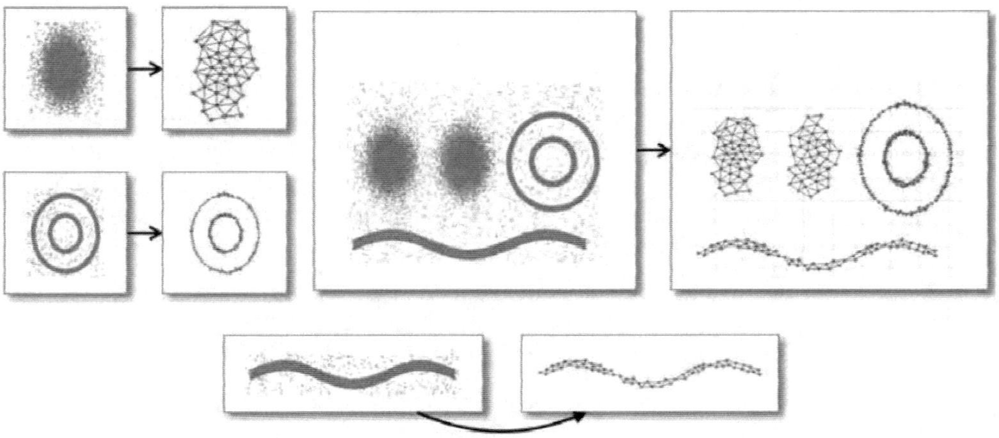

**Figure 2.** Effect demonstration on artificial data set

## 3. Training process of SOINN algorithm

From the point of view of specific algorithm, the training process of SOINN is mainly divided into four steps: neuron distribution, dynamic node adjustment, topology representation and denoising process. Here are the basic ideas of each part.

### *3.1. Distribution of neurons*

Using a certain number of representative data to approximate the original complete data set, and then making decisions on the new data based on these representative points is the basic idea of the prototype based learning algorithm. These representative data are called prototypes. They usually reflect the distribution information of the original data in some way. For example, each clustering center of K-means is a typical prototype. SOINN is also a prototype based algorithm. Its prototype consists of neurons distributed in the input space. Each neuron represents the most similar input pattern around it. From this point of view, distributed neurons are essentially a vector quantization process.

Vector quantization (VQ) [4] was originally used to compress and code the digital signal, so that the signal can be transmitted with a lower bandwidth. In short, the process of VQ is to use a set of fixed number of codebook vectors $w_i$ $(i = 1, 2, ..., K)$ the original signal x is encoded to achieve the purpose of data compression, so as to minimize the expected error value $E(w)$ of signal reconstruction at the decoding end to:

$$E(w) = \int \|x - w_{i(x)} - \|_2^2 p(x) \mathrm{d}x \tag{1}$$

Where $p(x)$ is the probability density function of signal $x$. $E(w)$ usually has no closed optimal solution, so it needs to be optimized by iterative algorithm, among which LBG algorithm based on Lloyd-Max condition is the most widely used one. The formula (1) is optimized by using random gradient descent to get the following updated formula, but it usually needs to traverse multiple training data to make the algorithm converge to the local optimal solution:

$$w_i^{(t+1)} \leftarrow w_i^{(t)} + \alpha(t)(x^{(t)} - w_i^{(t)}) \tag{2}$$

On the basis of formula (2), a neighbourhood function is added to control the sensitivity of different codebook vectors to the input, which can form a mapping with spatial order. The formula is as follows:

$$w_i^{(t+1)} \leftarrow w_i^{(t)} + \alpha(t) h(x^{(t)}, W, \sigma)(x^{(t)} - w_i^{(t)}) \tag{3}$$

In general, the existence of $h$ enables only those codebook vectors closest to the input mode to compete and move to the input data, which may be adjacent in the input space. The result of network convergence is that the output codebook vector and the input pattern form a spatial ordered mapping relationship.

For SOINN, its neuron is equivalent to codebook vector in vector quantization. Every time the data is input, only the nearest neuron and its neighbour nodes in the topology have the chance to move, that is, the topology neighbourhood. SOINN gives each neuron a different learning rate parameter $(t)$, which has the same effect as that in VQ. This makes the movement of neurons eventually tend to be stable and eventually converge without continuous oscillation. In general, the form of $(t)$ needs to meet certain constraints to ensure the final learning effect:

$$\sum_{t=1}^{\infty} \varepsilon(t) = \infty, \sum_{t=1}^{\infty} \varepsilon^2(t) < \infty \tag{4}$$

The learning rate of formula (4) is satisfied, which ensures that each node always maintains a certain learning ability when it is gradually stable.

### 3.2. *Dynamic adjustment*

Dynamic adjustment is the key to realize self-organization and incremental learning of SOINN. It makes the weight vector of neuron and topological structure of network dynamically adjust with the arrival of input mode to optimize the expression accuracy of input data. At the same time, it can adapt to the input mode that has not been learned before without affecting the previous learning results.

SOINN defines intra class node insertion and inter class node insertion respectively to achieve these two purposes. In order to reduce the quantization error of neurons adaptively and approximate the distribution of the original data as accurately as possible, the node insertion in the class is mainly used. It records the cumulative quantization error of each neuron, finds out the two nodes with the largest cumulative quantization error among all nodes, inserts a new node between them, updates their cumulative quantization error value by interpolation, and constantly increases the number of nodes. Therefore, SOINN will judge whether the insertion has significantly reduced the quantization error after each node insertion in the class. If not, the insertion will be cancelled.

### 3.3. *Topological representation*

Dynamic adjustment of neurons is for incremental learning of data. In addition, SOINN uses competitive Hebb learning rules to establish connections between neurons. At the same time, when the neurons are dense enough, the topological neighborhood of the original data can be perfectly maintained.

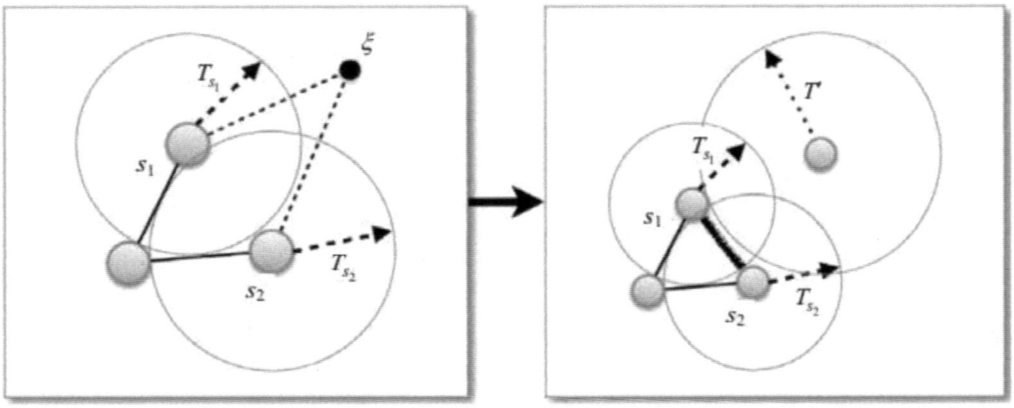


(a) (b)

**Figure 3.** Interclass node insertion

For each input data sample, two neurons most similar to the input pattern are found in the feature space, and then they are connected. The bold connection in Figure 3 (b) is the new connection established according to ChL. Considering that the position of the nodes will be adjusted during the running process, the two nodes that are relatively close before may not be activated by an input signal at the end of the algorithm at the same time, and there is no connection between them. Reset the parameter to 0 every time two endpoints of the edge connection are activated at the same time; otherwise, the parameter will be deleted if it exceeds a certain threshold.

### 3.4. *Network denoising*

In practical application, random input sequence will also make the results of SOINN unstable, and there is often noise in the input data, and unnecessary nodes are generated. In order to alleviate this problem, we should find and delete some nodes that may not reflect the distribution information of the original data, they are often in the low density area.

So how to judge? In SOINN, the number of times a node becomes the winner and the number of its neighbors are used to determine whether it is in a low data density area. If a node has no or only one neighbor node, and the number of times it becomes the winner is lower than the average number of current neurons, a certain proportion will be deleted, and all the edges associated with it will be removed.

### *3.5. Simulation test*

The SOINN algorithm is realized by MATLAB software, and the corresponding machine learning model is established. In order to test the performance of the model, the artificial data set is used to simulate and test the machine learning model. Using the rand function of MATLAB software, 50000 random data points are generated and divided into six groups for simulation. The self-organizing incremental learning neural network algorithm can extract the features of sample data and replace the original redundant data with fewer nodes and topology. When the amount of sample data is small, the number of nodes is large, and there is no obvious advantage. However, with the increase of data volume, the number of nodes does not show a linear growth, and the data compression ratio increases gradually. When the two-dimensional data points are 50000, there are only 85 nodes after learning, and the data compression ratio is as high as 588, that is, a large number of nodes can be used to represent the distribution characteristics of data.

The three-dimensional data set generated manually is input into the machine learning model and simulated. The simulation result is shown in Figure 4, which is the simulation result of the three-dimensional data. In the figure, the gray bubble points represent the distribution of the original data, the white red circle represents the nodes learned by the SOINN algorithm, and the blue line represents the topological relationship between them. It can be seen that the model has a good ability to extract features from 3D data. Using the rand function and radon function of MATLAB software, 50000 artificial data sets with noise points are generated, which are divided into six groups for simulation.

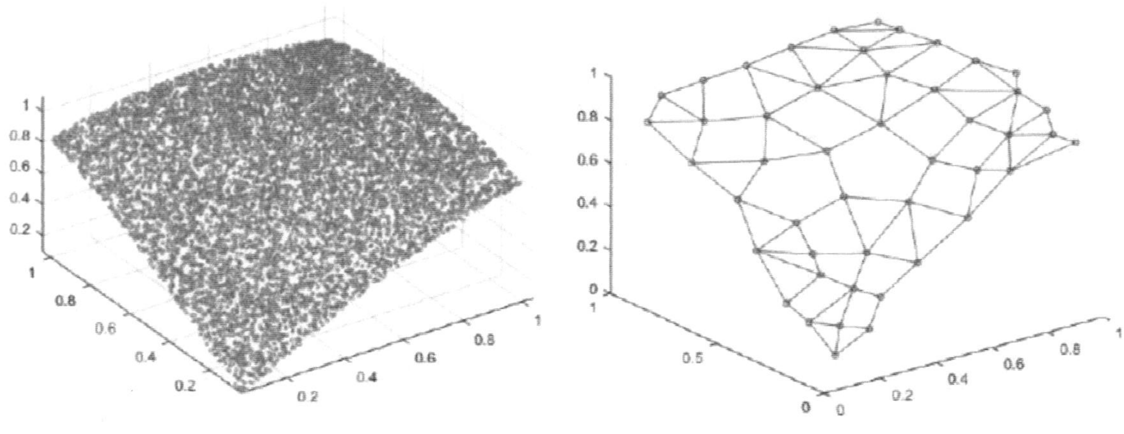


**Figure 4.** Learning simulation results of 10000 sample points

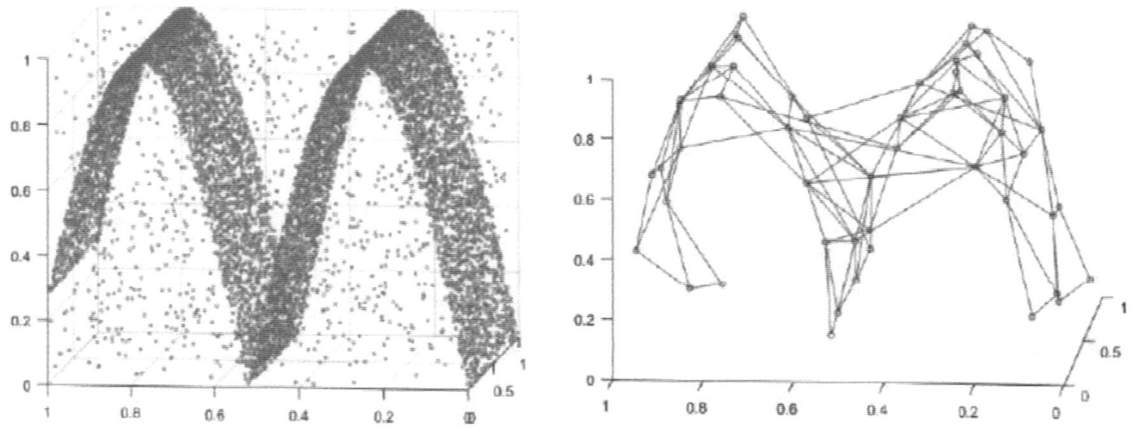

**Figure 5.** Simulation results of 10000 samples + 1000 noise data set

From the simulation results, the neural network model based on self-organizing incremental learning has a very good data extraction ability for three-dimensional data, that is, it can replace the distribution characteristics of a large number of redundant data with fewer nodes and network topology, and has a certain anti-interference ability.

## 4. Conclusion

SOINN has been applied in many fields, including medical expert system, computer vision, anomaly detection, robot intelligence, etc. For example, in the aspect of artificial intelligence, the robot's learning system can describe the current physical scene with some simple natural languages through the multi-modal training process such as image display and voice guidance, and at the same time, it can communicate with people with simple vocabulary; in the aspect of medical diagnosis, the hybrid expert system includes data processing, classification and knowledge extraction To assist doctors in the diagnosis process. This shows that SOINN makes robots have certain autonomy and learning ability in the face of constantly changing external environment.

In general, self-organizing incremental learning is a competitive neural network which can cluster data online and represent topology, and robot intelligence is its main application direction. The ability of incremental learning is very suitable for the training process of robot intelligence. It does not need too many people's intervention, and does not need to specify the number of clusters. As a result, SOINN can get the approximate density distribution and topology information of data without prior knowledge.

## References

[1] Shen F, Hasegawa O. An incremental network for on-line unsupervised classification and topology learning. Neural Networks, 2006, 19 (1): 90 - 106.

[2] Shen F, Ogura T, Hasegawa O. An enhanced self-organizing incremental neural network for online unsupervised learning. Neural Networks, 2007, 20 (8): 893 - 903.

[3] Shen F, Hasegawa O. Self-Organizing incremental neural network and its application. In: Proc. of the Int'l Conf. on Artificial Neural Networks (ICANN 2010). Berlin, Heidelberg: Springer-Verlag, 2010. 535 - 540.

[4] Gray RM. Vector quantization. IEEE ASSP Magazine, 1984, 1 (2): 4 - 29.

IOP Conf. Series: Materials Science and Engineering **914** (2020) 012029 doi:10.1088/1757-899X/914/1/012029

# Case-based Analysis of Discrimination in Police Surveillance Scene Regarding Facial Recognition

**Tingyao Chen**[a]

Law School, the South China University of Technology, Guangzhou 510006, China

[a]13824450795@163.com

**Abstract**. The application of facial recognition technology in police surveillance scene is justified, but it will result in discrimination. In this paper, the first case of facial recognition in the world, R (Bridges) v Chief Constable of South Wales Police, is taken as the starting point, and the problems of algorithmic discrimination and indirect discrimination reflected in this case are discussed while analyzing the case, and the author will predict what kind of method the plaintiff may take in the appeal.

## 1. Context

Facial recognition technologies, or face id, are biometric systems that combine the advantages of the humans' amazing skill of perception with the benefits of immense processing power and memory capacity of computers [1]. In the case, plaintiff R (Bridges) accused South Wales Police (SWP) of improper use of automatic facial recognition technology (AFR), which automatically processes images containing faces to verification, identification or surveillance, and it mainly uses the biological information feature of the human face to distinguish individuals [2]. Verification is the process of verifying whether a person is a match with a specific individual stored in the system, and it is a one-to-one process in face information matching [3]. Identification can be divided into the closed-set identification and the open-set identification. The former is the information of the object to be identified which has been stored in the database and designated and it is only necessary to compare the individual face information with the information to be recognized; the latter can be understood simply as the recognizer does not know whether the face information to be recognized is already stored in the database [4]. Surveillance is usually a many-to-many process, in which the AFR collects facial information from a non-specific individual to compare it with non-specific individuals in the database who may be suspected of committing a crime, and the determination of the suspect standard will be determined by the recognizer according to the specific circumstances or regulations to set up different algorithms.

The monitoring process of SWP using AFR Locate for facial recognition belongs to a many-to-many process, and it is roughly divided into the following steps: First, to establish a facial information database or watch list to store the information of individuals with illegal records or criminal suspects; Second, the CCTV real-time monitoring function is used to capture suspects' facial information in moving crowds, which is the main process of facial information collection. The system then extracts images from the video to determine whether it contains human faces and categorizes the images that contain different faces. The system then extracts facial features from the images, and the images will be compared with the individuals on the watch list. Finally, based on matching similarity SWP will determine whether the identified facial information is on the watch list. The construction of SWP's facial information database is based on the needs of previous police activities and investigation of specific cases, and the corresponding threshold is set for the system in each recognition to avoid the false-positive rate of the results. When the system has matched, both images are first reviewed by an AFR operator with the identity of a police officer, who confirms the match is accurate and then sends it to the intervention officers for review. SWP has developed the 'traffic light' system to help the examiner determine the urgency of the match, but not interfering with the examiner's opinion. AFR Locate will delete the face data if the result shows no match.

SWP consults the Surveillance Camera Commissioner on the use of Surveillance cameras and chooses the location of each AFR deployment based on the location of a particular crime and the need to maximize the collection of facial information. Despite this, many people who have their faces captured by AFR claim that they do not aware of this fact when AFR is used.

## 2. Facts

In this case, the plaintiff Mr. Bridges sued SWP because SWP was illegally using AFR, for the reason that the plaintiff's facial information was captured by SWP respectively at Queen Street Cardiff on 21 December 2017 and in Defence Procurement, Research, Technology and Exportability Exhibition on 27 March 2018 without the plaintiff's consent, according to which the plaintiff argued that the defendant had violated Article 8 of the European Convention on Human Rights, the Data Protection Act 1998 (DPA 1998) and the Data Protection Act 2018 (DPA 2018), and the s149(1) of Equality Act 2010.

## 3. Issue

In response to Article 8 of the European Convention on Human Rights, the plaintiff found that the defendant had violated the Right to respect for private and family life by using AFR to collect information on human faces without his consent. In the case of the s4(4) DPA 1998 and s35 of DPA 2018, the plaintiff held that the defendant had failed to meet the requirements of the first Data Protection principle and had violated the obligation of assessment under the s64 DPA 2018. The plaintiff also claimed that the defendant failed to consider the possibility of a false positive result during the early evaluation phase of the adoption of the NeoFace system, which resulted in a violation of the "promotion of equal relations" requirement of the s149(1) Equality Act 2010. In this case, SWP recognized the plaintiff's status in the judicial review process and recognized his eligibility as a victim under section 7 of the Human Rights Act 1998, and relieved the plaintiff of the burden of proofing that the image of his face had been collected actually, for the court to pass judgment on the substantive issues involved in the case.

## 4. Analysis

In response to Article 8 of the European Convention on Human Rights, the court invokes R. Wood V. The commission of Police of the Metropolis and believes that the core values of Article 8 are "physical and psychological integrity" and "physical and social identity". It also supports the claim that the SWP's collection of facial information without the plaintiff's consent and plaintiff did not have any prior criminal record, violates Article 8, and constitutes an interference with individual autonomy. The use of AFR by the defendant without the plaintiff's consent falls within the scope of the right to privacy protected by this article, and the instantaneous and transient nature of the data processing does not affect the infringement judgment, because the fact that the AFR processes biological information is in itself a sufficient violation of Article 8 rights, and SWP's use of AFR is not a "predictable and unsurprising" simple photograph. The plaintiff claimed that the use of AFR Locate had to have a real legal basis, and the court reasoned that it was reasonable to place AFR Locate under the common law framework. The common law gives the police the power to police and punishes crime, including the power to make public the portraits of suspects and the power to use personal portraits to prevent crime. The AFR's access to information is not physically intrusive, and the police have the power to use personal portraits fairly obtained through AFR for crime prevention purposes, and there is no need to create new legal instruments for each new technology. Based on the determination of the common law hearing, the court

IOP Conf. Series: Materials Science and Engineering **914** (2020) 012029 doi:10.1088/1757-899X/914/1/012029

established three frameworks for the analysis of the legitimacy of the AFR, namely, the primary legislation, secondary legislative instruments, and SWP's local policies. Based on the infringement, the court analyzed the justification of SWP's use of AFR Locate and borrowed the four-part test from Bank Mellat v Her Majesty's Treasury. Since a reasonable connection between the proportionality test and conduct and objectives can be demonstrated by police practice, the main consideration is whether the SWP's conduct is less intrusive and whether it strikes a balance between individual rights and community interests. The court found that AFR was used for a specific time, place, and purpose, with limited coverage and obvious benefits from its use. Besides, the court denied the plaintiff's claims that the use of AFR was not necessary, that it was not targeted in its application and that it did not improve the SWP's ability to detect crimes, and therefore the court could consider that both of these criteria were met, namely that the violation of Article 8 rights by AFR was justified.

Secondly, in respect to the plaintiff's claim that SWP did not fulfill its obligations as the data controller under DPA 1998 s.4(4), the court's examination of whether SWP's conduct was consistent with the first data protection principle was premised on an understanding of concepts such as "personal data". The court referred to the definition in Directive 95/46/EC and held that the definition should be defined in terms of both indirect identification and personalization. The problem in the indirect identification path is that if the information collected by AFR Locate does not conform to the definition of personal data, then the SWP will belong to personal data according to the matching result obtained by the information in the future. Drawing on the broad interpretation of indirect recognition given by CJEU in the Breyer v Bundesrepublik Deutschland, the court stated that all reasonable means used by the controller to identify an individual should be included in indirect recognition unless it falls within the scope of the risk of identification "appears in reality, to be significant". The second path to personal data is the individual, which is the ability to fully identify individuals in a crowd by analysing the facial information collected. The court decided that the individual path was more appropriate based on the AFR pattern and further considered whether SWP's actions were consistent with the first data protection principle. In cases where both the plaintiff and the SWP agree that the facial information does not belong to sensitive information, the court only needs to examine whether the SWP's behaviour complies with Schedule 2 and finds that Paragraph6's analysis of the data processing necessity is the focus of the review. The court uses the same reasoning as Article 8 to explain the legality of the SWP's behavior, i.e., there is no violation of DPA 1998.

The plaintiff's facial information was collected twice before DPA 2018 came into effect, but both parties argued that it is necessary for the court to review whether the defendant's use of AFR conforms to DPA 2018. The plaintiff claimed that the defendant had violated s. 35 (3) sensitive treatment relates to the interpretation of "biometric data". The court upheld the plaintiff's claim that s. 35 (3) collection of "biometric data" covered both individuals on the watch list and the general public, and that the collection of "biometric data" by the public had to meet s.35 requirements. The court recognized both "the Processing is strictly necessary for the Law Enforcement purpose" and the Processing must meet at least one of the conditions in Schedule 8, but questioned the appropriateness of the "Policy on Sensitive Processing for Law Enforcement Purposes" formulated by the SWP and indicated a need for Information Commissioner guidance.

The plaintiffs also relied on DPA 2018 s.64 to assert that SWP had not conducted a Data Protection Impact Assessment before using AFR Locate, but SWP had provided a "Privacy Impact Assessment" of 12th February 2018 to demonstrate that it had fulfilled its assessment.

Obligations. The court found that the assessment issued by the SWP was clear, took into account the impact on the public's right to privacy, identified the legal risks of AFR Locate's use and the possibility of violation in practice, and was consistent with Article 8 of the European Convention on Human Rights.

The plaintiffs also argued that SWP's use of facial recognition software, called NeoFace Watch, resulted in indirect discrimination based on the s149(1) Equality Act 2010 that did not take into account false positive results, i.e., AFR Locate has a higher error rate in identifying women and minorities. The court found that there was no evidence that the SWP had known or should have known in its earlier assessment that indirect discrimination would result and that it had not been established that the use of

AFR Locate would necessarily result in indirect discrimination. While the plaintiff's expert witness, Dr. Anil Jain, cannot collect information from the AFR Locate database to determine whether the system may have caused indirect discrimination, it contends that SWP also can not accurately analyse the information in the database, it is therefore argued that the SWP should not use AFR Locate where there is a risk of discrimination. As a representative of SWP, Dominic Edgell proposed that the high false-positive rate of women in the matching results was due to the system's recognition of female facial features. The court held there is no conclusive evidence at this time that the AFR has an indirectly discriminatory effect, although whether the AFR Locate has an indirectly discriminatory effect at this time needs to be further determined. However, this does not mean that in the past the SWP failed to comply with its obligations under the s149 (1) Equality Act 2010, and the court dismissed the plaintiff's claim.

## 5. Comments

The discrimination consequences of matching results seem to be a common feature of AFR Systems, which involves the discussion of algorithmic discrimination. AI Now's 2019 Report introduces some cases of algorithmic discrimination [5]. David Heine Meier Hansson, a prominent software engineer, found that Apple Card gave him 20 times as much credit as her wife, and Janet Hill as the wife of Apple co-founder Steve Wozniak, has only a tenth of her husband's credit limit. Although the design of the algorithm does not contain discriminatory factors at the code level, the data in the training set of the system are biased to different groups, such as race and age distribution, which leads to discriminatory results [6]. An article on the ICO blog analyses the gender bias involved in loan quotas, meaning that the algorithm, when applied to a loan assessment, builds a statistical model that requires a sample to be filled in to assess an individual's credit limit [7]. If the sample is male-dominated, the system will not be able to accurately judge the ability of women to repay loans, and then reduce the number of women's loans, which in the results show indirect discrimination against women. Technology companies have taken notice of algorithmic discrimination, but the protection of citizens'Anti-Discrimination Rights has given way to huge commercial interests. Google Chairman Eric Schmidt has said publicly that companies are aware of algorithmic discrimination and there is no need for too much public attention to algorithmic discrimination. Companies such as Amazon and Facebook have also spoken out against the regulation of facial recognition, arguing that it would affect the development of the industry, which actually to maintain the lucrative business of facial recognition. AFR's algorithm plays an important role in the process of facial recognition, and the discrimination result caused by AFR error recognition belongs to the category of algorithmic discrimination in essence.

In the present case, the plaintiff claims that SWP constitutes indirect discrimination on the basis that SWP violated its evaluation obligation under the Equality Act 2010, and that non-discrimination has different sources in European Union Law. For example, Article 14 of the European Convention on Human Rights (ECHR) and Article 21 of Charter of Fundamental Rights of The European Union (The Charter) both stipulate non-discrimination. The Charter's anti-discrimination provisions are similar to ECHR, but broader and set out as a separate right, making its protection not subject to other provisions. There are also specific directives or ordinances regulating the situations of discrimination enumerated in the Charter, such as Gender Equality Directive (Recast, 2006/54/EC), Gender Equal Access Directive (2004/113/EC) and Racial Equality Directive (2000/43/EC) in the aspect of race and ethnicity. The definition of indirect discrimination in Racial Equality Directive (2000/43/EC) is that which shall be taken to occupy where a neutral vision, criterion or practice would put persons of a racial or ethnic origin at a particular advantage compared with other persons. Indirect discrimination is difficult to avoid in the statistical model built by the algorithm since the matching result is often affected by the number of images and the skin color of the people collected by AFR, since the accuracy of skin color recognition has large differences in different light conditions, and these differences are usually difficult to be improved through mathematical or technical methods. The AFR's black box structure makes it impossible to know the factors that influence the results of the specific processes of facial information collection, recognition, and matching. Therefore, even if AFR is found to have indirect discrimination results, people can also sue AFR users who violate their fundamental rights under different regulations, and this situation cannot be effectively improved in a short period.

The admission of SWP in this case relieved the plaintiff of the burden of proof that his face information had actually been collected, but the court stated in its judgment that the plaintiff's substantive rights had not been violated. The plaintiff's allegation that the SWP's use of AFR Locate resulted in indirect discrimination was also denied by the court. It is worth noting that the plaintiff accused SWP of constituting indirect discrimination rather than direct discrimination. However, from the perspective of the probability of victory, once SWP constitutes direct discrimination, it must constitute a violation of the Equality Act. In other words, if SWP carries out direct discrimination, there is no room for defense in law, and SWP shall directly bear the illegal consequences. However, even if the SWP constitutes indirect discrimination, it does not necessarily have to bear responsibility, as for example, Directive 2000/43/EC provides that as long as the suspected act of indirect discrimination is objectively proved to have a legitimate aim and the manner of achieving that aim is appropriate and necessary, then the act does not amount to indirect discrimination. So when the plaintiff sued SWP for indirect discrimination, SWP could claim that the use of SWP was following the exception clause, which increased the burden of proof of the plaintiff. A reasonable explanation is that SWP's use of AFR Locate does not qualify as direct discrimination, so the plaintiff turns to indirect discrimination. According to the BBC, Lord Justice Singh granted the plaintiff's appeal in this case [8]. At the appeal stage, the author thinks that the plaintiff can continue to insist that SWP constitutes indirect discrimination but also needs strong evidence to show that SWP's indirect discrimination has had an adverse effect or absorb the victims who have suffered actual losses into the appeal procedure. Black, Asian, and minority ethnic (BAME) groups showed greater attention to the AFR's prior consent, meaning that Bame was more likely to suffer harm in the AFR's implementation scenario [9]. If Bridges can persuade the actual victims of indirect racial discrimination -- those who suffer inequality treatment because of AFR Locate's use and their race -- to join the appeal, SWP could be sued under Directive 2000/43/EC for violation of the s.2 (1) prohibition of indirect discrimination, increasing the probability of success while also reducing the risk of AFR Locate continuing to produce discriminatory results until the risk assessment is fully completed.

**References**

[1] Welinder, Yana. (2012) A face tells more than a thousand posts: Developing face recognition privacy in social networks. Harv. JL & Tech. 26: 165.
[2] Selvadurai, Niloufer. (2015) Not just a face in the crowd: addressing the intrusive potential of the online application of face recognition technologies. International Journal of Law and Information Technology 23.3: 187 - 218.
[3] Lucas D. Introna and Helen Nissenbaum, Centre for Catastrophe Preparedness and Response. (2020) Facial Recognition Technology: A Survey of Policy and Implementation Issues., 4 - 5.
[4] Ibid., 3.
[5] AI Now. (2019) AI NOW 2019 REPORT. https://ainowinstitute.org/AI_Now_2019_Report.pdf.
[6] Yavuz C. (2019) Machine Bias: Artificial Intelligence and Discrimination [J]. SSRN 3439702: 21 - 22.
[7] ICO. (2019) Human bias and discrimination in AI systems. https://ico.org.uk/about-the-ico/news-and-events/ai-blog-human-bias-and-discrimination-in-ai-systems.
[8] Nick Dermody. (2019) Facial recognition technology: Ed Bridges appeals human rights ruling. https://www.bbc.com/news/uk-wales-50495575.
[9] Ada Lovelace Institute, (2019) Beyond face value: public attitudes to facial recognition technology., 8 - 18.

IOP Conf. Series: Materials Science and Engineering **914** (2020) 012030 doi:10.1088/1757-899X/914/1/012030

# Data-parallel clustering algorithm based on mutual information mining of joint condition

**Changjiang Huang**

Sontan College Guangzhou University, Guangdong Guangzhou, 511370, China

33946676@qq.com

**Abstract.** In order to improve the reliability service ability of the cloud storage database, the data parallel clustering process is carried out, and the data parallel clustering algorithm based on the mutual information mining of the joint condition is proposed. a large data configuration structure model of a cloud environment virtual resource is constructed, data compression and characteristic reconstruction are carried out by adopting an online dictionary learning method, a regression analysis of the cloud environment virtual resource configuration data and a point cloud structure recombination are carried out in combination with a non-linear statistical sequence analysis method, the parallel characteristic scheduling of the large data of the virtual resources of the cloud environment is realized, the mutual information feature quantity of the joint condition is mined, the characteristic quantity of the mining is subjected to the characteristic filtering and the attribute set merging processing by adopting the fuzzy C-means clustering algorithm, by using the self-adaptive optimization algorithm, the automatic retrieval of the fuzzy clustering center is carried out, and the parallel clustering optimization of the large data is realized. The simulation results show that the classification performance of the large data clustering of the cloud environment virtual resources is good, the property classification and fusion capability is high, and the error rate is lower.

## 1. Introduction

With the development of cloud computing and large data information processing technology, the cloud storage technology is used for virtual resource storage, the parallel storage and information management capability of the data is improved, constructing a feature extraction and parallel mining model of the cloud storage virtual resource large data, carrying out feature extraction and mining on the large data of the cloud storage virtual resource by using a joint association rule mining method, and constructing a parallel clustering model of large data, In order to improve the retrieval and information management ability of the cloud storage virtual resource big data, the relevant data clustering algorithm research has been greatly concerned by people [1].

The clustering processing of cloud storage virtual resource big data is based on the feature extraction and classification identification of data, and the statistical feature quantity of cloud storage virtual resource big data is extracted, and the relevant classification algorithm is adopted for clustering analysis. In the traditional method, the big data clustering analysis method of cloud environment virtual resources mainly includes fuzzy C mean clustering method, K mean clustering method, grid clustering method and irregular triangular network clustering method [2], the detection statistical analysis model of cloud environment virtual resources is constructed, and the correlation rule attribute set of extraction data is realized to realize the optimization clustering of cloud environment virtual resource big data clustering. In the document [3], the parallel clustering algorithm of cloud environment virtual resource configuration data based on depth learning is proposed in the document [4], and the phase space reconstruction method is adopted to reconstructing the feature reconstruction of the cloud environment virtual resource configuration data of cloud environment virtual resource configuration data, and the method is based on cloud environment virtual resource configuration data, but the calculation cost of the calculation cost of cloud environment is not good. In the literature [4], the method of virtual resource allocation data clustering for cloud environment based on association rule mining and fuzzy C mean value is proposed. The feature extraction is carried out by using principal component analysis to optimize clustering of cloud environment virtual resource data. Aiming at the above problems, this paper presents a parallel clustering algorithm based on joint conditional information mining. Construct big data configuration structure model of cloud environment virtual resources, adopt online dictionary learning method for data compression and feature reconstruction, realize parallel feature scheduling of big data of cloud environment virtual resources, mining joint condition mutual information feature quantity, use fuzzy C mean clustering algorithm for feature selection and attribute collection processing for the feature quantity of mining, and carry out automatic retrieval of fuzzy clustering center in combination with adaptive optimization algorithm to realize data clustering optimization. Finally, simulation experiment analysis is carried out, which shows the advantages of the method in improving data clustering capability.

## 2. Analysis of Storage configuration structure and characteristics of data

### *2.1. Big data configuration structure Model*

In order to realize that optimization design of the parallel clustering algorithm for the virtual resource allocation data of the cloud environment, the data storage structure analysis is carried out in combination with the distributed structure recombination method of the cloud environment virtual resource configuration data storage node, and a cloud environment virtual resource configuration data architecture model is established, The distributed mobile social network (DSNs) is used to design the cloud environment virtual resource allocation structure model, and the optimal distribution model of the cloud environment virtual resource configuration data storage node is constructed by combining the vector quantization analysis method [5]. The graph model structure of the cloud environment virtual resource configuration data is represented by a binary digraph $G=(V,E)$, wherein, $V$ is a vertex set deployed in the cloud environment virtual resource configuration map model distribution node, and the $M_1$ , $M_2$ $\cdots$ $M_N$ is a set of all edges of the cloud environment virtual resource configuration data in the finite field distribution area *G*. assuming that the $W=\{u,w_1,w_2,\cdots,w_k\}$ is the Sink node of the cloud environment virtual resource configuration data, the Euclidean distance is used to represent the phase track pitch of the data transmission node of the virtual resource allocation data transmission node of the cloud environment, and the information coverage area of the cloud environment virtual resource allocation data, It is assumed that the transmission link layer data of the virtual resource allocation of the M cloud environments is $x_s=\left[x(\eta_1),\dots,x(\eta_N)\right]^T$ , and the characteristic value of the initial position $x(k-1),\dots,x(k-M)$ of the resource allocation data is as follows:

$$\hat{x}_s = W_s^T y \tag{1}$$

Mining association rule attribute set for virtual resource allocation data for cloud environment:

$$\max_{x_{a,b,d,p}} \sum_{a\in A}\sum_{b\in B}\sum_{d\in D}\sum_{p\in P} x_{a,b,d,p} V_p \tag{2}$$

$$\text{s.t.} \quad \sum_{a\in A}\sum_{d\in D}\sum_{p\in P} x_{a,b,d,p} R_p^{bw} \le K_b^{bw}(S), b\in B \tag{3}$$

According to the priority attribute scheduling method, the load model for constructing the cloud environment virtual resource allocation is as follows:

$$r(t)=\sum_{i}\sum_{j=0}^{N_f-1}\sum_{l=0}^{L-1} b_i\alpha_l p(t-iT_s-jT_f-c_jT_c-\tau_l)+\omega(t)=\sum_{i}\sum_{j=0}^{N_f-1} b_i p_h(t-iT_s-jT_f-c_jT_c-\tau_0)+\omega(t) \tag{4}$$

Wherein

$$p_h(t)=\sum_{l=0}^{L-1}\alpha_l p(t-\tau_{l,0}) \tag{5}$$

In addition, the $\omega(t)$ is the data dimension of the virtual node, and $p_h(t)$ is the load between the cloud environment virtual resource configuration data Source and the Sink node. The virtual resource distribution structure model of the cloud environment is as follows:

$$X=[s_1,s_2,\cdots s_K] = \begin{bmatrix} x_1 & x_2 & \cdots & x_K \\ x_{1+\tau} & x_{2+\tau} & \cdots & x_{K+\tau} \\ \cdots & \cdots & \cdots & \cdots \\ x_{1+(m-1)\tau} & x_{2+(m-1)\tau} & \cdots & x_{M+(m-1)\tau} \end{bmatrix} \tag{6}$$

Where $K=N-(m-1)\tau$ , representing the embedded dimension of the cloud environment virtual resource allocation data search feature space, sf is the delay. according to the above analysis, a big data configuration structure model of cloud environment virtual resources is constructed. online dictionary learning is used for data compression and feature reconstruction [6].

### 2.2. *Data feature analysis*

Assuming that the statistical distribution sequence $\{x_1,x_2,\cdots x_N\}$ , of virtual resource allocation data flow in cloud environment to be parallel clustering makes $x(n)$ a set of regression analysis features, the sparse evacuation point cloud mapping of cloud environment virtual resource allocation data is carried out in $m$ dimension reconstruction space, and the distributed reorganization structure of cloud environment virtual resource allocation data is obtained as follows:

$$\boldsymbol{X}(n)=\{x(n),x(n+\tau),\cdots,x(n+(m-1)\tau)\} \quad n=1,2,\cdots,N \tag{7}$$

Among them, $\tau$ represents the embedding delay of cloud environment virtual resource allocation data in high-dimensional phase space, and establishes the state transfer model [7]. The characteristic evaluation concept expression of cloud environment virtual resource allocation data is as follows:

$$p(y\mid\alpha,\theta)=\sum_{k=1}^{K}\alpha_k p_k(y\mid\mu_k,\textstyle\sum_k) \tag{8}$$

Using the method of cloud sparse evacuation random point structure reorganization, the scattered point set of the first cloud environment virtual resource allocation data is $P_i=(p_{i1},p_{i2},\cdots p_{iD})$, in which:

$$j\in N_i(k), N_i(k)=\{\|x_j(k)-x_i(k)\|<r_d(k)\} \tag{9}$$

The regression analysis of the cloud environment virtual resource configuration data and the point cloud structure recombination are carried out in combination with the non-linear statistical sequence analysis method [8], and the trust relation of the cloud environment virtual resource configuration data is expressed as $A \to B$, $B \to C$, and the regression analysis model is expressed as follows:

$$MSD_{a\to b} = 1 - \frac{\sum_{i=1}^{|I_{a,b}|} \sqrt{(d_{a,i} - \bar{d}_a)^2 + (d_{b,i} - \bar{d}_b)^2}}{|I_{a,b}| \times \sum_{i=1}^{|I_{a,b}|} \left[ \sqrt{(d_{a,i} - \bar{d}_a)^2} + \sqrt{(d_{b,i} - \bar{d}_b)^2} \right]} \tag{10}$$

According to the attribute set of virtual resource allocation in cloud environment, the vector quantitative decomposition is carried out, and the fuzzy domain of distributed big data retrieval of virtual resource allocation in cloud environment is recorded as $a_t(\theta)$, that is, the fuzzy domain of distributed retrieval of virtual resource allocation in cloud environment is as follows:

$$a_t(\theta) = T^H a(\theta) \tag{11}$$

The frequency domain and time domain feature analysis of data in fuzzy domain is carried out, and the time-frequency distribution structure model of feature distribution in cloud environment virtual resource allocation is obtained, and the data parallel clustering analysis is carried out under the control of fusion clustering constraints [9].

## 3. Optimization of parallel clustering algorithm for data

### *3.1. Joint conditional mutual information mining*

Based on the construction of big data configuration structure model of virtual resources in cloud environment, the online dictionary learning method is used for data compression and feature reconstruction and reconstruction [10]. In this paper, a data parallel clustering algorithm based on joint conditional mutual information mining is proposed. In the cloud environment virtual resource allocation big data association scheduling set, the fuzzy scheduling point set satisfies $U^T = U^{-1}$, and $V^T = V^{-1}$. The fusion clustering output of parallel allocation of virtual resources in cloud environment is as follows:

$$x_{n,G} = x_{n,G} + \Delta x_i \tag{12}$$

The cloud environment virtual resource allocation query model is constructed, the interference component is $p(e_k \mid v_k)$, and the cloud environment virtual resource allocation big data optimal clustering characteristic component:

$$\mu_{ik} = \frac{1}{\sum_{j=1}^{c} \left(\frac{d_{ik}}{d_{jk}}\right)^{\frac{2}{m-1}}} \tag{13}$$

$$V_i = \frac{\sum_{k=1}^{m} (\mu_{ik})^m x_k}{\sum_{k=1}^{n} (\mu_{ik})^m} \tag{14}$$

The detection statistic characteristic value $w_{BLCMV}$ of the distributed large data in the cloud environment virtual resource allocation is expressed as follows:

$$w_{BLCMV} = \tilde{R}_y^{-1}[a_t(\theta_0),C]([a(\theta_0),C]^H \tilde{R}_y^{-1}[a_t(\theta_0),C])^{-1} \tag{15}$$

Where, $C=\left[c_1,c_2,\cdots,c_g\right]$ represents the cloud environment virtual resource configuration big data's block matching set, the description is shown below:

$$U_{v_i} = \beta_{v_i} \times \log(1+\partial_{v_i} \times \sum_{j=1}^{K} S_{v_i} e_j^T \frac{R_{C_j}}{n_{C_j}}), v_i \in v, C_j \in C \tag{16}$$

Among them, $n_{C_j} = \sum_{j=1}^{K} S_{v_i} e_j^T$ , $e_j^T = (0,...,1,...,0)$ , calculates the fuzzy clustering center of virtual resource allocation in cloud environment, and the mutual information of joint conditions is extracted as follows:

$$I_{SM} = A_{SM} H\left(\sin\left(\frac{2\pi t}{\rho_{SM}}\right)\right)\left(1-H\left(\sin\left(\frac{2\pi t+D_{SM}}{\rho_{SM}}\right)\right)\right) \tag{17}$$

Wherein, $A_{SM}$ is the weighted output amplitude of big data parallel programming for virtual resource allocation in cloud environment, $\rho_{SM}$ is the adaptive adjustment parameter of big data parallel programming for virtual resource allocation in cloud environment, and $D_{SM}$ is an inequality constraint. According to the results of feature mining, the optimal design of data parallel clustering is carried out [11].

### 3.2. Implementation of clustering algorithm

The feature extraction technology is adopted to extract the average mutual information feature quantity of the cloud environment virtual resource configuration data, and the mutual information of the attribute distribution of the output cloud environment virtual resource configuration data is as follows:

$$I(Q,S) = H(Q) - H(Q|S) \tag{18}$$

Wherein:

$$H(Q|s_i) = -\sum_j \left[ p_{sq}(s_i,q_j) \big/ p_s(s_i) \right] \log_2 \left[ p_{sq}(s_i,q_j) \big/ p_s(s_i) \right] \tag{19}$$

The fuzzy C-means clustering method is adopted to realize the parallel clustering on the extracted feature quantity, and the fuzzy C-mean region distribution set of the large data of the cloud environment virtual resource allocation is obtained as follows:

$$\Omega_i(t) = \frac{\gamma_{th}\sigma^2}{h_i[G-(N(l)-1)\gamma_{th}]} \tag{20}$$

The self-adaptive information fusion method is adopted to carry out the information fusion of the data output characteristics of the cloud environment virtual resource configuration data, and the judgment criteria of the data parallel clustering are satisfied:

Norm (1):

$$\sqrt{\frac{R_{(m+1)n}^2 - R_{mn}^2}{R_{(m+1)n}^2}} = \frac{\left|x_{\eta(n)+m\tau} - x_{n+m\tau}\right|}{R_{(m+1)n}} \geq R_{tol} \tag{21}$$

Norm (2):

$$\frac{R_{(m+1)n}}{\sqrt{\frac{1}{N}\sum_{k=1}^{N}\left[x_k-\frac{1}{N}\sum_{k=1}^{N}x_k\right]^2}} > A_{tol} \tag{22}$$

According to the fuzzy C-means judgment criterion of the cloud environment virtual resource configuration data, the main component analysis of the cloud environment virtual resource configuration data is carried out, the $\{u_1,...,u_N\}$ represents a non-trusted node set, $\{v_1,...,v_M\}$ represents a user behavior set of the cloud environment virtual resource configuration data, the self-adaptive optimization of the data clustering center is realized, and the iterative formula of the optimization of the data clustering center is as follows:

$$x_i(k+1)=x_i(k)+s\left(\frac{x_j(k)-x_i(k)}{\left\|x_j(k)-x_i(k)\right\|}\right) \tag{23}$$

Wherein, $\|\bar{x}\|$ represents the norm of $\bar{x}$. according to the above algorithm design, the parallel clustering processing of big data for virtual resource allocation in cloud environment is realized [12].

## 4. Simulation experiment and result analysis

In order to verify the application performance of this method in parallel clustering of virtual resource allocation data in cloud environment, the simulation results are carried out with Matlab and C programming software. The sampling sample database of virtual resource allocation data in cloud environment comes from cloud combination database Pearson Database, data set size 1200, training data set size 120, and the time length of collecting cloud environment virtual resource big data is 400. The dimension of feature reconstruction is 4, the sampling time delay is 0.25, and the convergence judgment threshold is 0.28. according to the above simulation environment and parameter setting, the virtual resource allocation data in cloud environment is clustering in parallel, and the time domain waveform of data distribution is shown in Fig. 1.

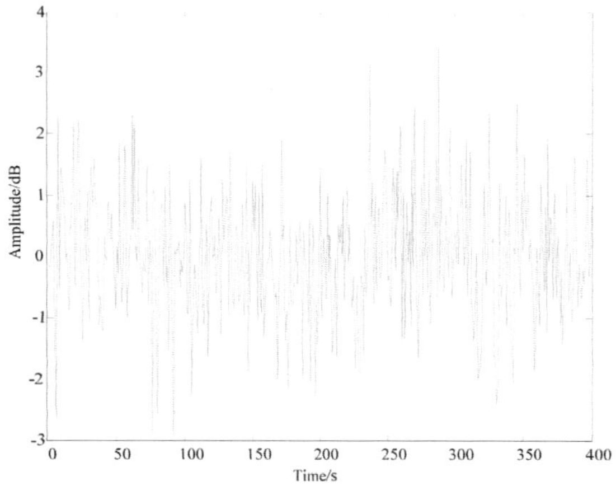


**Figure 1.** Data distribution time-domain waveform

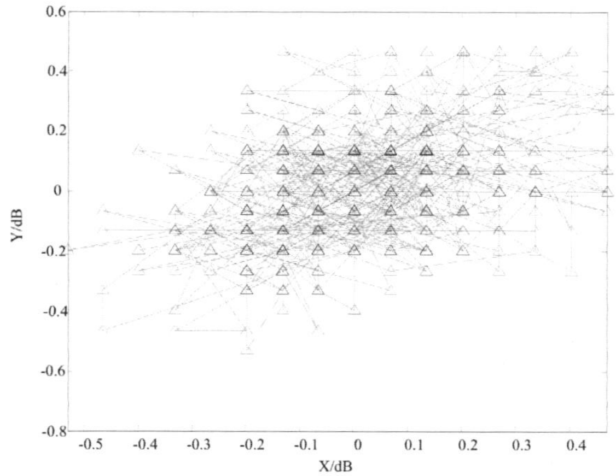


**Figure 2.** Data feature reconstruction

Taking the data of Fig.1 as input, digging the mutual information feature quantity of the joint condition, adopting a fuzzy C-means clustering algorithm to carry out characteristic filtering and attribute set merging processing on the extracted feature quantity, and obtaining the characteristic reconstruction output of the data as shown in Fig.2.

The analysis of Fig.2 shows that the clustering processing of the cloud environment virtual resource configuration data is carried out by the method, and the feature reconstruction capability is good and the discrete distribution is strong, so that the data optimization clustering is realized, and the clustering output is obtained as shown in Fig.3.

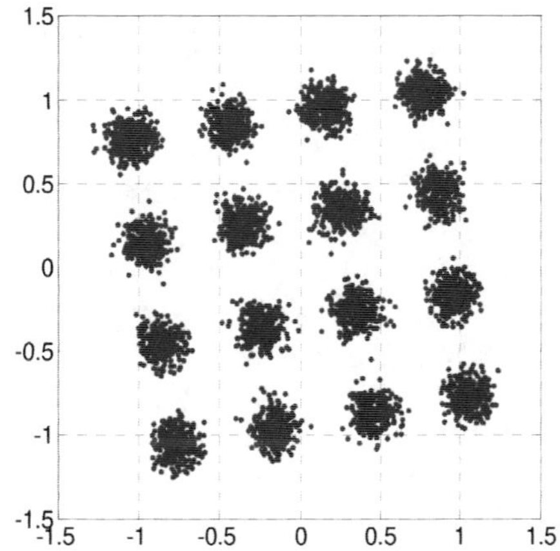


**Figure 3.** Data cluster output

Figure 3 shows that the classification performance of cloud environment virtual resource allocation data clustering is better by using this method. Test error score rate, get the contrast result as shown in Table 1.

**Table 1.** Comparison of error rates

| Iterations | Proposed method | K-means clustering | Particle cluster |
|---|---|---|---|
| 100 | 0.105 | 0.565 | 0.335 |
| 300 | 0.046 | 0.356 | 0.275 |
| 500 | 0.008 | 0.264 | 0.211 |
| 700 | 0 | 0.165 | 0.114 |

## 5. Conclusion

In the cloud storage system, the data of the cloud storage virtual resource needs to be optimized cluster analysis, and the data parallel clustering algorithm based on the mutual information mining of the joint condition is proposed in this paper. The optimal distribution model of the cloud environment virtual resource allocation data storage node is constructed, the sparse scattered point cloud mapping of the cloud environment virtual resource configuration data is carried out, and the regression analysis of the data and the reconstruction of the point cloud structure are realized. The feature extraction technology is adopted to extract the average mutual information feature quantity of the cloud environment virtual resource configuration data, and the fuzzy C-mean region distribution set of the large data of the cloud environment virtual resource allocation is obtained, so that the data optimization clustering is realized. The results show that the error rate of the parallel clustering of the data is small and the feature extraction capability is good.

## References

[1] DING J, LI A X, HU Z Q, et al. Accurate pulmonary nodule detection in computed tomography images using deep convolutional neural networks [C] //Proceedings of the 2017 International Conference on Medical Computing and Computer-Assisted Intervention. Berlin: Springer, 2017: 559-567.

[2] KAMNITSAS, K, LEDIG C, NEWCOMBE V F, et al. Efficient multi-scale 3D CNN with fully connected CRF for accurate brain lesion segmentation [J]. Medical Image Analysis, 2016, 36: 61-78.

[3] DOU Q, CHEN H, YU L Q, et al. Automatic detection of cerebral microbleeds from MR images via 3D convolutional neural networks [J]. IEEE Transactions on Medical Imaging, 2016, 35 (5): 1182-1195.

[4] RONNEBERGER O, FISCHER P, BROX T. U-Net:convolutional networks for biomedical image segmentation [C] //Proceedings of the 2015 International Conference on Medical Computing and Computer-Assisted Intervention. Berlin: Springer, 2015: 234-241.

[5] SU Kuanhao, HU Lingzhi, STEHNING C, et al. Generation of brain pseudo-CTs using an undersampled, single-acquisition UTE-mDixon pulse sequence and unsupervised clustering [J]. Medical Physics, 2015, 42 (8): 4974-4986.

[6] JOHANSSON A, KARLSSON M, NYHOLM T.CT substitute derived from MRI sequences with ultrashort echo time [J]. Medical Physics, 2011, 38 (5): 2708-2714.

[7] JOHANSSON A, KARLSSON M, YU Jun, et al. Voxel-wise uncertainty in CT substitute derived from MRI [J]. Medical Physics, 2012, 39 (6): 3283-3290.

[8] ANDREASEN D, VAN LEEMPUT K, EDMUNDJ M. A patch-based pseudo-CT approach for MRI-only radiotherapy in the pelvis [J]. Medical Physics, 2016, 43 (8): 4742-4752.

[9] ANDREASEN D, VAN LEEMPUT K, HANSEN R H, et al. Patch-based generation of a pseudo CT from conventional MRI sequences for MRI-only radiotherapy of the brain [J]. Medical Physics, 2015, 42 (4): 1596-1605.

[10] HAN Xiao.MR-based synthetic CT generation using a deep convolutional neural network method [J]. Medical Physics, 2017, 44 (4): 1408-1419.

[11] QIAN Pengjiang, SUN Shouwei, JIANG Yizhang,et al. Cross-domain,soft-partition clustering with diversity measure and knowledge reference [J]. Pattern Recognition, 2016, 50: 155-177.

[12] TU Guangsheng, YANG Xiaoyuan, ZHOU Tanping. Efficient identity-based multi-identity fully homomorphic encryption scheme[J]. Journal of Computer Applications, 2019, 39 (3): 750-755.

IOP Conf. Series: Materials Science and Engineering **914** (2020) 012031 doi:10.1088/1757-899X/914/1/012031

# The Optimal Environment Selection Strategy of Risk Model With Perturbed Diffusion

**Shumei Guo[1], Lili Zhang[2], Jie Guo[1], Xinna Li[1], Jiandong Yang[3]**

[1]Department of Basic, Information Engineering University ZhengZhou 450001, China
[2]Institute of Information Engineering, HangHe Science and Technology College ZhengZhou 450000, China
[3]China Satellite Maritime Tracking and Control Department, Jiangyin, 214431, China

**Abstract**. In this paper, the authors consider the optimal investment of the risk model with perturbed diffusion. Insurance companies invest the surplus in risky asset and risk-free asset. This paper discusses the problem of minimizing the ruin probability of insurance company. By solving the corresponding Hamilton-Jacobi-Bellman equations, the optimal investment portfolio and the upper bound of Lundberg of the minimal ruin probability are obtained. Especially,when the claim distribution is exponential distribution, asymptotic optimality and asymptotic uniqueness of strategy $A^*$ and $R$ are obtained.

## 1. Introduction

In order to ensure the interests of policyholders, insurance companies should minimize the ruin probability. So, finding the optimal investment is a hot issue in risk theory. Browne[1] first considered this issue, obtained the display solutions of the optimal investment strategy and the minimal ruin probability under the terms of surplus process was perturbed Brownian motion, risky asset was only one kind, claims followed a geometric Brownian motion and no reinsurance.Taksarand Markus-sen [2],LUO [3]reconsidered the model in Browne[1] in the case of making proportional reinsurance for compensation,also obtained clear expressions of optimal strategies and optimal value function. Scheer N,Schmidli H [4] considered a classical risk model with dividend payments and capital injections in the presence of both fixed and proportionals administration costs. Liu X,Chen Z,Ming R [5]considered the dividend problems in the perturbed compound Poisson risk model. Hipp and Plum [6] considered the traditional risk model and supposed surplus process could invest one kind of risky asset that the price followed a geometric Brownian motion,obtained the Hamilton-Jacobi-Bellman equations of minimal ruin probability in the case of none reinsurance, proved the existence of solutions and presented numerical results.LIU and YANG [7]reconsidered the model in Plum[6] in the case of surplus could invest no risky asset. Mataramvura S, ksendal,Bernt [8] considered the effect of demand on price where the objective of the pricing model is to maximize the expected utility of the insurer's terminal wealth.Mao H,Carson J M,Ostaszewski K M,et al [9]considered the effect of demand on price where the objective of the pricing model is to maximize the expected utility of the insurer's terminal wealth. YANG and ZHANG [10] considered the optimal investment problem of jump-diffusion model, discussed the numerical results of the optimal investment and the minimal ruin probability under the terms of claim settlement followed exponential distribution. and Gamma distribution. Yuen K,Lu Y,Wu R[11] considered the compound Poisson process perturbed by a diffusion with a threshold dividend strategy. This paper consider the classical risk model with perturbed diffusion and dividends, research the optimal investment problem, and obtain the optimal investment proportion of minimal ruin probability. When the claim distribution is exponential distribution, asymptotic optimality and asymptotic uniqueness of strategy $A^*$ and $R$ are obtained.

## 2. Model Introduction

Let $(\Omega, A, P)$ be a complete probability space containing all objects defined at the following.We consider the continuous time risk time:

$$U(t) = u + (c-d)t - \sum_{i=1}^{N(t)} X_i + \delta_1 W(t) \tag{1}$$

where $u$ is the initial reserve of the insurance company, the premiums are received continuously at a constant rate: $c$ ,per unit time;pre unit rate, $d$ is dividend payout ratio, $N(t)$ is Poisson process with intensity $\lambda$ , $\{X_i\}_{i=1}^{\infty}$ are non-negative, i.i.d., the distribution function of $Y(t) = \sum_{i=1}^{N(t)} X_i$ is $F(x)$ , $Y(t)$ and $\{X_i\}_{i=1}^{\infty}$ are i.i.d. $\delta_1$ is constant, $W(t)$ is a standard Brownian motion,which is independent with $Y(t)$ .

We consider the investment of insurance company when surplus is $X(t)$ at the moment of $t$ .We assume the risky asset is $(1-a)X(t)$ $(0 < a < 1)$ , rate of return is $r_0$ $(r_0 > 0)$ ,the percentage of the risk-free asset is $a$ , the situation invested in risk project $S(t)$ is

$$\mathrm{d}S(t) = \mu S(t)dt + \delta_2 S(t)\mathrm{d}B(t),\ t \ge 0 \tag{2}$$

where $\mu$ , $\delta_2 > 0$ are fixed constants; $\mu$ is the risk expectation; $\delta_2$ is volatility of risk; $B(t)$ is a standard Brownian motion, which is independent with $U(t)$ .

As a result

$$\mathrm{d}X(t) = \mathrm{d}U(t) + a\mu X(t)\mathrm{d}t + (1-a)r_0 X(t)\mathrm{d}t + a\delta_2 X(t)\mathrm{d}B(t) \tag{3}$$

Thus $X_0 = u$ , $u > 0$ , $t \ge 0$, the correlation between $B(t)$ and $W(t)$ is

$$\mathrm{d}W(t)\mathrm{d}B(t) = \rho dt \tag{4}$$

Definition of the ruin probability is

$$\Psi(x) = P(T < \infty \mid X_0 = x) \tag{5}$$

where $T = \inf\{t : X_t < 0\}$ is a stopping time.

There are many reserches on model (1), this paper will obtain the HJB equations of ruin probability by dynamic programming method in literature [12], then obtain the optimal risky investment and an upper bound of Lundberg for ruin probability.

## 3. Intgerodifferential Equation of Ruin Probability

Suppose ruin time of the surplus process that with initial surplus $x$ and investment strategy $A$ is

$$\tau(x, A) = \inf\{\, t \ge 0 : X(t, A) \le 0\} \tag{6}$$

ruin probability is

$$\Psi(x,A) = P(\tau(x,A) < \infty) \tag{7}$$

admissible set of research strategy is $A$ , $A \in \mathrm{A}$ , $A$ is $\sigma$ -manifold measurable.
We can obtain the HJB equations of ruin probability $\Psi(x)$ through formula and regularization method:

$$\inf\left\{\frac{1}{2}\left[\delta_1^2 + 2\rho\delta_1\delta_2 A + \delta_2^2 A^2\right]\Psi''(x) + [(\mu - r_0)A + c - d + r_0 u]\Psi'(x) + \lambda E[\Psi(u-x) - \Psi(u)]\right\} = 0 \tag{8}$$

the boundary conditions are $\Psi(0) = 1$ , $\Psi(\infty) = 0$ .
The optimal investment strategy can be got through equation (8):

$$A^*(u) = \frac{-(\mu - r_0)\Psi'(u) - \rho\delta_1\delta_2\Psi''(u)}{\delta_2^2\Psi''(u)} \tag{9}$$

## 4. Upper Bound of Lundberg

In order to discuss the Lundberg upper bound,we suppose the distribution function of the claim settlement followed is exponential decline, which means moment generating function:

$$M(r) = E[e^{ry}] = \int_0^\infty e^{ry} dF(y) \tag{10}$$

We assme that existing $r_\infty \in (0,\infty]$ , then $r \uparrow r_\infty$ , $M(r) \uparrow \infty$ .It is easy to know that $M(0) = 1$ and $M$ is a continuously increasing convex function.

Assuming the solution of equation (2) is $\Psi(x) = e^{-Rx}, x \geq 0$ ,and supposing $R(A)$ is the solution of equation $\frac{1}{2}(\delta_1^2 + 2\rho\delta_1\delta_2 A + \delta_2^2 A^2)r^2 - [(\mu - r_0)A + c - d + r_0 u]r + \lambda[M(r) - 1] = 0$ , then it is the definition of Lundberg index that with investment strategy $A$ (constant), where Lundberg index is defined as $R = \sup R(A)$ . If strategy $A$ is optimal, then the ruin probability achieve to a minimum under this strategy. For when Lundberg index is larger and initial funding increased, the speed of ruin probability decline becomes faster, so we can obtain the asymptotic optimal investment strategy through the method that find the largest Lundberg index.
Hence $R$ is the solution of equation below

$$\inf\{\frac{1}{2}(\delta_1^2 + 2\rho\delta_1\delta_2 A + \delta_2^2 A^2)r^2 - [(\mu - r_0)A + c - d + r_0 u]r + \lambda[M(r) - 1]\} = 0 \tag{11}$$

denotes the left of equal as

$$f(A,r) = \frac{1}{2}(\delta_1^2 + 2\rho\delta_1\delta_2 A + \delta_2^2 A^2)r^2 - [(\mu - r_0)A + c - d + r_0 u]r + \lambda[M(r) - 1] \tag{12}$$

Note: For any $A \geq 0$ , $f(A,r)$ is a convex function with variable $R$ . Since $R \geq R(A)$ , $f(A,r) \geq f(A,R(A)) = 0$ .Therefore optimal investment strategy $A^*$ ,corresponds to the largest

Lundberg index $r$ satisfies $f(A,r) \geq f(A^*(u),R)=0$ , that means $R$ is the solution of equation $\inf\{f(A,r)\}=0$.

Now what we need to do is to find $R$ and $A^*$. Assuming $R$ is known, then

$$f(A,R)=\frac{1}{2}(\delta_1^2+2\rho\delta_1\delta_2 A+\delta_2^2 A)R^2-[(\mu-r_0)A+c-d+r_0 u]R+\lambda[M(R)-1] \tag{13}$$

If and only if $A^*=\dfrac{\mu-\rho\delta_1\delta_2 R}{\delta_2^2 R^2}$ that $f(A,r)$ becomes minimum.

We put $A^*$ into $f(A,R)$, it shows that $R$ is the solution of the equation in the following:

$$\frac{1}{2}(1-\rho^2)\ \delta_1^2 r^2+\lambda M(r)=\lambda+cr+\frac{\mu^2}{2\delta_2^2}-\frac{\mu\rho\delta_1 r}{\delta_2} \tag{14}$$

**Lemma 1** Suppose $x \geq 0, \mu \neq 0, \delta_2 \neq 0$, then there exists a unique $r, 0<r<r_\infty$, satisfied equation below:

$$\frac{1}{2}(1-\rho^2)\ \delta_1^2 r^2+\lambda M(r)=\lambda+cr+\frac{\mu^2}{2\delta_2^2}-\frac{\mu\rho\delta_1 r}{\delta_2} \tag{15}$$

**Proof:** Denoting $h_1(r)=\frac{1}{2}(1-\rho^2)\ \delta_1^2 r^2+\lambda M(r),\ h_2(r)=\lambda+cr+\frac{\mu^2}{2\delta_2^2}-\frac{\mu\rho\delta_1 r}{\delta_2}$, then $h_1(0)=\lambda<h_2(0)=\lambda+\frac{\mu^2}{2\delta_2^2}$

Obviously, $h_1'(r)>0$, and $\lim_{r\to r_\infty} h_1(r)=\infty$. Therefore there exists a unique $r$ ,such that $h_1(r)=h_2(r)$. The proof completes.

**Lemma 2** $R$ and $A^*$ are defined as above, then $M_t := \exp\{-RX_t^{x,A^*}\}$ is martingale.

**Proof:** Obviously, $M_t$ adapts to $F_t$. On the other hand, for any $0<s\leq t<\infty$,

$$\begin{aligned} E[M_t|F_s] &= E[\exp\{-RX_t^{x,A^*}\}|F_s] \\ &= E[\exp\{-R[X_t^{x,A^*}-X_s^{x,A^*}]\}\exp\{-RX_s^{x,A^*}\}|F_s] \\ &= M_s\exp\{f(A^*,R)(t-s)\} \\ &= M_s \end{aligned} \tag{16}$$

The proof completes.

**Theorem2** for investment strategy $A^*(u)=\dfrac{\mu-\rho\delta_1\delta_2 R}{\delta_2^2 R^2}$ (constant), $\Psi(x)$ has an upper bound. $\Psi(x,\ A^*)\leq e^{-Rx}, x\geq 0$

**Proof:** Since $\{M_t\}_{t\geq 0}$ is Non-negative martingale, stopping time process $M^{\tau(x,A)^*}=M_{t\wedge r}(x,A^*)$ is martingale. Hence

$$e^{-Rx}=E[M^{\tau(x,A^*)}]=E[M^{\tau(x,A^*)}I_{\{\tau(x,A^*)\leq t\}}]+E[M^{\tau(x,A^*)}I_{\{\tau(x,A^*)>t\}}] \tag{17}$$

Where $I_C$ is the characteristic function of $C$. By monotone convergence theorem, we have

$$e^{-Rx} \geq E[M^{\tau(x,A^*)} \mid \tau(x,A^*) < \infty] P(\tau(x,A^*) < \infty) \tag{18}$$

then,

$$\Psi(x, A^*) = P(\tau(x,A^*) < \infty) \leq \frac{e^{-Rx}}{E[M^{\tau(x,A^*)} \mid \tau(x,A^*) < \infty]} \tag{19}$$

by the definition of $M^{\tau(x,A)^*}$, we can easily obtain the denominator

$$E[M^{\tau(x,A^*)} \mid \tau(x,A^*) < \infty] > 1 \tag{20}$$

The proof completes.

## 5. The Asymptotic Optimality and Asymptotic Uniqueness of Strategy $A^*$ And $R$

When claim distribution is exponential distribution, the asymptotic optimality and asymptotic uniqueness of strategy $A^*$ and $R$ are essential.

Definition1 Suppose $0 < r < r_\infty$, we say the variable $Q$ had uniform exponential moment in the tail distribution for $R$, in the event of $\sup_{y \geq 0} E[e^{-r(y-Q)} \mid Q > y] < \infty$

The risk ratio of claim $Q$:

$$h(y) = \frac{g_Q(y)}{1 - G_Q(y)} > 0 \text{ satisfies } \liminf_{y \to \infty} h(y) > R \tag{21}$$

$G_Q(y)$ is distribution function and $g_Q(y)$ is density function of random variable $Q$.

**Lemma** 3 Suppose $Q$ has uniform exponential moment in the tail distribution for $R$, so for any $A \in \mathrm{A}$, the process $\{e^{-(RY_t^{x,A})}\}_{t \geq 0}$ is uniformly integrable.

**Lemma** 4 If variable $Q$ has uniform exponential moment in the tail distribution for $R$, so for any $A \in \mathrm{A}$ and $x \in (0, \infty)$, when $t \to \infty$, the process $Y_t^{x,A}$ with set $\{\tau(x,A) = \infty\}$ almost everywhere converges to $+\infty$.

**Theorem** 3 Suppose $Q$ has uniform exponential moment in the tail distribution for $R$, so for any admissible strategy $A \in \mathrm{A}$, ruin probability satisfies $\Psi(x,K) \geq Le^{-Rx}$

$$\text{especially } L = \inf_{y \geq 0} \frac{\int_y^\infty dG_Q(u)}{\int_y^\infty e^{-R(y-u)} dG_Q(u)} = \frac{1}{\sup_{y \geq 0} E[e^{-r(y-q)} \mid q > y]} > 0 \tag{22}$$

**Lemma** 5 In theorem 3, if there is $\varepsilon > 0$ and $x_\varepsilon \geq 0$, such that,

$| A(x) - A^* | \geq \varepsilon$ with all $x > x_\varepsilon$ are established, then there exis $r_\varepsilon$ and $A_\varepsilon$ satisfied $\Psi(x,A) \geq A_\varepsilon e^{-r_\varepsilon x}$

**Theorem** 4 In theorem 3, suppose $A(x)$ is defined function of $A$. Then, $\lim_{x \to \infty} A(x) = A^*$.

**Proof**: Suppose $\lim_{x\to\infty} A(x) \neq A^*$ then there exists $\varepsilon$, $x_\varepsilon > 0$, such that $|A(x) - A^*| \geq \varepsilon$ with all $x > x_\varepsilon$. So by lemma 5, we can get $\Psi(x) \geq A_\varepsilon e^{-r_\varepsilon x}$ with one of $r_\varepsilon < R$, then we get the conclusion which contradicts with the optimality of $A^*$ $\lim_{x\to\infty} \frac{\Psi(x)}{e^{-Rx}} = \infty$.

The proof completes.

**Referencre**

[1] BROWNE S.Optimal investment policies for a firm with risk process:Exponential utility and minimizing the probability of ruin[J].Mathematics Methods Operator Research,1995,20:937-957.

[2] TAKSAR M,MARKUSSEN C.Optimal dynamic reinsurance policies for large insurance portfolios[J]Finance and Stochastics,2003,7:97-121.

[3] LUO Shangzhen,TAKSAR M,TSOI A.On reinsurance and investment for large insurance portfolios[J].Insurance:Mathematics and Economics,2008,42:434-444

[4] Scheer N,Schmidli H.Optimal dividend strategies in a Cramer–Lundberg model with capital injections and administration costs[J]. European Actuarial Journal, 2011, 1(1):57-92.

[5] Liu X,Chen Z,Ming R . The optimal dividend barrier in the perturbed compound Poisson risk model with randomized observation time[J]. Journal of Systems Science and Complexity,2015, 28(2):451-470.

[6] HIPP C,Plum M.Optimal investment for inverstors with state dependent income and for insurers[J].Finance and Stochastics,2003,7:299-321.

[7] LIU Chisang,YANG Hailiang. Optimal investment for an Insurance to minimize its probability of ruin[J]North American Actuarial Journal ,2004,8:11-31.

[8] Mataramvura S,ksendal, Bernt. Risk minimizing portfolios and HJBI equations for stochastic differential games[J]. Stochastics An International Journal of Probability and Stochastic Processes, 2008, 80(4):317-337.

[9] Mao H,Carson J M,Ostaszewski K M,et al.Optimal decision on dynamic insurance price and investment portfolio of an insurer[J]. Insurance: Mathematics and Economics, 2013, 52(2):359-369.

[10] YANG Hailiang,ZHANG Lihong.Optimal investment for Insurer with jump-diffusion risk process[J]. Insurance: Mathematics and Economics,2005,37:615 -634.

[11] Yuen K ,Lu Y,Wu R.The compound Poisson process perturbed by a diffusion with a threshold dividend strategy[J]. Applied Stochastic Models in Business & Industry, 2009, 25(1):73-93.

[12] FLEMING W H,SONER H M.Controlled Markov Processes and Viscosity Solutions[M].New York:Springer Verlag,2005.

IOP Conf. Series: Materials Science and Engineering **914** (2020) 012032 doi:10.1088/1757-899X/914/1/012032

# Study on the preparation technology of Sanqi Shangyao smeared films for percutaneous absorption

**Shi jun Liu[1,2,3,*], Pan Zhang[1,2,3], Zhi shu Tang[1,2,3], Yu Zhang[1,2,3], Jing Sun[1,2,3], Zhong xing Song[1,2,3], Chun li Cui[1,2,3], Hong bo Liu[1,2,3], Hong bo Xu[1,2,3]**

[1]Shaanxi University of Chinese Medicine/Shaanxi Collaborative Innovation Center of Chinese Medicinal Resource Industrialization, Xianyang, Shaanxi, 712083, China
[2]State Key Laboratory of Research & Development of Characteristic Qin Medicine Resources (Cultivation), Xianyang, Shaanxi, 712083, China
[3]Shaanxi Innovative Drug Research Center, Xianyang, Shaanxi,712083, China

[*]Corresponding author e-mail: 1181618@126.com

**Abstract**. Objective: The dosage form of Sanqi Shangyao tablets was reformed and made smeared films. Methods: PVA-124 was used as film forming material, glycerol as plasticizer, Twain 80 as surfactant and azone as penetration enhancer. On the basis of the single factor test, the orthogonal test $L_9$ ($3^4$) was used to screen the matrix prescription, and the optimum matrix ratio was obtained by investigating the fluidity, color, flexibility, film forming time and the appearance uniformity of the coating agent. Results: The best substrate ratio was Twain 80 0.5 g, PVA-124 4.0 g, glycerol 2.0 g, azone 0.3 g. Conclusion: The proportion of the smeared films prepared by orthogonal test is stable. The film forming time is short, and the preparation process is simple and feasible.

## 1. Introduction

The smeared film is a percutaneous absorption preparation [1], which refers to a liquid preparation made of raw materials and film-forming materials for external use and can form a film [2]. The smeared film can avoid the first-pass effect of the liver [3]. Its preparation process is simple, does not need backing material, does not need special equipment, and has a good effect on the prevention and treatment of some skin diseases.

Sanqi Shangyao tablet has the effect of relieving menstruation and activating blood circulation, dispelling blood stasis and relieving pain. At present, there are tablets, granules and capsules [4], and there are no literature reports on the smeared films, so the research group reformed the dosage form of Sanqi Shangyao tablet to make the smeared films, which not only avoids the first-pass effect of oral administration, but also can be administered in many directions. The curative effect is safe, which provides a certain basis for industrial production.

IOP Conf. Series: Materials Science and Engineering **914** (2020) 012032 doi:10.1088/1757-899X/914/1/012032

## 2. Methods and results

### *2.1. Prescription Optimization*

*2.1.1. Single factor test.* In order to determine the dosage range of each matrix in the orthogonal test, the research group did several groups of single factor experiments. The amount of PVA-124 is taken as a variable, and the dosage of Tween 80, glycerol and azone remains unchanged. It is found that the greater the proportion of film-forming materials, the worse the fluidity, but too little is not easy to smear. When the amount of water is 10 times, the fluidity is better and the film-forming time is less. Taking glycerol as a variable and other substrates as quantitative, it was found that the greater the content of the film, the more greasy the skin, but when the content was low, it was not easy to tear off, and the effect was better at 4 g. The situation of Tween 80 is similar to that of glycerol, the more azone, the worse the color of the matrix, and the better effect is at 0.6 g. The research group determined the dosage range of each matrix in the orthogonal test according to the results of single factor test.

*2.1.2. Orthogonal test design.* On the basis of pre-test, $L_9$ ($3^4$) test is used to design the level table of orthogonal factors. The levels and factors of orthogonal test were shown Table 1.

**Table 1.** Factors and levels of orthogonal test

| Levels | Factors | | | |
|---|---|---|---|---|
| | (A) Twain 80 /g | (B) PVA-124/g | (C) Glycerol/g | (D) Azone/g |
| 1 | 0.5 | 2.0(20mL water) | 2.0 | 0.3 |
| 2 | 1.0 | 3.0(30mL water) | 3.0 | 0.6 |
| 3 | 1.5 | 4.0(40mL water) | 4.0 | 0.9 |

*2.1.3. Appearance evaluation and scoring standards [5-6].* Flowability: The smeared film prepared by 1 mL was poured into the inclined glass plate under the same conditions, and the fluidity was observed. The smaller the fluidity, the higher the score, and the total score was 20.
Color and lustre: Observe the color transparency of the smeared film substrate, the better the transparency, the higher the score, and the total score was 20.

Flexibility: Observe the stretchable shape of the matrix after smearing, the more difficult it is to break, the higher the score, and the total score was 20.

Film forming time: Pour 2 mL of matrix onto the glass plate and observe the time it takes to form the film. The shorter the time is, the higher the score is, with a total score of 20.

Uniformity: Smear on the glass plate, observe the uniformity of the film, the more uniform, the higher the score, and the total score was 20.

The score is calculated on the basis of the average score of 10 students.

### 2.2. *Results of orthogonal test*

**Table 2.** Results of orthogonal test

| Serial number | A | B | C | D | Total score |
|---|---|---|---|---|---|
| 1 | 1 | 1 | 1 | 1 | 58.4 |
| 2 | 1 | 2 | 3 | 2 | 64.6 |
| 3 | 1 | 3 | 3 | 3 | 73.6 |
| 4 | 2 | 1 | 2 | 3 | 62.2 |
| 5 | 2 | 2 | 3 | 1 | 64.7 |
| 6 | 2 | 3 | 1 | 2 | 68.5 |
| 7 | 3 | 1 | 3 | 2 | 59.4 |
| 8 | 3 | 2 | 1 | 3 | 63.5 |
| 9 | 3 | 3 | 2 | 1 | 80.4 |
| $K_1$ | 196.6 | 180 | 190.4 | 203.5 | |
| $K_2$ | 195.4 | 192.8 | 207.2 | 192.5 | |
| $K_3$ | 203.3 | 222.5 | 197.7 | 199.3 | G=595.30 |
| $K_{1/3}$ | 65.533 | 60.000 | 63.467 | 67.833 | CT=39375.79 |
| $K_{2/3}$ | 65.133 | 64.267 | 69.067 | 64.167 | |
| $K_{3/3}$ | 67.767 | 74.167 | 65.900 | 66.433 | |
| R | 2.634 | 14.167 | 5.600 | 3.666 | |

### 2.3. *Analysis of variance results*

**Table 3.** Analysis of variance results

| Source of variance | Sum of squares from mean deviation | Degree of freedom | Variance | F | P |
|---|---|---|---|---|---|
| A | 12.082 | 2 | 6.041 | | |
| B | 316.909 | 2 | 158.455 | 26.230 | <0.05 |
| C | 47.309 | 2 | 23.655 | 3.916 | |
| D | 20.542 | 2 | 10.271 | 1.700 | |
| Error(e=A) | | | | | |

note: $F_{0.05}(2, 2) = 19.00$ $F_{0.01}(2, 2) = 99.00$

The comparison of the variance between the range R in Table 2 and the variance in Table 3 showed that the best matrix prescription was $A_3B_3C_2D_1$. The influence of the four factors on the test was B>C>D>A,the influence of factor B on the test results is significant ($P < 0.05$), and A, C and D had no significant effect on the test results. Therefore, the dosage of A, C and D factors can be prepared according to the minimum amount, that is, $A_1B_3C_1D_1$, Tween 80 0.5 g, PVA124 4.0 g, glycerol 2.0 g, azone 0.3g.

## 3. Verification test

Three batches of samples were prepared after the best matrix ratio obtained from the orthogonal test was mixed with 0.5g Sanqi Shangyao extract, and 10 students were asked to make a comprehensive score, and the results showed that the score of each batch was more than 80, indicating that the preparation process of the orthogonal test was feasible and repeatable.

## 4. Conclusion

The research group conducted an orthogonal experiment on the basis of the results of single factor pre-test, and optimized the dosage of Tween 80, PVA124, glycerol and azone by the method of appearance evaluation, and prepared Sanqi Shangyao smeared films after adding the Sanqi Shangyao extract. The experiment can provide some help to the development of the smeared films agent, but the mechanism of transdermal absorption is not clear, and needs to be further tested.

**Acknowledgments**
This work was supported by Shaanxi Project (Grant no. 2015KTCL03-14, 15JF001, ZYMS 025, 2018TD-005, 2018SF-285) and Shaanxi Education Finance ((2013)171); Key disciplines of traditional Chinese Medicine Pharmaceutical Engineering of Shaanxi Provincial Administration of traditional Chinese Medicine (2018).

## References
[1] Zhang ZW. Monograph on traditional Chinese medicine pharmaceutics [M]. Beijing. People's Health Publishing House, 2009:329-330.
[2] Fu CM, Liu W. Pharmaceutics of Chinese Materia Medica [M].Beijing. China Medical Science and Technology Press, 2014: 212-213.
[3] Ding JT, Zhang XM, Li H., Intestinal first-pass effect: research advances[J]. J Int Pharm Res, 2010, 37(4): 269- 272, 275.
[4] Chinese Pharmacopoeia Commission, Pharmacopoeia of the people's Republic of China (Part 1) [S]. Beijing. China Medical Science and Technology Press, 2015,482-485.
[5] Shi YJ, Cheng YQ, Li MC, et al. Optimization of material composition in compound kuhuang prescription coating agent by orthogonal test[J]. Pharm J Chin PLA, 2014, 30 (2):122-124.
[6] Sun HP, Han Y. Preparation of Jinhuang Plastics [J].Drugs & Clinic,2014,29(2):151-153.

IOP Conf. Series: Materials Science and Engineering **914** (2020) 012033 doi:10.1088/1757-899X/914/1/012033

# Acoustic target recognition algorithm based on particle swarm neural network

**Yalei Liu[1,*], Xiaohui Gu[2,a]**

[1]Department of mechanical and electrical, China maritime police academy of CAPF, Zhejiang, China
[2]School of Mechanical Engineering, Nanjing University of Science and Technology Jiangsu, China

*Corresponding author e-mail: liuyalei820@163.com, [a]Guxiaohui@njust.cn.com

**Abstract.** In order to improve the automatic recognition rate of acoustic targets, this paper conducts research on acoustic target recognition algorithms based on particle swarm neural network. Firstly, the mathematical description of the particle swarm optimization algorithm is described, and the initial parameters and algorithm flow of the particle swarm optimization algorithm in the experiments in this paper are given. Second, the design includes the central processor, power supply, signal conditioner, filter, trigger circuit, and state. Acoustic target recognition prototypes of display circuit, memory, target type indication circuit, serial port, crystal circuit, microphone and hardware interface circuit, etc. Finally, using the collected acoustic signals of tanks and helicopters, a semi-physical simulation experiment was designed to carry out target recognition. Experimental research and experimental results verify the effectiveness and stability of the acoustic target recognition system in this paper.

## 1. Introduction

Acoustic target recognition is an important task of intelligent transportation systems, which has important academic value, and will also generate huge social and economic benefits [1]. In the military field, how to classify and identify various wheeled vehicles and tracked vehicles is of great value for ground reconnaissance,battlefield situation awareness, threat assessment, command decision-making, and precise strike in modern warfare [2, 3].At present, for the detection and identification of civilian acoustic targets, optical methods are mainly used [4], and the detection and identification technologies for military acoustic targets include acoustic detection [5, 6], seismic wave detection [7, 8], and electromagnetic wave detection [9, 10] Three. Among them, sound detection technology has the advantages of good concealment, all-weather work, no radar and other working blind spots, simple system structure, low cost and small size, and is widely used. For the automatic recognition of acoustic targets under complex background conditions, Shanghai Jiaotong University designed an acoustic target recognition system based on fuzzy comprehensive evaluation [11]. Based on the characteristics of the acoustic signals of typical targets on the battlefield, Nanjing University of Science and Technology designed DSP-based and wavelet Fast battlefield sound target recognition system [12].With the development of electronic technology, the acoustic target recognition system tends to be miniaturized and fast. In this paper, an acoustic target recognition system based on particle swarm neural network algorithm and a prototype are designed, and the prototype is analyzed for tank and helicopter targets through experiments the validity and stability of identification.

## 2. Swarm Neural Network Optimization Algorithm

Particle swarm neural network is a neural network model in which particle swarm optimization algorithm and neural network are mixed with each other [13]. This paper introduces the mathematical description of the particle swarm optimization algorithm, and gives the initial parameters and algorithm flow of the particle swarm optimization algorithm in the experiments in this paper. At the same time, in the experiment, it is compared with the traditional neural network algorithm.

### *2.1. Mathematical description*

In the basic particle swarm optimization algorithm, a particle swarm consists of $m$ particles, and the position of each particle represents the potential solution of the optimization problem in the $n$ dimensional search space.The mathematical description of the particle swarm optimization algorithm is as follow. Assume that in a one-dimensional search space, a population of particles is $x=(x_1,x_2,\cdots,x_m)^T$, where the position of the first particle is $x_i=(x_{i,1},x_{i,2},\cdots,x_{i,n})^T$, and its speed is $v_i=(v_{i,1},v_{i,2},\cdots,v_{i,n})^T$. Its individual extremum is $p_i=(p_{i,1},p_{i,2},\cdots,p_{i,n})^T$, the global extremum of the population is $p_g=(p_{g,1},p_{g,2},\cdots,p_{g,n})^T$. After the particle finds the above two extreme values, it updates its speed and position according to the following two formulas:

$$v_{i,d}^{k+1}=v_{i,d}^{k}+c_1 rand()(p_{i,d}^{k}-x_{i,d}^{k})+c_2 rand()(p_{g,d}^{k}-x_{g,d}^{k}) \quad (1)$$

$$x_{i,d}^{k+1}=x_{i,d}^{k}+v_{i,d}^{k+1} \quad (2)$$

Where, $c_1$, $c_2$ called the learning factor, $rand()$ is a random number between (0, 1); and the velocity $v_{i,d}^k$ and position $x_{i,d}^k$ of the particle in the $d$ dimension in the $k$ iteration; and the position $p_{i,d}^k$ of the individual extreme value of the particle in the $d$ dimension; $p_{g,d}^k$ is the position of the global extremum of the population in the $d$ dimension. In order to reduce the possibility of particles leaving the search space during the evolution process, it is usually limited to a certain range, that is $v_{i,d}\in[-v_{\max},v_{\max}]$, if the search space of the problem is limited at $[-x_{\max},x_{\max}]$, it can be set as $v_{\max}=kx_{\max}, 0\le k\le 1$.

### *2.2. Initial parameter settings*

One of the biggest advantages of the original particle swarm optimization algorithm is that it does not need to adjust too many parameters, but a few parameters in the algorithm directly affect the performance and convergence of the algorithm. The parameter settings of the particle swarm optimization algorithm involved in this article are as follows:

1) Number of particles. The larger the number of particles, the larger the spatial range of the algorithm search, and it is easier to search for the global optimal solution. However, the increase in the calculation amount increases the search time, which is not conducive to the timeliness of the system. Based on the analysis of previous experiments, the number of particles is 30.

2) Particle length. The particle length is the length of the problem, which is determined by the specific optimization problem.

3) Particle range. The particle range is determined by the specific optimization problem. The particle range in the experiments in this paper is the value range of the problem parameters.

4) Maximum particle velocity. The maximum particle velocity determines the maximum distance a particle can move in a flight. In this paper's experimental analysis, the maximum particle velocity is set to $v_{\max} = kx_{\max}$, $k = 0.5$

5) Learning factors. $c_1$, $c_2$ indicates the degree to which particles are affected by social knowledge and individual cognition, so it set to the same value to give both the same weight, $c_1 = c_2 = 2$.

6) Algorithm termination conditions. Similar to the genetic algorithm, the termination condition of the particle swarm optimization algorithm can generally be set to reach the maximum number of iterations or meet a certain error criterion.

7) Fitness function. In this paper's experimental analysis, the objective function is directly used as the fitness function.

### *2.3. Particle Swarm Optimization Algorithm Flow*

The algorithm flow of the particle swarm optimization algorithm is as follows:

1) Initialize the random position and velocity of the particle swarm according to the initialization process;

2) Calculate the fitness value of each particle;

3) For each particle, compare its fitness value with the fitness value of the best position it has experienced. If it is better, use it as the current best position;

4) For each particle, compared its fitness value with the fitness value of the best position experienced globally, and if it is the best, used it as the current global best position;

5) Evolve the speed and position of the particles according to the iterative equations (1) and (2);

6) If the end condition is not reached (usually a good enough adaptation value or a preset maximum number of generations is reached), then return to step (2).

## 3. Experimental analysis

### *3.1. Prototype of acoustic recognition system*

As shown in Figure 1, it is a sound identification system test prototype, which includes the above-mentioned peripheral hardware circuits, where 1 is a central processing unit, 2 is a power source, 3 is a signal conditioner, 4 is a filter, and 5 is a trigger circuit. 6 is a status display circuit, 7 is a memory, 8 is a target type indicating circuit, 9 is a serial port, 10 is a crystal oscillator circuit, and 11 is an interface between a microphone and a hardware circuit.

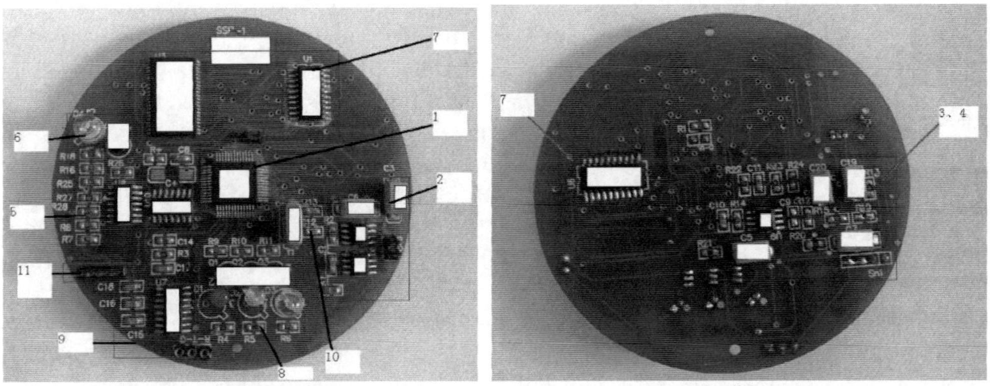


(a) (b)

(a) The front side of the acoustic recognition hardware system test prototype
(b) The back side of the acoustic recognition hardware system test prototype

**Figure 1.** Prototype of acoustic target recognition hardware system

IOP Conf. Series: Materials Science and Engineering **914** (2020) 012033 doi:10.1088/1757-899X/914/1/012033

### *3.2. Acoustic target recognition signal acquisition test*

This test uses a quaternary acoustic sensor to form a circular array with a diameter of 1m. During the test, the sampling frequency of the signal is 62.5KHz and the range is 10V. The acoustic signals of tanks and helicopters were collected in the early stage of this research group. This experiment uses acoustic playback to simulate acoustic targets. The location of the test device is shown in Figure 2. The specific test steps are as follows:

1) The microphone array is placed in a non-same wave-front, and the relative position of the microphone and the sound source is shown in Figure 3;

2) Acquire the signal at a fixed position of the sound source 1 # ~ 11 # opposite the static sound array;

3) Move the sound source at a uniform speed at positions 1 # ~ 11 # opposite to the static sound array to collect signals;

4) Randomly move the sound source at positions 1 # ~ 11 # opposite to the static sound array, with different speeds, and collect signals.

**Figure 2.** Location of the device

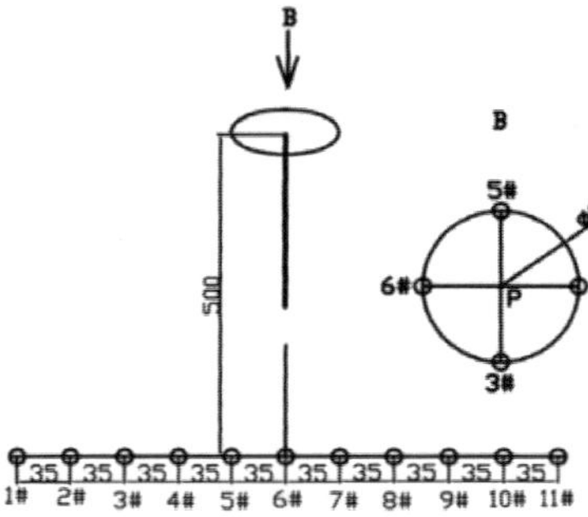


**Figure 3.** Relative position of the test

### *3.3. Analysis of acoustic target recognition results*

Based on the particle swarm neural network optimization algorithm, the total number of acoustic signal samples selected in the stationary state, uniform motion state, and maneuvering state were: 20, 25, 30, 35, 40, 45, 50, and the tank and helicopter targets were identified respectively. Test. The test results are counted based on the target type indicator signal in the prototype. At the same time, in order to verify the effectiveness of the particle swarm neural network optimization algorithm, it is compared with the traditional neural network algorithm. Table 1 shows the recognition results of helicopter and tank target test prototypes.

(1) Analysis of the effectiveness of the test prototype of the acoustic recognition system. It can be seen from Table 1 that whether the traditional neural network algorithm or the particle swarm neural network optimization algorithm is applied in this test prototype, the recognition rate of the tank remains above 66.7%, and the recognition rate of the helicopter target remains at 73.3% Above, especially in combination with the particle swarm neural network optimization algorithm of this article, the recognition rate of the tank has remained above 75%, and the recognition rate of the helicopter target has remained above 84%, thus confirming whether it is based on traditional theoretical algorithms or based on In the improved algorithm, the experimental prototype of the acoustic recognition system designed in this paper can effectively complete the recognition of acoustic targets, and can ensure the effectiveness of the recognition results.

(2) Stability analysis of test prototype of acoustic recognition system. As can be seen from Table 1, based on the traditional neural network algorithm, for tank targets, as the total number of samples increases, the recognition rate remains at 66.7% ~ 72%, the average recognition rate is 68.8%, and the recognition rate variance is 1.75%. As the total number of samples increases, the recognition rate stays at 73.3% ~ 80%, the average recognition rate is 75.3%, and the variance of the recognition rate is 2.2%; it can be seen that the prototype of the acoustic target recognition test is based on a traditional neural

network algorithm, The degree of fluctuation of the recognition rate is less, and the system is stable. In the particle swarm neural network optimization algorithm, for tank targets, as the total number of samples increases, the recognition rate remains at 75% to 88.6%, the average recognition rate is 83.5%, and the variance of the recognition rate is 4.12%; for helicopter targets, With the increase of the total number of samples, the recognition rate remained between 84% and 91.1%, the average recognition rate was 87.3%, and the variance of the recognition rate was 2.35%. By the same token, the recognition rate of the test prototype has less fluctuation and the system is stable. However, from the analysis of the degree of fluctuation of the recognition rate, the optimization algorithm based on the particle swarm neural network is more complicated and computationally intensive than the traditional neural network algorithm. Therefore, the degree of fluctuation of the recognition rate of the corresponding experimental prototype also increases.

(3) Analysis of recognition rate of different targets. From Table 1, it can be seen that, whether based on the traditional neural network algorithm or the particle swarm neural network optimization algorithm, the recognition rate of the acoustic target recognition test prototype for the helicopter target is higher than that for the tank target. Related. From the analysis of the sound signal propagation, compared with the helicopter target, the distance from the tank target to the sound array is much lower than the distance from the helicopter to the sound array. Therefore, the helicopter sound source is more satisfied with the point sound source. From the collection and analysis of acoustic signals, the reflection of the tank acoustic signals through the ground and the interference of environmental noise are strong, and the signal-to-noise ratio of the collected sound source signals is low. Therefore, the recognition rate of the tank target is low. Analyze from another angle to improve the sound source signal. The signal-to-noise ratio can improve the recognition rate of acoustic targets.

**Table 1.** Identification results of helicopter and tank test prototypes

| sample | Neural network algorithm | | | | Particle Swarm Neural Network Optimization Algorithm | | | |
|---|---|---|---|---|---|---|---|---|
| | Tank target | | Helicopter target | | Tank target | | Helicopter target | |
| | Identify correctly | Recognition rate% | Identify correctly | Recognition rate% | Identify correctly | Recognition rate% | Identify correctly | Recognition rate% |
| 20 | 14 | 70.0 | 15 | 75.0 | 15 | 75.0 | 17 | 85.0 |
| 25 | 18 | 72.0 | 20 | 80.0 | 21 | 84.0 | 21 | 84.0 |
| 30 | 20 | 66.7 | 22 | 73.3 | 25 | 83.3 | 26 | 86.7 |
| 35 | 24 | 68.6 | 26 | 74.3 | 31 | 88.6 | 31 | 88.6 |
| 40 | 27 | 67.5 | 30 | 75.0 | 34 | 85.0 | 35 | 87.5 |
| 45 | 31 | 68.9 | 34 | 75.6 | 38 | 84.4 | 41 | 91.1 |
| 50 | 34 | 68.0 | 37 | 74.0 | 42 | 84.0 | 44 | 88.0 |

## 4. Conclusion

All manuscripts must be in English, also the table and figure texts, otherwise we cannot publish your Based on the software and hardware design, this paper completes the assembly and commissioning of the prototype of the acoustic target recognition system based on the particle swarm neural network algorithm. The effectiveness and stability of the test prototype are verified by taking tank and helicopter targets as the target. Based on the test results, it is known that the recognition rate of the tank target remains at 75% to 88.6%, the average recognition rate is 83.5%, and the variance of the recognition rate is 4.12%; for the helicopter target, the recognition rate is maintained at 84% to 91.1%, and the average recognition is The rate is 87.3%, and the variance of the recognition rate is 2.35%. The experimental results verify the effectiveness and stability of the acoustic target recognition system designed in this paper.

IOP Conf. Series: Materials Science and Engineering **914** (2020) 012033 doi:10.1088/1757-899X/914/1/012033

## References


[1] Jian Zhao, Weiwen Su, Jian Jia.Research on depression detection algorithm combine acoustic rhythm with sparse face recognition [J]. Cluster Computing: The Journal of Networks, Software Tools and Applications, 2019, 22 (12): 7873-7884.

[2] Sung-Hwan Shin, Takeo Hashimoto, Shigeko Hatano.Automatic Detection System for Cough Sounds as a Symptom of Abnormal Health Condition [J]. IEEE Transactions on Information Technology in Biomedicine, 2009, 13 (4): 486-493.

[3] Ravichandran B, Gandlhe Avinash, etal. Robust automatic target recognition using learning classifier systems [J]. Information Fusion, 2007, 8 (3): 252-265.

[4] YANG S, CAO J, WANG J, et al. Linear prediction of one sided autocorrelation sequence for noisy acoustics recognition of excavation equipments [C]. Intelligent Control and Automation, IEEE, 2016: 924-928.

[5] Gaetano,C., Ivan, R. and Pierluigi, T., Acoustic system for aircraft detection and tracking based on passive microphones arrays, 148th Meeting of the Acoustical Society of America, San Diego, November, 2004: 1-14.

[6] Liu Yalei. Study for the tracking theory of 3D dynamic acoustic array for acoustic. Nanjing: Nanjing University of Science and Theology. 2013

[7] Lan Jinhui, Nahavandi Saeid etal. Research on seismic signals for vehicle targets and recognition by data rusion [A]. Proceedings of the 4th International Conference on Control and Automation [C]. Montreal, Canada: IEEE, 2003: 733-736.

[8] ZHONG Xionghu, HOPGOOD J R. Particle filtering for TDOA based acoustic source tracking Nonconcurrent multiple talkers [J]. Signal processing, 2014, 96: 382-394.

[9] Liu Yalei, Gu Xiaohui. Optimal arrangement of four-sensor dynamic acoustic array. Sensor review, 2012: 32 (4): 288-295

[10] LIU Ya-lei, GU Xiao-hui.Tracking algorithm study of CEFMDVDAKF in brainpower submunition for sound-source bearing [J]. Chinese Journal of Scientific Instrument, 2011, 32 (4): 748-755.

[11] ZHONG Xionghu, PREMKUMAR A B. Multiple wideband source detection and tracking using a distributed acoustic vector sensor array: A random finite set approach [J]. Signal processing, 2014, 94: 583-594.

[12] HUANG G, HUANG G B, SONG S, et al. Trends in extreme learning machines: a review [J]. Neural Networks: the Official Journal of the International Neural Network Society, 2015, 61 (C): 32.

[13] HUANG G B, ZHU Q Y, SIEW C K. Extreme learning machine: a new learning scheme of feed forward neural networks [J]. Proc Int Joint Conf Neural Netw, 2004 (2): 985-990.

IOP Conf. Series: Materials Science and Engineering **914** (2020) 012034 doi:10.1088/1757-899X/914/1/012034

# Design and Research on Temperature Control System of Electric Furnace Based on PLC Touch Screen Integrated Machine

**Haiyu Zhao[1], Jiashuo Zheng[2]**

[1]School of Electrical and Electronic Engineering, North China Electric Power University, Beijing, 102206
[2]School of Mathematics and Physics, North China Electric Power University, Beijing, 102206

**Abstract.** With the development of science and technology in China, temperature control system has been widely used in many fields. The emergence of PLC effectively improves the shortcomings of PID control system, optimizes the performance of PID control system, and plays a very important role in promoting industrial production and construction. In view of the design of temperature control system of heating furnace, configuration software and PLC can be used the combination of technology can achieve the goal of furnace temperature control, and it can also monitor the furnace temperature in real time and comprehensively, and then adjust the furnace temperature according to the application requirements. The investment in the process is less, which can reduce the cost to a certain extent, and also has the application advantage of high accuracy. The temperature control system of electric heating furnace is designed in this paper. PLC is mainly used, and the temperature system is controlled intelligently with touch screen. According to the actual situation and cost budget, the controller of this paper chooses Mitsubishi FX2N-48MR series PLC. In order to achieve intelligent control, it needs to use digital PID control algorithm, which is convenient for the staff to operate Gt1040-qbbd-c series touch screen is used to realize human-computer interaction. The system can be started and stopped. The touch screen for temperature setting can be very simple to operate. The temperature change can also be visually displayed and monitored, and the whole system can be automatically set.

## 1. Introduction

The use of resistance furnaces in industry began in 1920. With the rapid development of science and technology in the world, resistance furnaces are very common in industrial production. In many production processes, whether the working environment of factory staff is safe, whether the products produced meet the market quality requirements, whether the production efficiency can be improved, and whether energy can reduce waste are closely related to the technology and economy of temperature control [1]. The accuracy of resistance furnace temperature control, whether the system can operate stably, whether the system is reliable and other requirements are the key indicators in various fields. The improvement and optimization of temperature control technology has become a hot issue for researchers all over the world. The theoretical technology of temperature control has roughly experienced three stages from the constant value switch control to PID control and then to the current popular intelligent control [2-3].

(1) Constant value switch control: if the actual temperature value measured by the system is lower than the given value, the system needs to raise the temperature, then start the control switch to heat the system, and when the measured temperature value is higher than the given value, the control switch turns off heating. But in the actual operation of the system, the temperature control error will be very high.

(2) PID control temperature: temperature control with PID algorithm. Proportion, integral and differential are the core contents of the design and they determine the quality of the system. The working principle of the resistance furnace depends on the electric heating of the resistance wire to increase the temperature. The temperature of the resistance furnace is reduced naturally due to the influence of the external temperature. If the actual temperature of the resistance furnace exceeds the maximum limit, the control means can no longer operate normally. In the large-scale industrial production process, the biggest problem is that using PID controller for temperature control can not be very accurate [4]. Instability may occur when disturbed, which is a thorny problem at home.

(3) Intelligent control: in short, it can be thought and operated by computer instead of human brain without human participation. It can get rid of the technician's operation and automatically command the machine control object to realize the expected function. The rapid progress of scientific theory in modern countries has made a new breakthrough in the field of temperature control, in which intelligent calculation should be used. For multi-objective can be optimized, in a complex environment can learn to adapt.

The object of study is tunnel electric heating furnace. It is 5.2m long, 1.5m wide and 1.8m high. It is heated by resistance heating tube. The resistance heating pipe is divided into three groups, one group is composed of 10 electric heating pipes of 1 meter long (37.4 ohm each, with a power of about 1.3KW) in parallel, the other two groups are composed of 1.5 meter long (45 ohm each, with a power of about 1.075kw each) in parallel, with a total power of about 34.5kw. The original electric heating and insulation furnace is controlled by temperature control meter, with temperature control range of 180℃~ 220℃. In view of the design of temperature control system of heating furnace, configuration software and PLC technology can be used Combined with the way to achieve the goal of furnace temperature control, it can also monitor the furnace temperature in real time and comprehensively, and then adjust the furnace temperature according to the application requirements. In the process, less capital is invested, which can reduce the cost to a certain extent, but also has the application advantage of high accuracy. In the industrial production, the equipment operation is relatively stable and safe, and the energy waste is greatly reduced.

## 2. Design scheme of temperature control system of electric furnace based on PLC

### *2.1. Overall design scheme*

The overall plan is mainly composed of the following parts:

(1) Touch screen: it is mainly used to set and display the temperature value of the corresponding electric heating furnace, stop and start, set the holding time, and display the temperature curve in real time

(2) PLC: it mainly completes PID adjustment function, data transformation, field equipment and process control.

(3) Temperature measurement circuit: the thermocouple is connected with PLC, and PLC processes the temperature measurement signal and outputs the control signal. The heating resistance bar is connected with the thyristor according to the principle of three-phase symmetry.

(4) Output regulating circuit: the main function is to pass the control signal sent by PLC processing operation through the pulse width

The modulation device outputs pulse signal to the thyristor and controls the conduction time of the thyristor to control the heating of the heating tube. The thermocouple is connected with PLC, and the deviation between the constant temperature setting value $Q_0$ and the actual heating pipe temperature $Q_e$.

### 2.2. *Qorking principle*

(k) According to the integral separation PID control algorithm, the output control quantity $U_{(k)}$ is obtained, the conduction time of the thyristor is controlled, and the temperature of the heating tube is adjusted so that it is consistent with the given constant temperature value to achieve the purpose of constant temperature control. When the constant temperature value is reached, the constant temperature indicator light will be on, so that the number of trigger pulses is 0, and the thyristor is not on, at this time, the system will not heat the resistance furnace. In case of input error or system failure, the system will send an alarm signal. At the same time, GT1040-QBBD-C screen is used for real-time display of heating tube temperature and input of temperature setting value.

## 3. Hardware design of temperature control system of electric furnace based on PLC

### 3.1. *PLC module*

PLC is generally divided into integral type and modular type, but its logical structure is basically the same. In terms of hardware structure, PLC is composed of CPU, memory, I/O interface unit, extension interface and extension part, peripheral interface and peripheral and power supply, etc. all parts are connected by system bus.

Mitsubishi FX2N series PLC is small in size, moderate in price, powerful in performance and high in cost performance. Compared with other PLCs, Mitsubishi programming is easy and convenient. Mitsubishi PLC provides many expansion modules to meet the needs of different users. In line with the practical and cost-effective situation, Mitsubishi FX2N Small PLC is selected. There are many kinds of Mitsubishi PLC, among which FX2N series is ahead of other models and the processor has powerful performance. If users have other special process requirements, they can also add special function modules for most occasions. Most of FX2N consists of three parts: basic unit, expansion module and expansion unit. The user memory capacity can be expanded to 16K steps. I/O points can be expanded up to 256. It has 27 basic instructions. Compared with other large-scale PLC FX2N, it is far ahead in basic instruction operation speed. Mitsubishi FX2N-48MR is Thermocouple Input relay output type, with 24 input and output points,

The number of points available for the expandable module is 48-64, with 8000 steps of ram. The interrupt number of the timer is as follows:

**Table 1.** Interrupt number of timer

| Enter number | Interrupt period (ms) | Interrupt inhibit special auxiliary relay |
|---|---|---|
| I6XX | In the XX part of the pointer name, enter an integer of 10-99. I610 executes timer interrupt every 10ms | M8056 |
| I7XX | | M8057 |
| I8XX | | M8058 |

### 3.2. *Analog module*

Because the Mitsubishi FX2N PLC selected in this paper needs to consider the matching between the selected analog module and Mitsubishi PLC in determining the analog quantity. FX2N-2AD has two analog quantity inputs, FX2N-4AD has four analog quantity inputs, and FX2N-8AD has eight analog quantity inputs. Combined with practical experience, FX2N-4AD is compared with FX2N-4AD It is more convenient to operate, cost-effective and has more spare parts. Finally, it is decided to use FX2N-4AD. It's an analog input module matched with Mitsubishi PLC. The analog input module consists of four analog input channels (CH1, CH2, CH3, and CH4). Any channel can transform the analog into digital. The resolution of FX2N-4AD is 12 bits.

### *3.3. Touch screen module*

In this paper, the constant temperature setting value of the resistance furnace adopts GT1040-QBBD-C series Mitsubishi touch screen and switch key to control the input, and the software (GT designer 3) is used to edit the touch screen interface, as shown in the figure below, including the temperature setting input of the resistance furnace, real-time display of the touch screen temperature, start and stop control buttons of the touch screen, constant temperature and alarm indicators. In order to enable the user to input the given temperature value of the system through multiple ports, 10 switch buttons are added to achieve the above functions. This port can receive decimal number. Touch screen program and switch key circuit are shown in the figure below. The given temperature value shall not be more than 100℃ and less than 0℃. If the input value is greater than or less than the above range due to misoperation, the red alarm light will start to alarm (red light is on). The system temperature setting input is realized by touch screen: (the temperature setting input is the number in the first row of the right figure):

## 4. The circuit design of the temperature control system of electric furnace based on PLC

### *4.1. Temperature detection circuit*

In this paper, the function of temperature detection in electric heating furnace is realized by thermocouple (the temperature of heating tube is measured by the hot end of thermocouple, and the cold end of thermocouple is placed at 0℃). The most commonly used temperature sensor is thermocouple. Generally, the thermocouple used in industry will be equipped with display instrument, and the recording instrument and electronic regulator will be added according to the situation. Thermocouple temperature measurement range can be from zero to one thousand three hundred degrees, and the industry often needs to measure the object of liquid steam, gas medium or solid surface temperature can be directly measured by thermocouple.

Thermocouple structure is simple, it is a passive sensor that can directly measure temperature and is a primary instrument. Its working principle is to transform the temperature signal into millivolt signal. The medium temperature to be measured can be obtained by secondary instrument.

Thermocouple temperature measurement is made of thermoelectric phenomenon. Two kinds of different wires are welded together to form the working end, and the two ends are connected with the measuring instrument to form the circuit. Put the working end at the temperature to be measured, because of the temperature difference between this end and the free end, electromotive force will be generated, and current will pass through the circuit. Through the measurement of electrical quantity, the temperature at the other end can be measured by using the known temperature. It can be used between two substances with large temperature difference. Some thermocouples can measure the high temperature of 3000℃ and others can measure the low temperature close to absolute zero.

In order to prevent adding the third metal conductor to change the thermoelectric potential generated by the couple and reduce the accuracy, it is necessary to ensure that the temperature at both ends of the third metal conductor is the same, so that the thermoelectric potential generated by the thermocouple circuit will not be affected by the newly added metal. Therefore, the temperature of the target medium can be calculated by connecting the measuring instrument with the thermocouple and obtaining the thermal potential.

### *4.2. Zero crossing detection circuit*

The trigger needs to input a synchronous signal, which can be provided by a pulse generated by the zero crossing detection circuit at the beginning of each power cycle. The voltage is divided and filtered by resistance, and a zero crossing comparison circuit is formed by four differential comparators to obtain the square wave signal synchronized with the input voltage signal of the power grid. The GND is the ground LM339, which is the four differential comparator. There are four independent voltage comparators in the LM339 integrated block, the common mode range is very large, from zero voltage to power voltage minus one to two volts; the differential input voltage range is also very large, even equal to VCC. Four way differential comparator uses diode to protect its input. The working waveform of the design is shown in Figure 1.

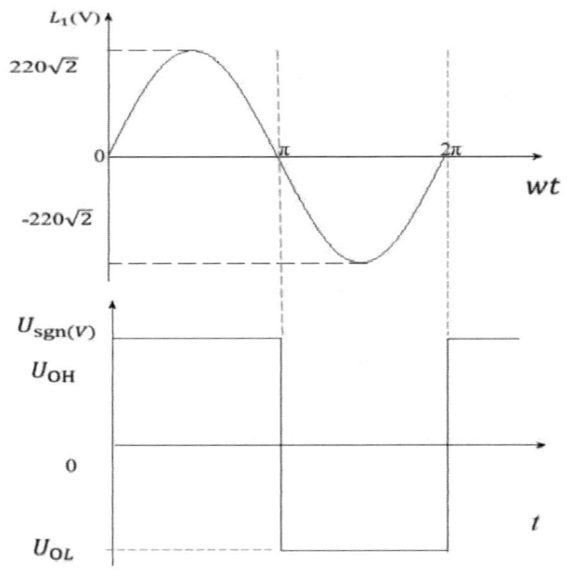


**Figure 1.** Working waveform of zero crossing detection circuit

*4.3. PID control circuit*

In this design, PID control algorithm of integral separation is used to realize constant temperature automatic control of resistance furnace. In practical application, engineering tuning is generally used to determine the parameters of PID. Proper increase of $K_P$ can speed up the adjustment and reduce the error. However, $T_i$ and $T_d$ continue to optimize the control level when $K_P$ regulation is not ideal. The increase of $K_P$ will decrease the stability of the system, increase the overshoot and even produce unstable oscillation. If it is very difficult to adjust only $K_P$ to achieve the desired control effect, then $T_i$ and $T_d$ are needed to help adjust.

$T_i$ is introduced to eliminate the static error of the control system and improve the accuracy of the system control. However, the addition of $T_i$ will cause some problems. For example, when the constant temperature value of the system is set at the beginning and the end or before and after a very large difference, the system may have a large error due to handling this change. At this time, the deviation has exceeded the adjustment range of the controller, but PID It is still in operation. With the accumulation of time, the system will have a large overshoot, or even an unstable oscillation. The introduction of $T_d$ can reduce overshoot and regulation time, but it can amplify noise interference.

## 5. Conclusion

The temperature control system of electric heating furnace is designed in this paper. The main hardware used is PLC (programmable controller) and touch screen to control the temperature system intelligently. According to the actual situation and cost budget, the controller of this paper chooses Mitsubishi FX2N-48MR series PLC. In order to achieve intelligent control, digital PID control algorithm is needed, and GT1040-QBBD-C is used for the convenience of staff operation Series touch screen to achieve human-computer interaction, the start and stop of the system, temperature setting with the touch screen can be very simple operation, temperature change can also be intuitive display monitoring and the whole system can be automatically set. In the industrial production, the human-computer interaction interface of the control system designed in this paper is easy to learn and operate, and it is very convenient. The whole temperature control system sclf-tuning shows great advantages.

IOP Conf. Series: Materials Science and Engineering **914** (2020) 012034 doi:10.1088/1757-899X/914/1/012034

## References

[1] Li, Sheng J. A Research on Information Collection Method for High-Precision Temperature Based on PLC Control [J]. Applied Mechanics and Materials, 2014, 529:148-152.
[2] MENG Fan-xing, LI Li-juan, CHENG Guo-qing. Research and Development of PLC Temperature Control System of Heating-sealing Device [J]. Packaging Engineering, 2016.
[3] Gao Y F, Li G P, Jiang C D, et al. Design of the Wire Rope Cleaning and Detection Line Control System Based on Touch Screen and PLC [J]. Applied Mechanics and Materials, 2013, 455: 495-501.
[4] L Pin. Application of Control System Composed of PLC and Touch Screen [J]. Process Automation Instrumentation, 2010, 31 (8): 45-47, 51

IOP Conf. Series: Materials Science and Engineering **914** (2020) 012035 doi:10.1088/1757-899X/914/1/012035

# Electronic circuit design of stepping motor based on 51 single chip microcomputer

**Jianwei Jia[1], Bo Xue[2], Yimeng Wang[3]**

[1]Nankai University Department of Microelectronic Engineering, Tianjin, China
[2]Physical college, University of Electronic Science and Technology of China, Chengdu, China
[3]College of the Arts, Guangxi University, Nanning, Guangxi, 530000

**Abstract.** With the development of automatic control system and the improvement of high-precision control requirements, stepping motor plays an increasingly important role in various automation equipment. The stepping motor control system designed in this paper takes STC89C51 as the core, LCM12864 as the display module, and eight independent keys as the control function keys. It realizes the start and stop state, speed and running distance, forward and backward control of the stepping motor, and adds Hall sensor to measure the speed of the stepping motor. Using SCM as the main control chip not only simplifies the hardware circuit, but also reduces the system cost. The system increases the interactive experience between people and machines, and maximizes the economic benefits, saves resources, and can be better applied to the actual daily life.

## 1. Design background

In our life, stepping motor is also called pulse motor. In 1920, step motor was invented. Thirty years later, the invention of transistors led to the use of stepping motors. Digital control becomes easier. Stepping motor has been widely used in a variety of mechanical control systems which require high performance of controllable random change. Stepping motor can be divided into hybrid type, variable reluctance type and permanent magnet type.

In real life, we can find the application of stepping motor everywhere. In particular, stepping motor is often used to control in various occasions which need the control of speed and position and the flexibility of completing various functions accurately. For example, stepping motor is one of the essential control components for 3D printers widely used in technology and manufacturing. Generally, there are four stepper motors installed on the printer, one for each x, y, Z axis and feeder. How to print out the desired shape and style requires accurate control of the speed, position and other factors affecting the stepper motor, which requires higher requirements for the control system of the stepper motor. The household fan can also be regarded as a stepping motor control system. By pressing the function button on the fan, you can control the start-up, customization, swing direction or fixed direction of the stepper motor, and the rotation speed of the fan blade determines the size of the wind speed. From this we can see that the normal operation of all parts of our life is inseparable from the stepping motor. Therefore, the stepping motor control system designed has excellent performance, simple operation and simple hardware circuit design, which are essential requirements.

## 2. Basic Requirements

This paper presents the design of a stepping motor control system with STC89C51 as the core chip. It is used to control the start rotation, stop, forward and backward, acceleration and deceleration of the stepping motor, and to set the running distance of the stepping motor. LCM12864 is used to display the start and stop of the motor, the forward and reverse rotation, the rotation speed of the motor, the display status of the preset distance and the actual rotation distance from the stepping motor. In this paper, 28byj-48 motor is selected, and the method of stepping motor drive is controlled by single chip microcomputer. Among them, the stepping motor is a controllable machine which converts the received electrical signal into angular displacement output.

When the driver receives the pulse signal from the main control chip, it drives the stepping motor to rotate at a fixed angle (called "step angle") in the set direction. And gradually turn in the same direction at the same angle. Therefore, the control of the running distance of the stepping motor can be realized by controlling the number of pulse groups sent by the main control chip. Using the time interval between each pulse signal, the speed of stepping motor can be changed. In particular, LCM12864 is added as the display, which can display the relevant rotation status and data of the stepping motor, and increase the visibility of the operating system.

## 3. Electronic circuit of stepping motor based on 51 single chip microcomputer

### *3.1. General design*

The system is a stepping motor control system based on STC89C51. It is mainly composed of power circuit, independent key control circuit, Hall sensor speed measurement module, LCM12864 display module and stepping motor driver module. When pressing the external independent key, the MCU sends a pulse signal. The stepping motor driver drives the motor to rotate according to the function of the corresponding key, and displays the function of the current running state of the stepping motor on LCM12864. The rotation of stepping motor will also affect the output of Hall sensor, so as to calculate the running speed of motor.

### *3.2. Singlechip module*

Selection of this paper STC89C51 MCU is the core control chip. Fifty-one Single chip microcomputer is Eight 32 bit CPU I/O Port, two programmable T0 and T1 , interrupt control system with five interrupt sources, single+ 5V Power supply. In the stepping motor control system designed in this paper, the P1 The interface is connected with an external independent key for function control; the step motor is connected to the P2 Port. In particular, the timer and external interrupt of single chip microcomputer are used to count the timer to determine which task blocks are executed to achieve the corresponding tasks.

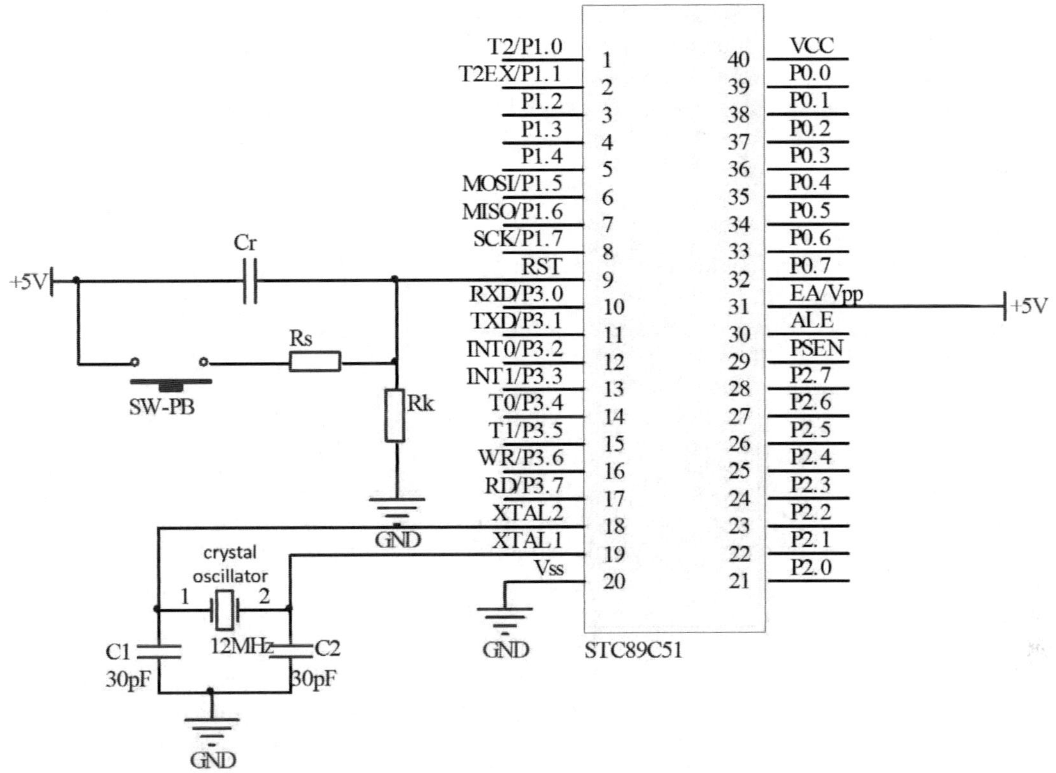


**Figure 1.** Circuit diagram of the minimum system of single chip microcomputer

*3.3. Sensor module*

Hall sensor is selected as the sensor module to measure the speed of stepping motor, as shown in the figure below. Use the 3144E Hall switch IC as shown in Figure 2.2. VCC is connected to the positive pole of the power supply, GND is connected to the negative pole of the power supply, do-ttl digital output is connected to the p3.3 port of the single chip microcomputer. When the hall sensor 3144E senses the magnetic field, the do sends an interrupt signal to the external interrupt 1 once. The integrated circuit uses Hall effect.

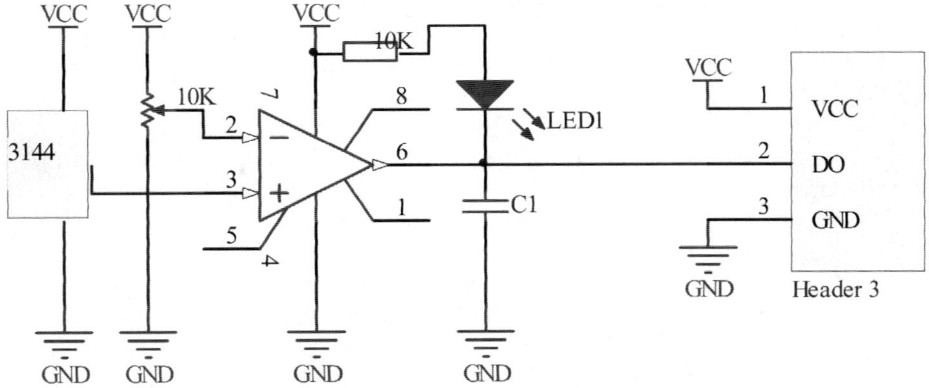


**Figure 2.** 3144E Hall switch IC

The magnetic sensing circuit includes voltage regulator, Hall voltage generator, differential amplifier, Schmitt trigger, temperature compensation circuit and open collector output stage. The input is the magnetic induction intensity and the output is the digital voltage signal. When the sensor senses the magnetic field, the digital output is low and the signal lamp is on; if the magnetic field is not sensed, the digital output is high and the signal lamp is not on. The sensor has the characteristics of small volume, high sensitivity and fast response.

### *3.4. Display module*
LCM12864 is used for LCD module. If the third foot V0 is not connected, the display is equipped with a potentiometer. In this case, the potentiometer on the back of the display must be adjusted to R1. The 18 pin Vout of LCM12864 can be connected or not in the suspended state.

LCM12864 with Chinese character library is selected as the display module. Because LCM12864 is a 128 * 64 lattice LCD module, each line can display 4 lines and 8 columns, 32 16 * 16 lattice Chinese characters in total. Each display RAM can display one Chinese character or two 16 * 8 dot matrix full height ASCII characters, that is, each screen can display up to 32 Chinese characters, or 64 American information exchange standard codes. The display module needs to display the start and stop status, rotation direction, speed gear, rotation speed, preset distance and actual running distance of the motor. Every time there is a new data change, the whole screen will be refreshed and reloaded to prevent garbled phenomenon.

### *3.5. Driver module*
ULN2003 is used as the stepper motor driver in the stepper motor control system, and 28byj48 stepper motor with 1 / 64 reduction ratio is selected as the stepper motor. The stepping motor supplies + 5V power and is connected with the power supply of single chip microcomputer. The standard interface of stepper motor is equipped on the drive board of ULN2003. The A, B, C and D LEDs on the drive board indicate the working state of the four phase stepper motor. The frequency of LED flashing represents the speed of stepping motor drive. The in1, in2, in3 and in4 of the ULN2003 driver are respectively connected with the p2.0, p2.1, p2.2 and p2.3 ports of the single chip microcomputer, which sends pulse signals to control the rotation of the stepping motor.

### *3.6. Key module*
The motor control system uses 8-bit independent keys to control the external functions. The key is connected with P1 port of single chip microcomputer, which is characterized by occupying I/O port line by itself, and each operation state button does not affect other I/O port lines.

Compared with matrix keyboard, the hardware configuration of independent key is flexible and the software architecture is simple. K1 is the motor start stop control key. The initial state of the motor is stop after power on. Press K1 to start the motor rotation, and then press the motor stop again. K2 is the motor rotation direction control key. After the motor starts to rotate, it is in the reverse state. Press K2, the motor starts to rotate forward, and then press the motor to reverse. K3 is the control key of adding and subtracting function, which controls whether the adding and subtracting function is to use and adjust the speed range, set the high position of the running distance or set the position of the running distance. K5 and K7 are function buttons. K6 and K8 are minus function buttons.

## **4. Simulation and experimental results**
First of all, LCM12864, timer 0, timer 1 and external interrupt 1 are initialized. In the whole step motor control system, the realization of each function is regarded as a task block. In the main function, only the timer timing is used to decide which task block to execute. The counting time of timer 0 is 2ms, and the corresponding tasks of 10ms, 20ms, 50ms and 100ms can be executed after counting 5, 10, 25 and 50 times; the counting time of timer 1 is 50ms, and the corresponding tasks of 200ms, 500ms and 1000ms can be executed after counting 4, 10 and 20 times. Among them, the task of 20ms is to scan each key to see if there is a key press, and if there is a key press, to complete the corresponding task

function; the task of 500ms is to calculate the actual running distance of the stepping motor; the task of 1000ms is to count the rotation speed of the stepping motor.

The actual running distance is displayed by serial port debugging. The setting distance is 179mm. When the actual running distance reaches 179mm, the stepping motor stops rotating.

**Table 1.** running distance test table

| | Preset running distance (mm) | Actual displayed running distance(mm) | Absolute error (mm) | Relative error |
|---|---|---|---|---|
| 1 | 176 | 177 | 1 | 0.50% |
| 2 | 59 | 60 | 1 | 1.69% |
| 3 | 99 | 100 | 1 | 1.01% |
| 4 | 1099 | 1099 | 0 | 0% |
| 5 | 600 | 604 | 4 | 0.67% |
| 6 | 255 | 255 | 0 | 0% |
| 7 | 256 | 259 | 3 | 1.17% |

According to the table, the average relative error is 0.72%, less than 5%. Because the program is designed to refresh LCM12864's fixed UI every time the data is updated. Therefore, the reason for the error is that the screen swiping takes too much time, which leads to the difference between the two distances is not timely. Therefore, the error occurs when the actual display running distance is greater than the preset running distance.

## 5. Conclusion

This design system is mainly based on STC89C51 single chip microcomputer to control the start, stop, forward and backward, acceleration and deceleration control and running distance of stepping motor. LCM12864 is used to display the start/stop and forward/reverse state of the motor, the speed of the motor, and the preset distance and actual running distance of the stepping motor. Eight independent keys are used as function keys.

After the MCU is powered on and the program is downloaded, LCM12864 displays the fixed UI interface. At this time, the content displayed is the initial state of the stepping motor set in the program. When key K1 is pressed, the motor will run according to "on, reverse and speed 05" displayed in the LCD. Among them, "speed 05" represents that the initial speed of the set motor is five gears. Press the start / stop key again, LCM12864 shows "off", and the motor stops. Press the K2 key when the disc is running, the motor will change from the original initial reverse state to forward rotation, and LCM12864 will display "forward rotation". The speed shift must be changed when the motor is stopped. First press K3 key to adjust to 01 state, and then press k5-k8 function plus / minus key to change the rotation speed.

The setting of running distance can only be changed when the motor is stopped. Press key K3 to 02, and then press key k5-k8 to set the high position of the distance. Press K3 to 03, press k5-k8 to set the low position of the distance, and then press K1 to make the motor start to rotate. When the actual running distance is equal to the preset running distance, the motor stops rotating. It can also modify the rotation direction, rotation speed and running distance of the motor according to the above method after the single chip computer is powered on.

## Reference:

[1] Cao Minggang. Research on stepping motor control system based on single chip microcomputer [J]. Digital technology and application, 2017,06:20 + 22

[2] Deng Lu. Research on linear motor control system based on Hall sensor [D]. Southeast University, 2017

IOP Conf. Series: Materials Science and Engineering **914** (2020) 012035 doi:10.1088/1757-899X/914/1/012035

[3] Alexandru Morar. Driver for 5-phase stepper motor pentagon connection with dedicated ICs[J]. Procedia Manufacturing,2018,22.
[4] YH Lee.Stepper motor motion control system design[J].Equipment Manufacturing Technology,2015,2(6):31-41.

IOP Conf. Series: Materials Science and Engineering **914** (2020) 012036 doi:10.1088/1757-899X/914/1/012036

# Influence of Irrigation Canal Construction on the Stability of A Natural High and Steep Slope

**Yanhui Guo*, Jin He**

College of Civil and Architectural Engineering, Yunnan Agricultural University, Kunming 650201, Yunnan, China

*Corresponding author e-mail: guoyanhui0818@163.com

**Abstract**. The excavation and construction of the irrigation canal on a high and steep slope usually change the stress state of the original slope, which may have a serious impact on the stability of the slope. Taking the construction of irrigation canal on a high and steep slope as the research background, the stability of this high and steep slope under natural condition, construction platform excavation and construction of irrigation channel is calculated by Midas GTS NX. The results show that the stability of slope is better in natural state. The excavation of the construction platform of the irrigation canal will cause a certain displacement of the original slope, but will not affect the overall stability of the slope. With the construction of the irrigation canal, the displacement of the slope increases a lot, and the stability of the slope decreases significantly.

## 1. Introduction

In recent years, China has been building a large number of irrigation systems to meet the needs of agricultural production. However, due to the influence of topography, geological conditions and irrigated areas, a large number of irrigation channels have to be built on high and steep slopes with complex geological conditions [1-2]. The excavation and construction of the irrigation canal on the high and steep slope will often change the stress state of the original slope, which has a serious impact on the stability of the slope. Moreover, the seepage of the irrigation canal will reduce the geotechnical mechanics parameters of the original slope, induce more and more landslide disasters, and may cause huge casualties and economic losses [3-4]. Therefore, the excavation and construction of the irrigation canal on the high and steep slope has important social benefits and practical significance for the study of the stability of the original slope [5]. In this paper, based on the research background of the construction of a high and steep slope, Midas GTS NX, a large-scale three-dimensional numerical calculation software, is used to calculate the stability of the slope under the natural state, the excavation of the irrigation channel platform, the construction of the irrigation channel, etc. The research results are of great significance to the construction safety and subsequent safety guarantee of irrigation canal project.

## 2. Engineering overview

The construction area of an irrigation canal project is a high and steep, with complex geological conditions, narrow access roads and many bends, which makes the construction difficult and slow. The original design and construction of the first section of the project is 3923m long. Up to now, more than

1000 m of irrigation channel platform excavation has been completed. Due to the steep mountains, developed valleys, complex geological conditions and fragile surface ecology in the route area of the first section of the irrigation canal, the surface soil mainly consists of Quaternary silty clay mixed with gravel, gravel mixed with silty clay, etc., with low mechanical strength of rock and soil. Slope gully development, debris flow and other geological disasters occur frequently. In the process of construction, local collapse of irrigation channel platform occurs consistently.

## 3. The calculation model and mechanical parameters

### *3.1. Calcuation model*

The whole slope angle of the no.1 section is 60°and the irrigation channel construction platform is 80°, which is selected as the research object. The numerical simulation of slope stability under the above three conditions is carried out. Based on the field survey, the possible influence scope of landslide is analyzed, and combined with relevant data, the slope excavation model is established, which is 173m long and 300m high. The whole slope model has 8918 units and 8854 nodes. The free boundary of the model is the surface, which is constrained by normal direction and fixed at the bottom. The numerical calculation model is shown in figure 1.

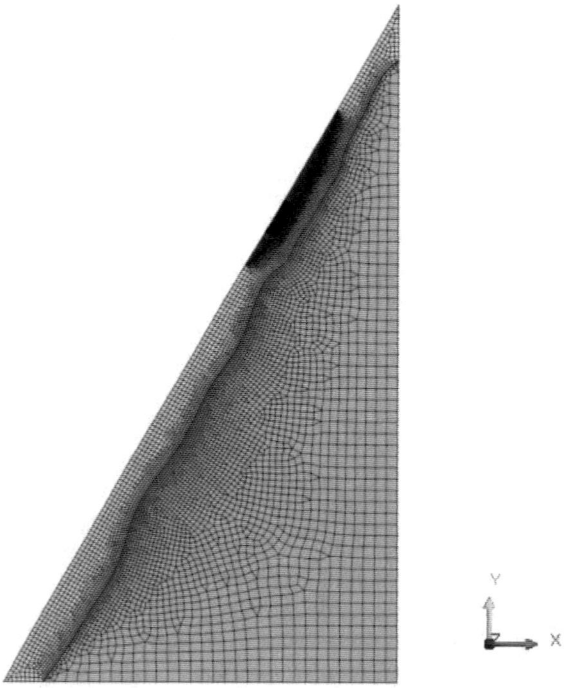


**Figure 1.** Computational mechanics model

### *3.2. Recommended values of mechanical parameters of rock and soil mass*

The Quaternary overburden on site is mainly composed of silty clay with gravel, gravel with silty clay and sand gravel. According to the test results of the feasibility study stage of the project, and referring to relevant specifications and similar engineering data, the recommended values of main physical and mechanical parameters of each soil layer are proposed, as shown in table 1. Mohr Coulomb elastic-plastic constitutive model is used for calculation [6].

IOP Conf. Series: Materials Science and Engineering **914** (2020) 012036 doi:10.1088/1757-899X/914/1/012036

**Table 1.** recommended values of main physical and mechanical parameters of overburden Table

| Lithology | Cohesion (kPa) | Internal friction angle (°) | Deformation modulus (MPa) | Saturation weight (kN/m$^3$) | Bearing capacity eigenvalue(kPa) | Base friction coefficient |
|---|---|---|---|---|---|---|
| Silty clay with | 30 | 21° | 4~5 | 20 | 200 | 0.3 |
| Crushed stone with silty clay | 15 | 28° | 20~23 | 21.5 | 280 | 0.45 |
| Crushed stone, block stone | 70 | 30° | 25~28 | 22 | 310 | 0.50 |
| Sandy cobble | 90 | 31° | 30~40 | 23 | 400 | 0.50 |

### *3.3. Analysis content and calculation steps*
The analysis content and calculation steps are as follows: (1) Based on the mechanical parameters in the engineering investigation, the initial stress field of the area is constructed; (2) The stability of irrigation canal slope under natural conditions (before excavation); (3) The stability of slope after excavation of construction platform of irrigation canal; (4) The stability of the slope after the excavation of the irrigation canal.

## 4. Numerical simulation analysis of slope stability of irrigation channel

### *4.1. Slope stability analysis in natural state*
In the natural state, the distribution of the maximum principal stress field of the slope is generally controlled by the influence of the gravitational field. The maximum principal stress of the slope generally appears as compressive stress, which gradually increases from surface to interiorthe of the slope. The maximum value can reach 4.061 MPa, and the maximum principal compressive stress is located in the lower right corner of the slope model. The minimum principal stress of the slope is still controlled by the gravity field. The minimum principal stress of the slope surface and below a certain depth range is expressed as tensile stress. Tensile stress concentration occurs in weakly weathered basalt, resulting in a maximum tensile stress of about 0.165 MPa. As the depth increases, the minimum principal stress gradually changes from tensile stress to compressive stress, and the maximum compressive stress is approximately 1.938 MPa.

### *4.2. Slope stability analysis after the excavation of the irrigation canal construction platform*
After the excavation of the irrigation channel construction platform, the bottom of the platform rebounded due to the excavation and unloading of the platform. The maximum rebound volume was within the range of 3.207 to 3.665 cm. The main displacement of the slope occurred in the rock and soil within a certain range near the excavation. The maximum total displacement reached approximately 5.498 cm and appeared in the quaternary overburden above the excavation site. The total displacement generally shows a trend of decreasing from the front to the back of the slope. Based on the distribution characteristics of the total displacement, it shows that the impact of excavation on the cover layer is greater, and the impact on the stability of the slope is mainly caused by the deformation of the front and local deduced stability. Figures 2 and 3 are the lateral and vertical displacements of the slope after the excavation of the construction platform. It can be found that the horizontal and vertical displacements of the deep bedrock are almost zero. The maximum horizontal displacement is 4.093cm, the maximum

vertical displacement is 3.717cm, the distribution positions are close, and the vertical displacement mainly occurs at the bottom of the platform.

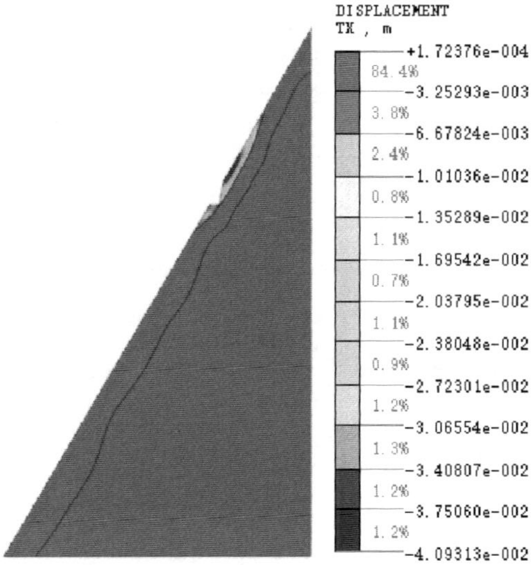


**Figure 2.** Lateral displacement of the slope

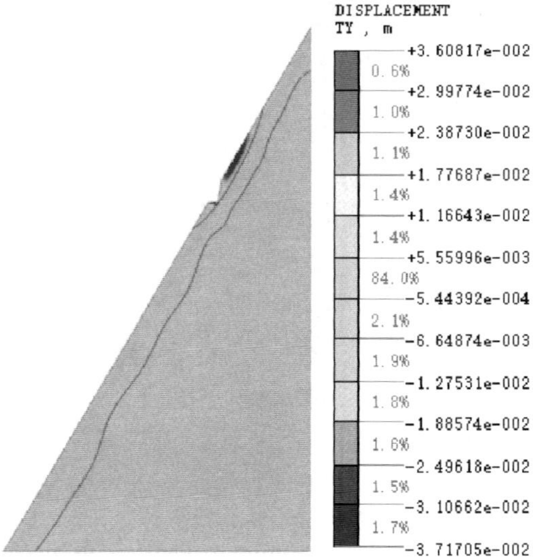


**Figure 3.** Vertical displacement of the slope

*4.3. Slope stability analysis after completion of irrigation canal construction*

Figure 4 shows the lateral displacement of the slope after the completion of the irrigation canal construction, and Figure 5 shows the vertical displacement of the slope after the completion of the irrigation canal. The main displacement of the slope occurred in the overburden slope of the upper part of the irrigation channel, and the total displacement of the rock and soil near the excavation showed a trend of increasing first and then decreasing from the slope surface to interior of the slope. The maximum total displacement of the slope was 2.0m. Compared with the end of the excavation stage of the platform, which Indicates that the excavation of irrigation channel has a greater impact on slope stability. Excavation of irrigation channel mainly causes the slope to generate horizontal displacement and subsidence along the free surface of the slope. The maximum horizontal displacement is 1.756m and the maximum vertical displacement is 1.089m, the distribution position is relatively close.

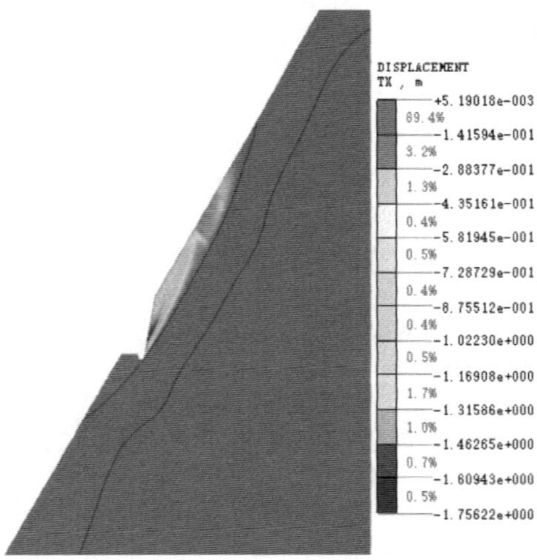


**Figure 4.** Lateral displacement of the slope

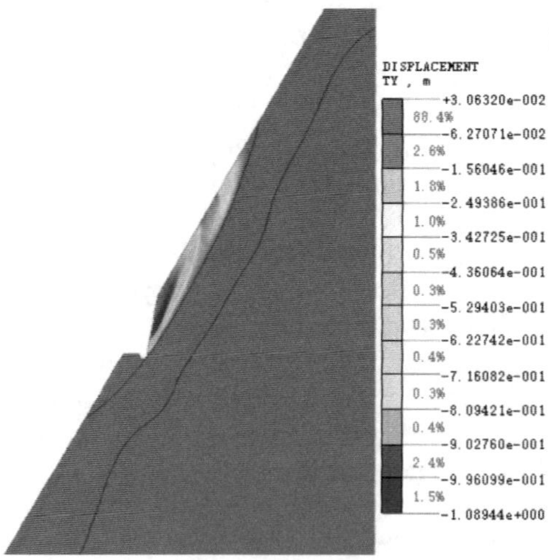


**Figure 5.** Vertical displacement of the slope

Irrigation canal construction after completion, filling presence slope canal upper body cover layer obvious maximum shear strain, the relatively high value of maximum shear strain mainly distributed to the portion of the rock and soil, a maximum value of 1.890m ,to form a more internet excavation stage At the same time, the maximum value of the maximum shear strain increases significantly, and at the same time, a sliding surface is formed at the interface between the overburden and the bedrock. The cloud diagram of maximum shear strain after excavation of the canal is shown in figure 6 .

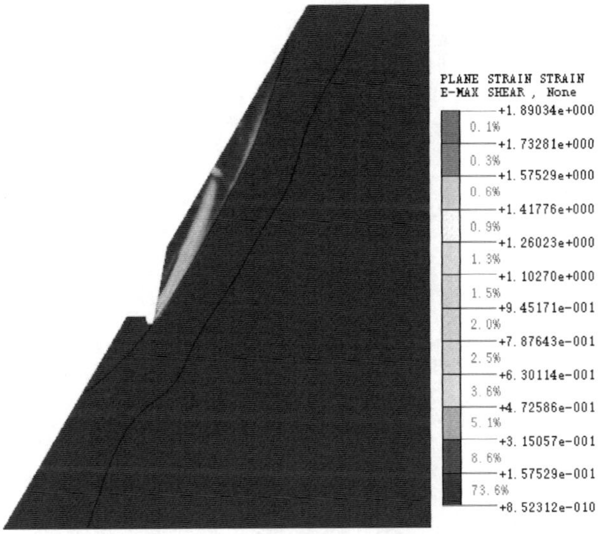


**Figure 6.** Cloud diagram of maximum shear strain

## 5. Conclusions

According to the numerical simulation results of the whole process of the irrigation canal construction on high and steep slope, the slope stability is better under natural condition. Excavation of the construction platform will cause a certain displacement of the original slope, but it will not affect the overall stability of the slope. With the excavation of the irrigation canal, the displacement of the slope increases significantly, and the stability of the slope decreases significantly. The excavation of the irrigation canal will have a significant adverse impact on the slope. During the operation of the irrigation canal, it is easy to get clogged and leak due to the influence of the surrounding rock and soil mass of the upper slope. The water in the irrigation canal leaks into the slope can further reduce the shear strength of the rock and soil mass, which may cause landslides and mud flow disasters.

## Acknowledgments

This work was financially supported by Applied Basic Research Project of Yunnan Science and Technology Department (NO.2018FB075) and The Postdoctoral Science Foundation of China (NO.2017M620433).

## References

[1] WANG Jiading, HUI Zanghe. Landslides in Crows Induced by Irrigated Waterin in Loess Area[J]. Scientia Geograpica Sinica, 2002, 22 (3): 305-310.

[2] DONG Ying, JIA jun, ZHANG Maosheng, et al. An Analysisis of Inducing Effects of Irrigation and the Responses of Loess Landslide Hessian Heifantai Area of Yongjing, Gansu[J]. Geological Bulletin of China, 2013, 32 (6): 893-898.

[3] WANG Zhirong, WU Wenjiang, ZHOU Ziqiang. Landslide Induced by Over-irrigation in Loess Plat-form Areas in Gansu Province. The Chinese Journal of Geological Hazard and Control, 2004. 15(3)43-46,54

IOP Conf. Series: Materials Science and Engineering **914** (2020) 012036 doi:10.1088/1757-899X/914/1/012036

[4] Griffiths D V, Lane P A. Slope stability analysis by finite element[J]. Geotechnique, 1999, 49(3):387-403.
[5] Dawson E M, Roth W H, Drescher A. Slope stability analysis by strength reduction[J]. Geotechnique, 1999,49(6): 835-840.
[6] Phogat V, Malik R S, Kumar S. Modelling the effect of channel bed elevation on seepage and water table rise in a sand box filled with loamy soil[J]. Irrigation Science,2008,27(3):191-200.

IOP Conf. Series: Materials Science and Engineering **914** (2020) 012037 doi:10.1088/1757-899X/914/1/012037

# A Comparison of Ride Performance of Hydro-Pneumatic Suspension System with Those of Rubber and Leaf Suspension Systems

**Le Xuan Long[a], Le Van Quynh[*], Nguyen Van Liem[b], Bui Van Cuong[c], and Hoang Anh Tan[d]**

Faculty of Automotive and Power Machinery Engineering, Thai Nguyen University of Technology, Thai Nguyen, Vietnam

[*]Corresponding author e-mail: lequynh@tnut.edu.vn, [a]lexuanlong@tnut.edu.vn, [b]xuanliem712@gmail.com, [c]cuongbui@tnut.edu.vn, [d]hoanganhtan86@gmail.com

**Abstract**. The objectives of this study is to compare the performance of the hydro-pneumatic suspension system of heavy truck on cab ride comfort with rubber and leaf springs of suspension systems. To compare the three types of suspension systems, a three-dimensional dynamic model of heavy truck with 15 degrees of freedom (DOF) is established based on Zhou Changfeng model under random excitation of road surface for simulation and analysis. The root mean square (r. m. s) acceleration of the vertical cabin, pitch angle, and roll angle of cabin is chosen as objective functions. Matlab/Simulink software is used to simulate and calculate the values of objective functions. The ride performance of the hydro-pneumatic suspension system on vehicle is analyze and compared to two other supension systems when vehicle operates under different operating conditions. The results show that the values of objective functions with the hydro-pneumatic suspension system are significantly reduced in comparison with the rubber and leaf suspension system. The ride performance of Hydro-pneumatic suspension system is much better than that of the rubber and leaf suspension systems.

## 1. Introduction

Hydro-pneumatic suspension system is becoming more and more widespread in the engineering vehicles because of its good nonlinear stiffness and vibration reduction characteristic. Optimizating the parameters of interconnected hydro-pneumatic suspension system of a three-axle vehicle for ride comfort and handling was devoloped by [1]. Analytical modelling, simulation and experimental study of a hydro-pneumatic suspension system to increase ride comfort and to obtain excellent performance in rough terrains was studied by [2]. The parameters of the hydro-pneumatic suspension system were optimized for ride comfort and the mathematical model of the hydro-pneumatic suspension system was derived and then incorporated into a quarter car vehicle model [3]. The new hydropneumatic suspension with the interconnection type about right and left was improved and the structure parameters and mathematical model of the new system was established by [4]. A multibody cosimulation approach was used to investigate the effects of hydropneumatic parameters on the ride safety and aid with design optimization and tuning of the suspension system was devoloped by [5]. A hydro-pneumatic suspension model built with the advantage of fractional order in viscoelastic material modeling considering the mechanics property of multiphase medium of hydropneumatic suspension system was researched by [6]. The performance of the hydro-pneumatic suspension system of heavy truck on the ride quality of road surfaces was analyzed by [7]. The dynamic models of the traditional and new air suspension systems to compare the performance of the air suspension systems for reducing the negative impacts on the road surface using a 3D dynamic model with 14 DOF was proposed by [9]. The dynamics models of a heavy truck using Zhou Changfeng model was developed for analyzing and evaluating the effects of suspension systems on road surface friendliness [12]. A three-dimensional nonlinear dynamical model of a typical heavy truck with 16-DOF (degree of freedom) was established to analyze and evaluate the performance of the air suspension system of heavy trucks with semi-active fuzzy control [15]. A 3-D dynamic model with 13 DOF was established to control the cab's isolation system of heavy truck using Fuzzy logic controller [16]. The acceleration responses and dynamic load coefficient (DLC) were chosen as an objective function for analyzing the effects of suspension design parameters of a semi-trailer truck on vehicle ride comfort as well as road surface friendliness [17].

In this paper, the three-dimensional nonlinear dynamic model of heavy truck with 15 DOF is established based on Zhou Changfeng model [8] and the dynamic models of three types of suspension systems such as the hydro- pneumatic, leaf and rubber springs of suspension systems are proposed to compare their effect on vehicle ride comfort. The performance of the hydro-pneumatic suspension system on vehicle ride comfort through three objective functions according to the international standard ISO 2631-1:1997[13] is analyzed and compared to the rubber and leaf springs of suspension systems when vehicle is operated under the different operating conditions.

## 2. Vehicle dynamic model

### *2.1. Full vehicle dynamic model.*

The dynamic model of AD250 articulated dump truck in Fig.1 consists of 15 degrees of freedom.

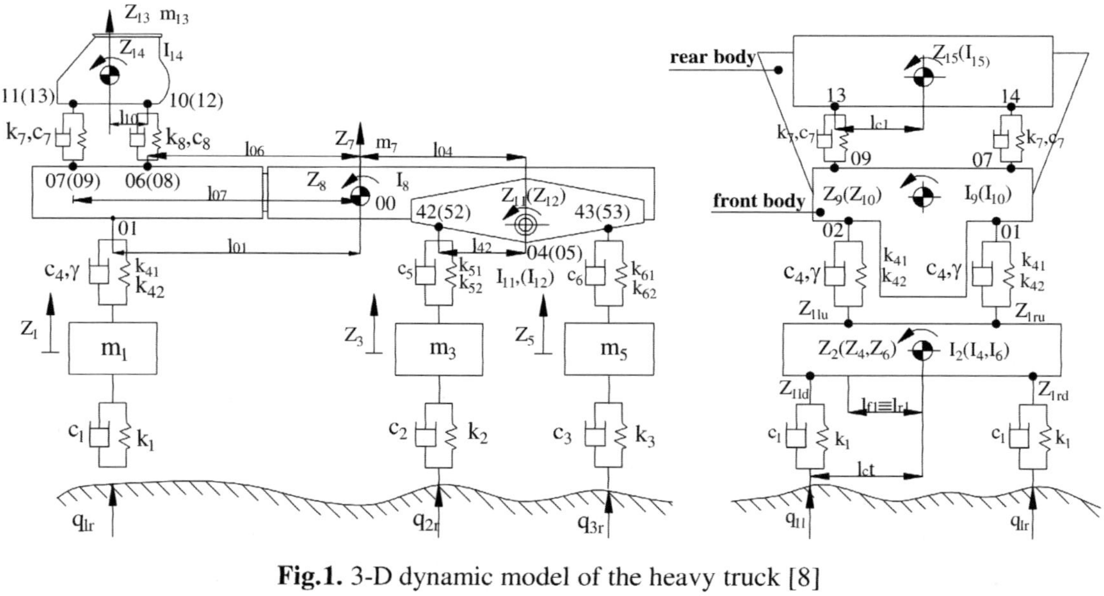


**Fig.1.** 3-D dynamic model of the heavy truck [8]

From the full vehicle dynamic model as shown in Fig.1, the motion equations of unspung mass, sprung mass and cab mass are written using Newton's second law. The general dynamic differential equation for the 3-axle heavy truck is given by the following matrix form:

$$[M]\{\ddot{z}\}+[C]\{\dot{z}\}+[K]\{z\}=[C_t]\{\dot{q}\}+[K_t]\{q\}. \tag{1}$$

where [M], [C], and [K] are vehicle mass, damping, and stiffness matrix of the suspension system; [$C_t$] and [$K_t$] are damping and stiffness matrix of the wheel system; {z} is the vector of displacements; {q} is the vector of excitation of road surfaces.

### *2.2. Mathematic model of the suspension systems.*

To analyze the performance of the hydro-pneumatic suspension system of heavy truck on the ride quality, the lumped model of three different types of suspension systems is established to determine the vertical forces of the suspension systems, as shown in Fig.2.

### *2.3. The rubber suspension system.*

From Fig.2 (a), the vertical nonlinear dynamic force of the rubber spring suspension system [8] is determined by

$$F_r = k_{r1}s + k_{r2}s^3 + \left[k_{r1}\left(z_b - z_a + s\right) + k_{r2}\left(z_b - z_a + s\right)^3\right] + c\left(\dot{z}_b - \dot{z}_a\right)\left(\frac{1+\text{sgn}\left(\dot{z}_b - \dot{z}_a\right)}{2}\right)$$
$$+ c\gamma\left(\dot{z}_b - \dot{z}_a\right)\left(\frac{1+\text{sgn}\left(\dot{z}_b - \dot{z}_a\right)}{2}\right). \tag{2}$$

where $k_{r1}$, $k_{r2}$ are stiffness of rubber spring, c is damping coefficient, s is the displacement of rubber spring when vehicle is static, $z_b$ and $z_a$ are vertical displacement of sprung and unsprung mass, $\gamma$ is the ratio of compression to tensile damping.

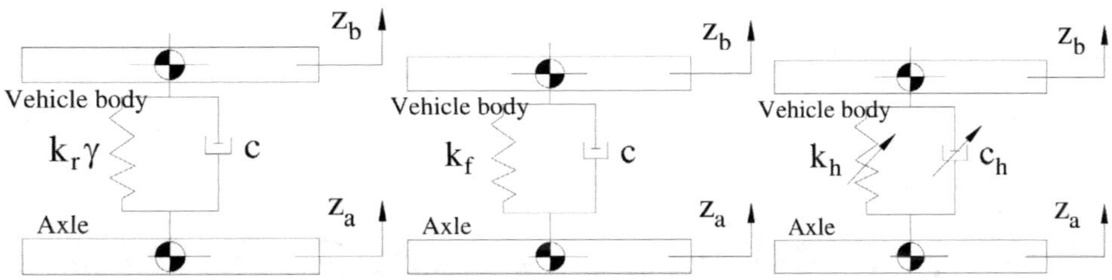


(a) Rubber suspension (b) Leaf suspension (c) hydro-pneumatic suspension

**Fig.2.** Lumped model of three different types of suspension systems

### *2.4. The hydro-pneumatic suspension system.*

From Fig.2 (b), the vertical dynamic force of hydro-pneumatic suspension system[7]can be computed by

$$F_h = -\frac{\rho\left(A_s - A_p\right)^3\left(z_b - z_a\right)^2}{2C_d^2A^2}\,\text{sign}\left(z_b - z_a\right) + p_0\left(\left(\frac{V_0}{V_0 + A_p\left(z_b - z_a\right)}\right)^n - 1\right) - m_D\left(\ddot{z}_b - \ddot{z}_a\right). \tag{3}$$

where, $C_d$ is the coefficient of discharge; A, $A_s$, and $A_p$ are the area of the orifice, cylinder and floating piston; $p_0$ and $V_0$ are the initial pressure and volume in air chamber; $m_D$, n are the mass and the polytrophic rate (1< k <1.4).

### *2.5. The leaf suspension system.*

From Fig.2 (c), the vertical dynamic force of the leaf suspension system is calculated by

IOP Conf. Series: Materials Science and Engineering **914** (2020) 012037 doi:10.1088/1757-899X/914/1/012037

$$F_l = k_l\left(z_b - z_a\right) + c\left(\dot{z}_b - \dot{z}_a\right). \tag{4}$$

where $k_l$ and c are the stiffness of the leaf spring and damping coefficient of the suspension system.

### *2.6. Road surface roughness.*

To analyze the effects on vehicle ride, the road surface roughness plays an important role. The road surface roughness is assumed to be a zero-mean stationary Gaussian random process. It can be generated through the inverse Fourier transformation [10], [18].

$$q(t) = \sum_{i=0}^{N} \sqrt{2\Delta n G_d\left(i\Delta n\right)} \cos\left(2\pi i \Delta n t + \varphi_i\right). \tag{5}$$

where, Δn is within a frequency band; $\varphi_i$ is the random phase uniformly distributed from 0 to 2π; $G_q(i\Delta n_0)$ is the road roughness coefficient which is defined for typical road classes from A to F according to ISO 8068(1995) [11].

## 3. Simulation and analysis results

Matlab/Simulink software is used to simulate with a set of parameters of vehicle and suspension parameters by the references [14] and compare the ride performance of hydro-pneumatic suspension system with leaf and rubber spring of suspension systems. The simulation results of the time domain responses of the cab vertical, pitch angular, roll angular accelerations ($a_z$, $a_{phi}$ and $a_{teta}$) when vehicle moves on the ISO class C road surface at v=40 km/h and full load are shown in Fig.3. From the results of Fig.3 we see that the peak acceleration amplitudes with hydro-pneumatic suspension systems respectively reduce in comparison with leaf and rubber springs of suspension systems. The performance of the hydro-pneumatic suspension will be continued to analyze and compare in the section below.

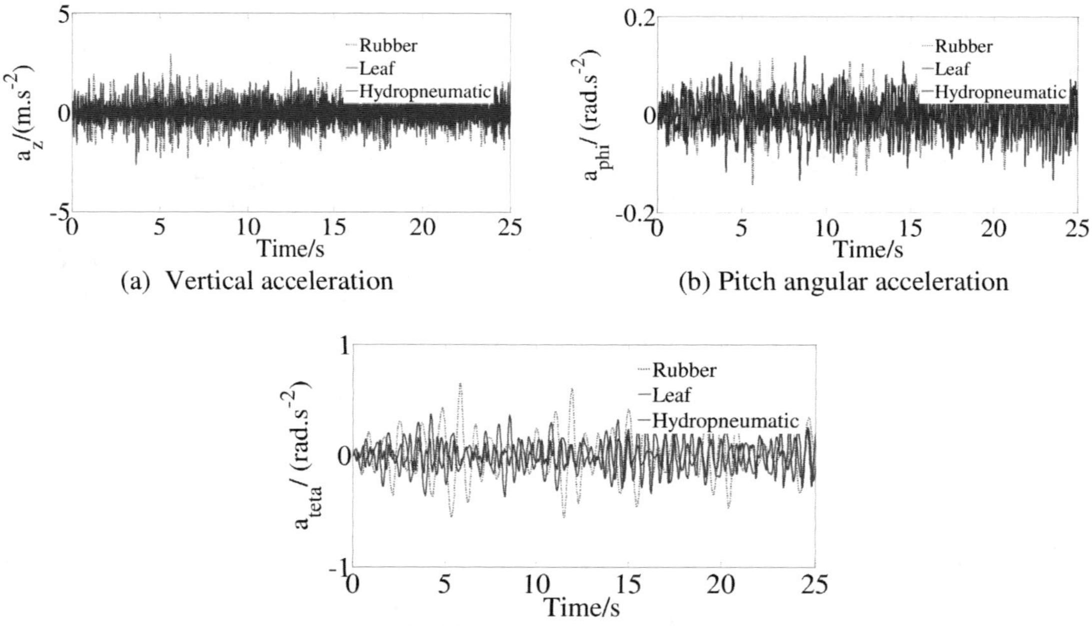


(a) Vertical acceleration (b) Pitch angular acceleration

(c) Roll angular acceleration

**Fig.3** The vehicle cab vertical, pitch angular, roll angular accelerations

To analyze the different road surface conditions on the ride performance of vehicle suspension systems, five road surface conditions from ISO class A (very good) to ISO class E (very poor) in ISO 8068[12] are used to compare when vehicle moves at the speed of 40km/h and full load. The values of the weighted RMS accelerations of the vertical, pitch and roll vehicle cab ($a_{wz}$, $a_{wphi}$ and $a_{wteta}$) according to the international standard ISO 2631-1 are shown in Fig.4.

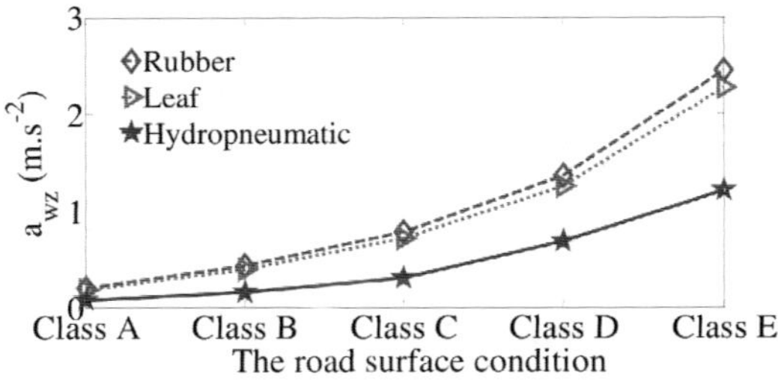


(a) Vertical motion of cab

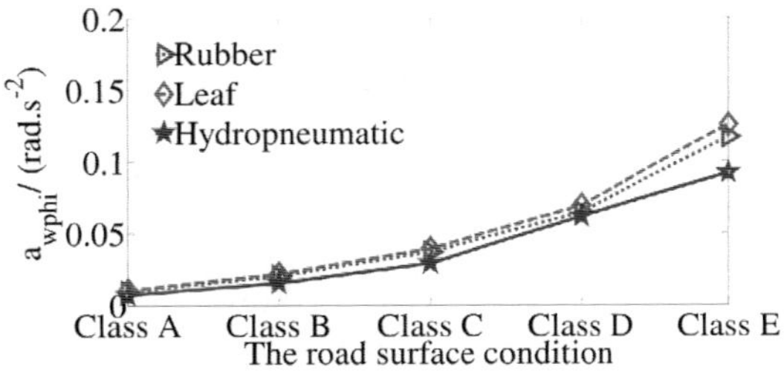


(b) Pitch motion of cab

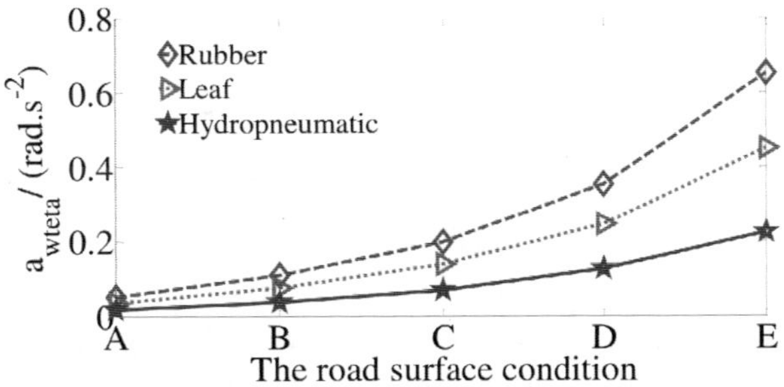


(c) Roll motion of vehicle body

**Fig.4.** The awz, $a_{wphi}$ and $a_{wteta}$ values with five different road suface conditions

Fig.4 shows that the $a_{wz}$, $a_{wphi}$ and $a_{wteta}$ values of three types of suspension systems quickly increase when vehicle moves on the different road surfaces, especially, when vehicle moves on bad road surface conditions such as ISO class C and ISO class E. The $a_{wz}$ values with the hydro-pneumatic suspension system reduce by 63.5%, 63.8%, 61.4%, 50.1%, and 50.8% in comparison with the rubber spring of the suspension system, and reduced by 60.1 %, 60.3%, 57.8%, 45.5%, and 46.7% in comparison with the leaf spring of the suspension system with the identical operating conditions and that is similar to the $a_{wphi}$ and $a_{wteta}$ values. Vehicle ride comfort has been a significant improvement the hydro-pneumatic suspension system.

To compare the ride performance of three types of the suspension systems, the vehicle speeds of 20 km/h, 30 km/h, 40 km/h, 50 km/h, 60 km/h, 70 km/h were considered when vehicle moves on the ISO class C and full load. When the vehicle moves at different speeds, the $a_{wz}$, $a_{wphi}$ and $a_{wteta}$ values with the hydro-pneumatic suspension system comparing with the rubber and leaf springs of the suspension systems are shown in Fig.5. Fig. 5 shows that the $a_{wz}$, $a_{wphi}$ and $a_{wteta}$ values with the hydro-pneumatic suspension system are much lower than those of the rubber and leaf springs of the suspension system when vehicle moves at the different speeds. The $a_{wz}$ values with the hydro-pneumatic suspension system reduce by 52.1%, 62.1%, 61.4%, 55.3%, 50.5%, and 44.9% in comparison with the rubber spring of the suspension system and reduce by 53%, 59.8%, 57.8%, 39.6%, 34.8%, and 34% in comparison with the leaf spring of the suspension system and that is similar to the $a_{wphi}$ and $a_{wteta}$ values.

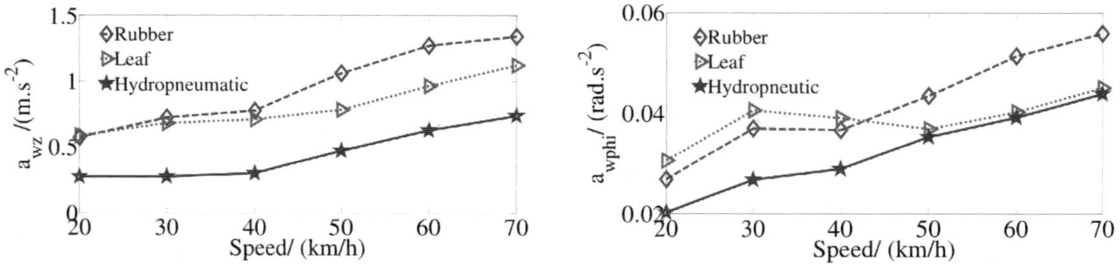


(a) Vertical motion of cab (b) Pitch motion of cab

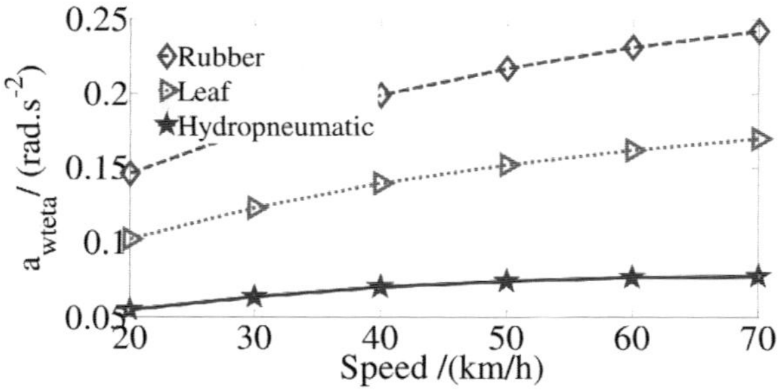


(c) Roll motion of vehicle body

**Fig.5.** The awz, $a_{wphi}$ and $a_{wteta}$ values at the different vehicle speeds

## 4. Conclusions

In this study, the three-dimensional nonlinear dynamic model of heavy truck with 15 DOF is established based on Zhou Changfeng model and the dynamic models of three types of suspension systems such as the hydro- pneumatic, leaf and rubber springs of suspension systems are proposed to compare their effect on vehicle ride comfort. From the analysis results, the conclusions may be drawn as follow: (i) The peak acceleration amplitudes with hydro-pneumatic suspension systems respectively reduce in comparison with leaf and rubber springs of suspension systems and (ii) The $a_{wz}$ $a_{wphi}$ and $a_{wteta}$ values with the hydro-pneumatic suspension system significantly reduce in comparison with the rubber and leaf spring of the suspension systems in all operating conditions of vehicle. Cab ride comfort has been a significant improvement the hydro-pneumatic suspension system.

## Acknowledgments

The work described in this paper was supported by Thai Nguyen University of Technology for a scientific project (T2018-B23)

## References

[1] Ferhat Saglam, Samim Unlusoy, Optimization of ride comfort for a three-axle vehicle equipped with interconnected hydro-pneumatic suspension system, Adv. in Auto. Eng., 1(2018) 1-20
[2] Ibrahim A. Badway, Mohamed Ib. Sokar, et al, Simulation and Control of a Hydro-pneumatic Suspension system, Inter. J. of Sci. & Eng. Resear. 8(2017) 930-935.
[3] Ferhat Saglam, Y. Samim Unlusoy, Analysis of interconnected hydro-pneumatic suspension system for load sharing among heavy vehicle axles. Proceed. of CDSR'16, Canada, 2016.
[4] Tian-hong Luo, Ting-qiong Cui, et al, A improved hydro-pneumatic suspension: mathematical modeling and simulation, Inter. J. of resear. In Mech. Engi. & Tech., IJRMET 5(2015) 37-42.
[5] Shousong Han, Zhiqiang Chao, et al, Research on the effects of hydropneumatic parameters on tracked vehicle ride safety based on cosimulation. Shoc. and Vibr. (2017) 1-10.
[6] Junwei Zhang, Sizhong Chen, et al, Research on modeling of hydropneumatic suspension based on fractional order. Math. Pro. in Engi. (2015) 1-11.

[7] Le Xuan Long, Tran Thi Hong, et al. Performance analysis of the hydro-pneumatic suspension system of heavy truck. Inter. J. of Mech. Engi. and Tech. 9(2018) 1128–1139.
[8] Zhou Changfeng, Dynamic modeling, optimization and experimental research of rubber suspension system of off-road vehicle. PhD thesis: Southeast University, Nanjing, China, 2007.
[9] Le V. Q, Comparing the performance of suspension system of semi-trailer truck with two air suspension systems. Vibr. Pro., 14(2017) 220-226.
[10] Le V. Q., Zhang J., et al, Nonlinear dynamic analysis of interaction between vehicle and road surfaces for 5-axle heavy truck. J. of South. Univer., 27(2011) 405-409.
[11] ISO 8068. Mechanical Vibration-Road Surface Profiles-Reporting of Measured Data. International Organization for Standardization, 1995.
[12] Van Cuong, B., Van Quynh, L., et al, Influence of heavy truck operating condition on dynamic load coefficient. ICERA (2018) 372–379.
[13] ISO 2631-1. Mechanical vibration and shock-evaluation of human exposure to whole-body vibration, Part I: General Requirements. The International Organization for Standardization, 1997
[14] Le Xuan Long, Performance analysis of the hydro-pneumatic suspension system on vehicle ride comfort. Scie. Proj. Thai Nguyen University of Technology. Thai Nguyen, Viet Nam, 2019.
[15] Nguyen Van Liem, Zhang Jianrun, et al. Performance analysis of air suspension system of heavy truck with semi-active fuzzy control, J. of South. Univer. 33(2017) 159-165.
[16] Nguyen Van Liem, Zhang Jianrun, et al, Study of fuzzy control for cab's isolation system of heavy truck. Vibro. Pro. 10(2016) 309-314.
[17] Le Van Quynh, Bui Van Cuong et al, Effects of suspension design parameters of a semi-trailer truck on vehicle ride comfort and road surface friendliness. Adva. In Engi. Re and App. (2019), 281-289.
[18] Le Van Quynh, Vu Thi Hien, et al, Influence of tire parameters of a semi-trailer truck on road surface friendliness. Inter. Re. J. of Engi. and Tech. (IRJET). 6(2019) 3674-3677.

IOP Conf. Series: Materials Science and Engineering **914** (2020) 012038 doi:10.1088/1757-899X/914/1/012038

# Status and Development Trend of Active Sub-arrays Structure Design in Active Phased Array Antenna

**Rui Xiao[*], Shouli Jiang[a], Jianfeng Zhong[b]**

Nanjing Institute of Electronic Technology, Nanjing, China

[*]Corresponding author e-mail: 250242323@qq.com, [a]jsl430@163.com, [b]zhongjianfeng712@163.com

**Abstract**. Active sub-arrays structure design is an important part of phased array radar design. The basic architecture and composition are introduced, especially the different between brick, tile and chip sub-arrays are discussed. Integrated design of radiating unit and sub-array, intelligent health detection system and electromagnetic compatibility, microelectronics and topology optimization design, application of new materials, which considered to be four key technologies, are also pointed out.

## 1. Introduction

With the development of weaponry technology, the speed and penetration capability of aircraft, missiles and other aircraft have been significantly enhanced. Radar, as the core component of battlefield sensing and detection, undertakes important tasks such as air defense and antimissile early warning. The traditional radar using a mechanical scanning system cannot adapt to the multi-warhead, multi-target, saturated attack combat mode of modern weapons. Phased array radar can take advantage of the characteristics of multiple beams without mechanical rotation to achieve tracking of multiple targets and complete a variety of tasks such as search and guidance [1, 2].

The reason why phased array radar can achieve this function is rely on the active sub-array of a large number of internal, arrayed and integrated with thousands of transceiver modules called T/R components. As the core component of a phased array antenna, the active sub-array is the basic unit for radar transmission and reception in terms of electrical performance; from a structural perspective, the purpose of improving the structural design of the active sub-array is to achieve lightweight and modular array maintain. Discussing the integration form of the active sub-array structure and the key technologies in the future, this will be of great significance for improving radar integration and realizing radar light-weighting.

## 2. Basic composition of active sub-array

Active sub-arrays belong to the radar antenna front sub-system. From the layout point of view, it is generally behind the antenna radiation unit. The active sub-array can be interconnected with the antenna radiating unit through the array winding layer according to the needs; at the same time, the back end of the active sub-array needs to be interconnected with digital modules, power supply networks, liquid supply networks, etc. to achieve control, radio frequency, power supply, thermal and other interaction.

The composition of the active sub-array includes T / R components, integrated networks, sub-array power supplies, delay components, and sub-array skeletons. Among them, the T / R component is usually a separately packaged micro-assembly integrated circuit (MMIC) module; the sub-array power supply includes primary and secondary power supplies; the sub-array skeleton often integrates liquid-cooled flow channels or air channels. The sub-array skeleton is not only the structural installation foundation of modules such as T/R components, power supplies, and time-delay components, but also the function of cooling and cooling. This is a typical structure-function integrated component.

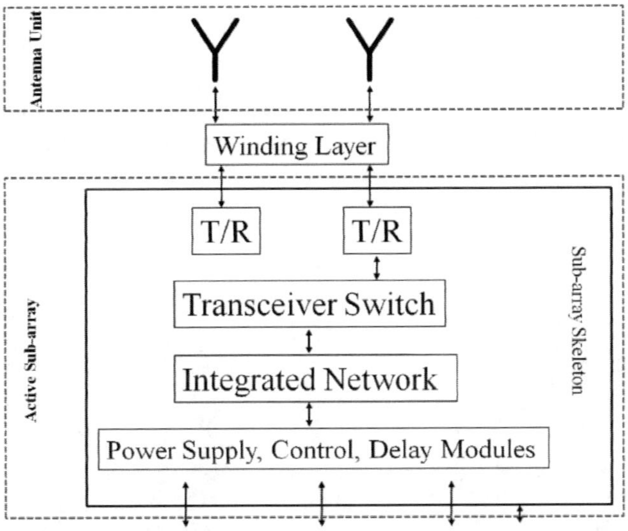


**Figure 1.** Schematic diagram of the composition and position of the active sub-array.

## 3. Basic classification of active sub-arrays

### *3.1. Classification by cooling method*

According to different classification standards, active sub-arrays can be classified differently. For example, according to the cooling method, active sub-arrays can be divided into liquid-cooled active sub-arrays, air-cooled active sub-arrays, and conductive active sub-arrays. Liquid-cooled active sub-arrays have the highest heat dissipation efficiency, but liquid cooling often requires a separate cooling unit and provides reliable front-surface runner channels. The advantages of air-cooled active sub-arrays and conductive natural cooling sub-arrays are that they do not need to be equipped with additional cooling units, and the heat dissipation effect is generally not as good as that of liquid cooling, which leads to increased system temperature and reduced reliability. Active modules such as components are derated, sacrificing some performance.

Many new cooling technologies have also appeared, such as two-phase flow and spray cooling. These technologies are still in the laboratory stage, and there is still a certain distance from engineering applications. The specific cooling method should be separately distinguished and designed according to the weight requirements of the front and the cooling environment conditions.

### *3.2. Classification by architecture system*

In addition to classification based on cooling methods, more sub-array classifications can be divided into brick-type active sub-arrays and tile-type active sub-arrays according to the architectural form. The typical difference between the two is the arrangement of the T / R components: in the brick sub-array, the T / R components are arranged perpendicular to the antenna array; in the tiled sub-array, the T / R components are arranged parallel to the antenna. Generally speaking, brick active sub-arrays can be subdivided into brick-type active sub-arrays based on two-dimensional directions [3] and blade linear arrays based on one-dimensional integration [4]. The number of sub-array channels is divided according to the requirements of the array. The general number of channels is 2n, that is, a sub-array often includes 16, 32, 64 channels. The one-dimensional array pattern after sub-array division is:

$$F(\theta)=(E\bullet A_{SA})\bullet A_P \quad (1)$$

in the formula: E represents the array element pattern;
$A_{SA}$ represents the array element sub-array pattern;
$A_P$ represents the array pattern of the terminal array behind the array [5].

Studies have shown that when the number of elements in the sub-array is too small, this is unfavorable to the weighting operation; if the number of elements is too large, increasing the pitch of the sub-array will cause excessively dense grating lobes at the element-level pattern. In the case of the same cell spacing, the number of channels in the sub-array directly determines the number of sub-arrays, the amount of internal equipment, the economy of the sub-array, and the size of the sub-array. Therefore, the determination of the internal channel number of the sub-array is the result of a joint iteration of telecommunication design and structural design.

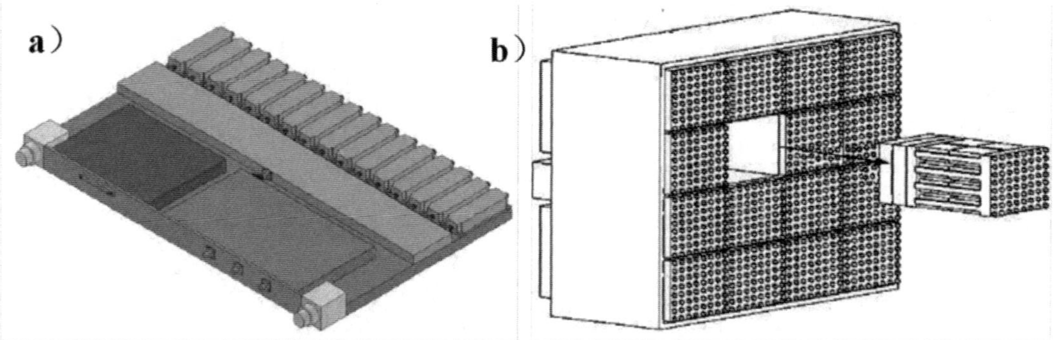


**Figure 2.** a) Blade sub-array based on one-dimensional direction [4]. b) Brick sub-array based on two-dimensional direction [7].

When the radar frequency band is increased to Ka frequency band and above, the antenna unit spacing is reduced, and the traditional brick sub-array size no longer meets the requirements. The tiled sub-array is placed on the stack by the internal modules such as T / R modules, which can effectively reduce the thickness of the front and reduce the weight of the front. It should be noted that although tile sub-arrays can partially realize the lightweight of sub-arrays, tile sub-arrays and chip sub-arrays are not the same concept: tile sub-arrays are still tiled stacks based on SOP packaging modules. The sub-array is a system integration based on the micro-system SIP independent packaging module, which has a higher degree of integration and integration. The realization of the chip sub-array depends on the maturity of the packaging process such as chip-based chip T / R components and TSV based on SIP packaging.

**Figure 3.** Tiled active sub-array [8].

## 4. Active Sub-array Development Trend

With the development of phased array radar in the past ten years, the internal composition of active sub-arrays has become more and more complex, and digital parts such as frequency conversion modules have been moved forward and integrated into the sub-arrays, which places higher requirements on structural design. The four key technologies of the active sub-array that need to be developed are discussed below.

### *4.1. Integrated design of radiating unit and active sub-array*

In a general sense, the sub-array and the antenna radiating unit are physically separated, and the middle is interconnected by a winding layer. With the development of 5G communication, the research of millimeter wave and terahertz radar has been deepened, and the loss and standing wave degradation caused by the interconnection between the sub-array and the antenna radiating unit through the winding layer are no longer accepted. The antenna radiating unit has the original form of a horn and a metal vibrator, and it is developed to a high-frequency printed board vibrator and a patch antenna. The antenna radiating unit and the sub-array need to realize the transition through integrated design integration and blind plug connectors. This design can reduce antenna weight and thickness. Active sub-array and radiating unit integrated design also need to consider the forward and backward maintenance in advance. Generally speaking, the spherical radome does not have high requirements for the forward or backward maintenance of the sub-array, and both can be implemented; the flat-type radome requires that the sub-array can be designed to perform rearward maintenance from the hatch.

During the integrated design of the radiating unit and the active sub-array, higher requirements are placed on the interconnection method within the sub-array. Conventional SMP, BMA and other blind plug forms, or SMA cable assembly link forms are gradually replaced by SSMP \ CSMP, and the use of wool buttons to achieve inter-plane interconnection is gradually being applied. No matter what kind of interconnection is mentioned above, interconnection reliability (especially reliability under comprehensive stress environment) needs to be taken seriously.

### *4.2. Intelligent health monitoring system and electromagnetic compatibility design*

The active sub-array integrates many sensitive modules such as T / R components. The reliability of these modules is largely dependent on temperature. These modules also need to avoid risk factors such as leakage. This means that the health monitoring of sub-arrays and the detection of hidden dangers to avoid greater losses need to be promoted to an increasingly important position. The first is the monitoring and feedback of the sub-array temperature. According to the classic theory of electronic modules, the reliability decreases by 50% during every 10 ° C increase in temperature. During the structural design of the active sub-array, the temperature sensor is usually placed in the water inlet and outlet of the sub-array hanger. The temperature and temperature difference of the water inlet and outlet are monitored to adjust the liquid supply situation of the entire array to improve the overall temperature consistency of the array. In addition to the external temperature sensor of the sub-array, the multi-function chip inside the T / R module sometimes integrates the temperature detection function as required. This integration can more accurately reflect the temperature information of the single channel inside the module, and the back-end controller can intuitively read it. Security situation of the entire sub-array and antenna.

In addition to temperature detection, multiple leak detection sensors will be integrated in the sub-arrays of the blindly inserted interconnects, which can detect early and detect the presence of water in the array in time to avoid short circuits caused by leaks or water. At present, leak detection is generally collected through leaks, and only when there is fluid accumulation. In this case, the sub-array leakage has occurred for some time and may have caused losses. Therefore, where to arrange in the sub-array and through what kind of structure to achieve rapid collection of leaks to reach the alarm threshold as soon as possible, this tests the experience and wisdom of the sub-array structural designer. Temperature and leakage are typical health monitoring items. Now when the radiating unit is integrated with the sub-array, in order to monitor the deformation of the antenna radiating unit, mechanical and deformation

IOP Conf. Series: Materials Science and Engineering **914** (2020) 012038 doi:10.1088/1757-899X/914/1/012038

monitoring sensors are also integrated inside the sub-array. It is common to attach strain gauges. Through the monitoring of stress and strain, it is timely found whether the sub-array has abnormal damage during the service process, and real-time electrical signal compensation based on the deformation of the array is realized through back-end control according to this [9].

### *4.3. Application of micro-system and topology optimization in sub-array design*

In recent years, micro-systems have received extensive attention in the design of phased array radars [10], and the core of which is the integrated implementation of 3D heterogeneous heterogeneity [11]. In the active sub-array, the module system-level package (SOP) such as T / R components is developed to a micro-system-based sub-array system-level package (SIP). The sub-array-level SIP system integration is different from the SOP package of the component: the SIP-based integration method is for the pursuit of functional and application integrity, which is consistent with the sub-array's complete telecommunications and structural function requirements, especially the sub-array and patch antenna units Integrated design.

The large number of applications of TSV technology in micro-system architecture put forward higher requirements for thermal stress modeling and topology optimization, and thermal simulation and optimization represented by micro-channels [12].

## 5. Conclusion

In this paper, the common architecture, classification, and composition of current phased array radar active sub-arrays are introduced, and the trend of subsequent sub-array structure design is proposed, and the matters needing attention are summarized. The structural design of the phased array radar active sub-array needs to be involved in the initial arrangement of the antenna array. It is determined together with the array scheme, and the optimal cooling form and layout scheme are selected. At the same time, the electromagnetic compatibility of each internal module is considered in advance. As the main hardware component of the antenna array, the active sub-array is not only a mechanical arrangement in the integrated design, but also needs to solve the problems of heat, force and electromagnetic compatibility. The design attention points and ideas mentioned in the article are not only applicable to the current brick, tile, or chip sub-arrays, but also have reference for the sub-array and front design of the highly integrated micro-system architecture.

## References

[1] Fan H T, Yan J. Development and outlook of active electronically scanned array guidance technology, Acta Aeronautica et Astronautica Sinica, 2015, 36(9):2807-2814.

[2] SHAO Chunsheng. Study Status and Development Trend of Phased Array Radar, Modern Radar,2016, 38(6):1-4+12.

[3] Tang Baofu, Zhong Jianfeng, Gu Yeqing. Design of Active Phased Array Radar Antenna, first, Xi'an, 2014( in Chinese).

[4] Cui Min, Zhang Juan, Design of a C-band Transmitting-receiving Sub-array System, Fire Control Radar Technology, 2014, 43(2):72-76.

[5] Arik D. Brown. Electronically ScannedArrays: MATLAB Modeling and Simulation, CRC Press: New York, 2010.

[6] CHEN Jian-feng, WU Hai. Research into Subarray Division Technology Based on Overlapped Subarray Architecture, Shipboard Electronic Countermeasure, 2016, 39(1): 95-98.

[7] Gu Ye-qing, Yao Ye, Wang Chao, Integration Design of Active Phased Array Antenna, Electro-Mechanical Engineering, 2016,32(6):29-32,40.

[8] Li Gang, Li Tao, Research on Smart Skin Technology of Tile Radar in Helicopter, Fire Control Radar Technology, 2018, 47(01): 28-31.

[9] Tang Baofu, Gu Yeqing, Wang Chao, A Study on the Applications of Intelligent Structure in Phased Array Antenna, Modern Radar, 2014,36(11): 8-11.

[10] Tang Weiqiang, Applications of Microsystem Technologies in Radar System and Related

Oversea Research Status, Electro-Mechanical Engineering, 2016,32(6):21-23+28.
[11] QIN H C,YAN Z W, SU D L, et al. Electromagnetic susceptibility analysis method for 3D TSV ICs,Journal of Beijing University of Aeronautics and Astronautics, 2017, 43(12):2406-2415.
[12] Ran Hong-lei, Peng Hao, Huang Jie,Research on TSV Technology in 3D Packaging Microsystem, Electronics Quality, 2018,12,111-115

IOP Conf. Series: Materials Science and Engineering **914** (2020) 012039 doi:10.1088/1757-899X/914/1/012039

# Health Management Information System Based On Regression Analysis

**Dong Wang[1, *], Xiujuan Chen[2, a], Honghe Wei[3, b]**

[1]Electrical and Information Engineering, Shanghai Urban Construction Vocational College, Shanghai, 201415, China
[2]Technical Section, Winning Health Technology Group Co., Ltd,Shanghai, 300253, China
[3]Shanghai Research and Development Center, ZTE Corporation, Shanghai, 201203, China

[*]Corresponding author e-mail: wd_lulu@163.com, [a]chenxj@winning.com.cn, [b]wei.honghe@zte.com.cn

**Abstract**. With the improvement of living standard, people's demand for health is increasing. However, the rapid pace of urban life makes it difficult for most people to stop and pay close attention to their bodies. Even after the hospital examination, it is difficult to achieve stable follow-up and health judgment. On the basis of regression analysis of data, an open source mobile terminal Android platform is used to design a health management information system (APP), which can realize the information sharing between residents and hospitals. The system allows people to know their physical condition instantly, and hospitals can track and advise users.

## 1. Introduction

With the continuous improvement of living standards and the rapid pace of society, health problems become a steady burden for the country and individuals. Health management information system is a software for comprehensive management of health data of individuals and people. The purpose is to share the health information of patients and healthy people with hospitals, so that patients and healthy people can have better health, and effectively reduce medical expenditure [1-2].

Most of the traditional health management information systems are based on the web page form of B/S architecture. It is not very convenient for users to view and understand their health information anytime and anywhere, which not only affects the user experience, but also lacks flexibility and practicability, and cannot achieve a good real-time communication between users and hospitals.

Nowadays, with the maturity of communication technology, mobile terminal is no longer limited to the terminal of communication network, but has become the terminal of Internet. Therefore, the application software and demand service of mobile terminal have a huge development space. The health management information system app based on regression analysis proposed in this paper is exactly based on Android system, adopting technologies such as separation of cloud storage and mobile client, introducing information sharing to realize seamless interaction between system server and client, and providing intelligent software for health information management [3-4].

IOP Conf. Series: Materials Science and Engineering **914** (2020) 012039 doi:10.1088/1757-899X/914/1/012039

## 2. System function

The health management information system based on regression analysis can automatically collect, control and process the data submitted by users, and then generate relevant health prompts through the judgment of hospital doctors. It mainly includes four modules: user login module, daily work module, statistical analysis module and quality control module.

For example, the incidence of hypertension is affected by the following five aspects: triglycerides, serum high-density lipoprotein, serum low-density lipoprotein, systolic blood pressure and diastolic blood pressure. Then when the user uploads the above five aspects of data, the system calculates the possibility of the disease through a specific algorithm. The system also classifies the incidence possibility into three levels: no hypertension, hypertension level 1, hypertension Level 2 and hypertension Level 3, and matches the symptoms with the doctor's suggestions one by one, and then sends them to the user, so as to effectively improve the people's healthy life of early treatment and early prevention.

This system takes android as the platform, uses web services, Java and other technologies to realize the function of information sharing between users and hospitals.

## 3. System implementation

### *3.1. User login*

After the Android client is started, the user is required to input the account and password. Because the software interacts with the cloud, when clicking the login button on the Internet, the system will automatically detect the account and password and send them to the cloud for encryption. If login fails, users will be given feedback of login failure; otherwise, users will be given feedback of login success.

### *3.2. Routine*

After logging into the system, the hospital operator and user can see two different interfaces respectively. As shown in Figure 1, the hospital operator interface has four functions: "Analysis of data ", "Userinfo", "Routine test" and "Suggest". On the user interface, there are four functions: "Upload data", "Health status", " Routine test" and "Doctor's advice".

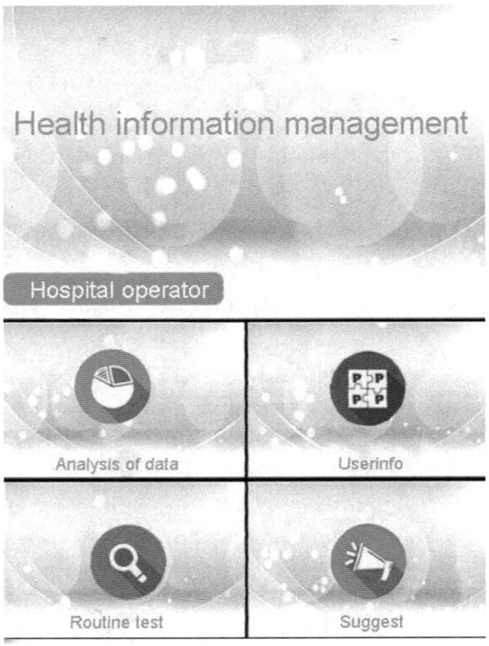


**Figure 1.** Hospital Operators' Interface

IOP Conf. Series: Materials Science and Engineering **914** (2020) 012039 doi:10.1088/1757-899X/914/1/012039

In the concrete implementation, we need to encapsulate the service address, namespace, method name and parameter value of the calling web service. Generate soap request information and call specific methods to get the return result. Some codes are as follows:

```
SoapObjiect jiemian=new SoapObjiect (nameSpace,methodName);
setProperties(jiemian);
envelope.bodyOut=jiemian;
Envelope.dotNet=true;
Envelope.setOutputSoapObjiect(jiemian);
```

During the data reading process, the access to MySQL database is applied. Some codes are as follows:

```
public class sqlconn
{
	public static void fetchdata(Connection conn)  // Function to read data
	{
		try
		{
			Statement statement = conn.createStatement();  // Using statement to execute SQL statement
			String sql = "select * from emp";  // Querying the contents of table emp with SQL
			ResultSet rs = statement.executeQuery(sql);  //Used to return results
			System.out.println("name" + "\t\t" + "pathogeny " + " physician-in-charge " + "upload date ");
			String date = null;
			while(rs.next()){
				job = rs.getString("job");

				System.out.println(id + "\t\t" + job + "\t" + date);
			}
			rs.close();
		}catch(SQLException e){
			e.printStackTrace();
		}catch (Exception e) {
			e.printStackTrace();
		}finally{
				System.out.println("Read data successfully!"+"\n");
		}
	}
```

### *3.3. Prediction of hypertension*

This system can monitor and predict hypertension. Hypertension is one of the most common chronic diseases. If this disease is not found and treated in time, it will easily cause cardiovascular and cerebrovascular diseases, with great harm [5].

There is a standard classification of blood pressure level in literature [1]. The classification is as follows: (1) normal blood pressure, systolic blood pressure < 120 mmHg and diastolic blood pressure < 80 mmHg; (2) normal high value, systolic blood pressure 120 ~ 139 mmHg and / or diastolic blood pressure 80 ~ 89 mmHg; (3) grade 1 hypertension (mild), systolic blood pressure 140 ~ 159 mmHg and / or diastolic blood pressure 90 ~ 99 mmHg; (4) grade 2 hypertension (moderate), systolic blood pressure 160 ~ 179 MmHg and / or diastolic blood pressure 100-109 mmHg; (5) grade 3 hypertension (severe), systolic blood pressure ≥ 180 mmHg and / or diastolic blood pressure ≥ 110 mmHg; (6) simple systolic hypertension, systolic blood pressure ≥ 140 mmHg, diastolic blood pressure < 90 mmHg.

According to the cardiovascular risk factors affecting the risk level of hypertension analyzed in literature [1], the following laboratory examination attributes are mainly concerned: triglyceride (TG), serum high-density lipoprotein (HDL), serum low-density lipoprotein (LDL), systolic blood pressure (sys), diastolic blood pressure (DLA).

In this paper, the laboratory examination attribute values of hypertension in the hospital are obtained for analysis. Here, only part of the data are taken for explanation, as shown in Figure 2.

| A | B | C | D | E | F |
|---|---|---|---|---|---|
| vid | SYS | DLA | TG | HDL | LDL |
| aad3e | 157 | 100 | 1.97 | 0.94 | 2.66 |
| 42189 | 103 | 65 | 1.1 | 1.46 | 4.18 |
| 9a7b5 | 146 | 80 | 2.12 | 1.07 | 2.89 |
| a238d | 152 | 90 | 2.15 | 1.22 | 1.87 |
| 6335a | 110 | 64 | 0.76 | 1.7 | 3.09 |
| 9c33d | 113 | 78 | 0.89 | 1.66 | 2.63 |
| 90983 | 128 | 69 | 1.43 | 1.22 | 2.31 |
| 472cd | 94 | 52 | 0.84 | 1.02 | 2.21 |
| 750a1 | 141 | 92 | 1.63 | 1.3 | 3.48 |
| d469d | 103 | 62 | 0.53 | 2.39 | 2.82 |
| 2fcdb | 143 | 87 | 1.49 | 1.61 | 3.87 |
| 01315 | 114 | 64 | 0.97 | 2.17 | 4.58 |
| 99906 | 145 | 81 | 1.78 | 1.42 | 3.02 |
| 566b5 | 119 | 79 | 1.04 | 1.21 | 2.9 |
| c83ed | 120 | 81 | 0.81 | 1.13 | 3.18 |
| de5d7 | 117 | 72 | 0.71 | 1.52 | 2.57 |
| a88bf | 108 | 70 | 0.62 | 1.88 | 2.4 |
| 4f09b | 140 | 90 | 1.92 | 1.06 | 2.95 |
| 25a6b | 121 | 83 | 1.36 | 1.35 | 2.3 |
| b5e85 | 110 | 66 | 0.67 | 1.53 | 1.42 |
| 97052 | 113 | 73 | 0.65 | 1.44 | 1.95 |
| a6522 | 115 | 66 | 0.98 | 1.63 | 2.33 |
| 6da4c | 166 | 108 | 2.11 | 1.64 | 5.19 |
| 357b1 | 106 | 68 | 0.94 | 1.25 | 2.69 |
| 3dd93 | 147 | 103 | 2.14 | 0.96 | 2.77 |
| d4e2b | 119 | 78 | 0.8 | 1.12 | 2.47 |
| 54a2c | 111 | 72 | 1.03 | 0.94 | 2.64 |
| 9f13d | 112 | 71 | 1.08 | 1.22 | 3.21 |
| 81b32 | 103 | 68 | 0.68 | 1.8 | 2.83 |
| e88b6 | 121 | 58 | 2.42 | 1.02 | 3.26 |

**Figure 2.** Partial Data on Factors Affecting Hypertension

Regression analysis can be used to deal with the data of influencing factors of hypertension, and calculate the relationship between systolic blood pressure and serum triglyceride. The calculation method of other factors is similar to it.

1.Calculate the value of $\bar{X}$, $\bar{Y}$, $lxx$, $lyy$, $lyy$, where X is TG and Y is SYS.

$$\begin{aligned}
&\bar{X}=1.25, \bar{Y}=123.23, \sum X=37.62, \\
&\sum Y=3697, \sum X^{2}=56.69, \sum Y^{2}=465653 \\
&\sum XY=4893.28 \\
&lxx=\sum X^{2}-\left(\sum X\right)^{2}/n=56.69-37.62^{2}/30=9.51 \\
&lyy=\sum Y^{2}-\left(\sum Y^{2}/n\right)=465653-(465653/30)=450131.23 \\
&lxy=\sum XY-\left(\sum Y*\sum X/n\right)=4893.28-(37.62*3697/30)=257.24
\end{aligned} \tag{1}$$

2.Calculate the regression coefficient b and intercept a.

$$\begin{aligned}
&b=\frac{lxy}{lxx}=\frac{257.24}{9.51}=27.05 \\
&a=\bar{Y}-b\bar{X}=123.23-27.05*1.25=89.42
\end{aligned} \tag{2}$$

3.List regression equations.

$$\hat{Y} = a + bX = 89.42 + 27.05X \tag{3}$$

4.Draw regression equation chart, as shown in Figure 3.

Figure 3 shows the relationship between systolic blood pressure and serum triglyceride. The abscissa is serum triglyceride and the ordinate is systolic blood pressure. It can be seen that systolic blood pressure and serum triglyceride are basically linear. It can be seen from the figure that with the increase of serum triglyceride, the systolic pressure also increases, so serum triglyceride has an impact on the incidence of hypertension.

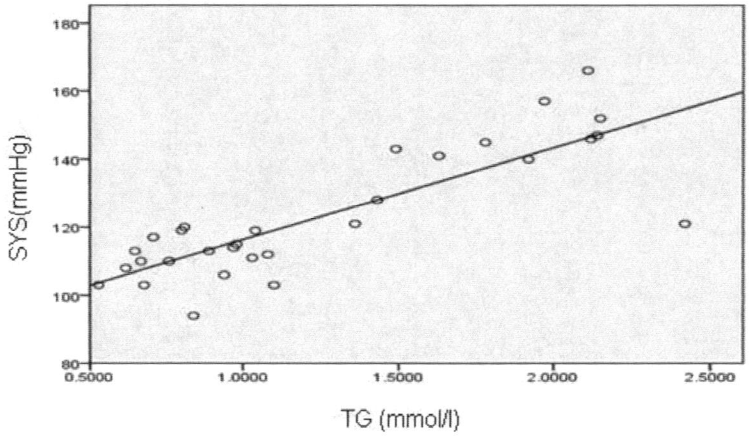


**Figure 3.** Relationship Between Systolic Blood Pressure and Serum Triglycerides

By calculating the relationship between examination attributes, we know that the influence of examination attributes on hypertension tends to be consistent. The higher the combination of all factors, the greater the possibility of suffering from hypertension. According to the possibility of different factor gradients, the system makes a comprehensive judgment, infers the grade of hypertension, matches the symptoms with the doctor's orders, and then transmits them to the client. At the same time of information sharing, the health goal of early medical treatment and early prevention should be achieved.

## 4. Conclusion

This system is mainly for information sharing between users and hospitals. Compared with the traditional mode, it is more conducive to information synchronization and interaction. The health management system based on regression analysis has a relatively large development and promotion space in the future, which is conducive to people's healthy life.

## Acknowledgments

This paper is sponsored by the research funds of Shanghai Urban Construction Vocational College to introduce high-level talents (No. CJXY-2018-011), and the Educational Science Research Project of Shanghai Urban Construction Vocational College in 2019 (No. cjyb201904).

## References

[1] China High Blood Pressure Prevention Guidelines Revision Committee. Guidelines for the prevention and treatment of hypertension 2010[J]. Chinese Journal of Hypertension, 2011,19(8):701-743.

[2] ZHU Hua-Jian, TAN Hai-Bo, ZHAO He, MA Zu-Zhang. Font End Operation Application of

Health Management Information System Based on Android Platform[J]. Computer Systems & Applications, 2015,9: 61-65
[3] ZHOU Guang-Qing, WU Shu-Yu, XUE Bing-Bing. et al. Design and Implementation of Community Health Management Platform Based on Internet of Things[J]. China Computer & Communication, 2014,35: 53-56
[4] DING Chong, ZHU Jie. Design and Implementation of the Chronic Disease Monitoring System Based on the Public Health Platform[J]. China Digital Medicine, 2017,4: 43-46
[5] LIN Ling-Ling. Research on Extracting Classification Rules of Hypertension Based on C4.5 Algorithm[D]. Taiyuan University of Technology, 2012

IOP Conf. Series: Materials Science and Engineering **914** (2020) 012040 doi:10.1088/1757-899X/914/1/012040

# Research on Design and Application based on Bridge Inspection and Maintenance Scheme

**Lingye Tan[1,*], Ziyang Zhang[1,a], Xin Wang[2,b]**

[1]Faculty of Engineering, University of new south wales, New South Wales, 2032, Australia
[2]School of Environment and Civil Engineering, Jiangnan University, Wuxi, 214000, China

[*]Corresponding author e-mail: tanlingyet@126.com, [a]zzy967230@gmial.com, [b]1101180128@stu.jiangnan.edu.cn

**Abstract**. Under the background of the rapid economic and social progress and development of the market economy in China, the cause of traffic construction has also been built and carried out like a raging fire. there is also a direct relationship between traffic engineering construction and the rapid development of science and technology and the national economy, which plays important technical support and driving role in the development of social and national scientific and technological and economic traffic engineering construction. Under the background of the current market economy with the rapid development of science and technology and the economy, a large number of large-scale bridge designs and buildings have been put into use and obtained the latest bridge scientific detection technology. in the process of bridge construction, the structural complexity of the detection technology and the span of the bridge have been improved to a great extent. This paper mainly discusses the bridge detection technology, specific maintenance, and bridge reinforcement technology in the process of design and construction of all kinds of large bridges, and makes a technical analysis on the maintenance life of all kinds of large bridges, to increase the safety and maintenance life of all kinds of large bridges and promote the realization of their greater economic and social benefits and value at the same time.

## 1. Introduction

Based on the continuous and rapid development of China's bridge construction, the number of bridges is increasing, which effectively relieves the pressure of highway traffic, promotes modernization, and improves the operation and management capabilities of vehicles. And, bridge inspection technology is also increasing. China has further strengthened the management process of highway and bridge construction. In order to better explore the application of bridge inspection technology.

Based on the analysis of the implementation of highway bridge building reinforcement and maintenance, this paper discusses the widely used detection technology of highway bridges, strengthens the reinforcement and maintenance evaluation of highway bridges, and puts forward more ideal design schemes, detection and Maintain highway bridges.

## 2. Analysis and Application of Bridge Detection Technology at present

### *2.1. Evaluation of bridge reliability*

After the completion of the construction of the main structure, the bridge project may be affected by the interaction of a variety of natural factors in the whole process of its operation. in many cases, the corrosion of bridge steel bar, the aging of bridge concrete and the fracture of bridge deck caused by an earthquake may occur directly. To effectively improve the reliability and service life of the main structure bridge, the reliability degree of the main bridge structure should be evaluated comprehensively under the guidance of the theory of modern bridge structure detection technology. to see whether the actual bearing capacity of the current bridge main structure is consistent with the technical requirements for bridge reliability design and whether the bridge main structure can fully meet the needs of normal bridge operation functions. And according to the specific operation status of the bridge to formulate and design an effective bridge reinforcement and maintenance plan. The design and formulation of the bridge maintenance and maintenance bridge reinforcement plan should be carried out according to the needs of the actual bridge structure evaluation. Under the guidance of the theory of the modern bridge reliability design scheme, the bridge structure and route should be re-planned and designed and the structure should be selected to minimize the damage caused by impact to the main structure bridge. The main consideration in the process of bridge design for reliability evaluation of bridge reliability is the reliable detection of bridge load and main structure resistance. According to the existing reliability evaluation and design methods and statistical standards of bridge engineering's main structure, the calculation formula of the random calculation process model of bridge main structure load and resistance is obtained. The structural material of the bridge is a uniform entity that exists in practical application, and each structural variable in the process of structural design is also fixed, so it is impossible to have any random variable change phenomenon. the calculation and acquisition of the values of each variable can be obtained by theoretical and experimental methods. Because the bridge materials selected and used in bridge structure design are non-uniform to a certain extent, they will be different in the structural properties of bridge materials to a certain extent. the specific uniformity is a process that increases and attenuates gradually with the strength and resistance of the bridge materials used to the structure [2].

### *2.2. Analysis of concrete content of Bridge Detection*

To provide the most basic transportation infrastructure and its construction guarantee for the sustained and steady development of China's socialist market economy and the deepening and development of reform and development in the new historical period. It provides a guarantee for the normal traffic travel of the people and the masses and maintaining the harmony and development security of the socialist economy, therefore, our country has further strengthened the research, establishment, and perfection of a modern transportation technology testing system in the construction and development of transportation in our country. The state strengthens the setting of bridge testing service content in our country, which includes two parts: regular testing and detailed testing. Regular inspection mainly refers to the regular maintenance of the basic situation and components of the bridge in the process of construction and use of the bridge at a certain date and time after the completion of the construction and use of large bridges in our country. Either one year or several years, professionals have been organized to repair and maintain the basic testing institutions, construction and use of large bridges in our country, as well as parts. Based on the comprehensive detection and analysis of the instrument and its basic structure and components, the main cause of the damage is accurately found and repaired. Secondly, the material is repaired in detail. in the process of inspection, it is found that the main causes of the damage of the bridge structure are comprehensively analyzed after the stability and damage degree of the bridge structure reaches a certain degree. the methods of drilling core and non-destructive testing are used to strengthen the risk assessment and detection of damaged bridge structures in the process of construction. In the aspect of evaluation and testing items, firstly, the overall damage of the material in the bridge and the structural stability of the material in the bridge are analyzed, and the specific damage and damage of the material in the bridge structure are obtained. the physical properties and dynamic properties of the bridge during

IOP Conf. Series: Materials Science and Engineering **914** (2020) 012040 doi:10.1088/1757-899X/914/1/012040

the construction of the material in the bridge structure are measured respectively. The chemical properties of the bridge structure of the damaged materials and the corrosion resistance mechanical properties of the bridge were tested respectively. After that, the overall damage performance of the material in the bridge structure and the stability of its operation in the bridge are analyzed. the specific contents include the stability and geometric shape of the material in the bridge structure, the inherent modal parameters of the bridge structure and the internal forces that directly affect the bridge members.

## 3. Analysis and Application of current Bridge maintenance and reinforcement Technology

As the society and the country attach great importance to the development of this new material technology and strengthen its environmental protection, in the structural and construction material design of concrete bridges, we use a new type of carbon fiber composite construction material with carbon fiber resin as reinforced plastics. The carbon-fiber structure of concrete bridge can not only play a strong stability and reinforcement role to the test beam but also effectively promote the renewal and reconstruction of concrete bridge engineering. This new type of carbon fiber composite is composed of resin and millions of new carbon fiber common materials, and the length and diameter of the fiber can generally reach 6-3 pm.

Therefore, under the premise that the FRP of carbon fiber has a certain tension compensation, the fiber can generally withstand a strong load-bearing force, which urges the carbon fiber resin to transfer the stress of the test beam directly to the neighboring fiber, and at the same time effectively control and protect the position and environment of the fiber itself, to prevent the fiber from being directly damaged by the surrounding harmful gases and ions. Under this ultimate stress condition, according to the stability and failure characteristics of the reinforcement test beam, the fiber of the steel bar under stress can yield easily, and the carbon fiber fracture phenomenon will occur at the same time of the stress failure phenomenon of the test beam, so in the current code for the structural design of concrete bridges, The bending stability and strength of the strengthened test beam are also effectively calculated by breaking the carbon fiber of the steel bar. To promote the national support and guidance for the improvement and promotion of bridge construction quality, it should be carried out strictly by the requirements of the construction process of carbon fiber cloth leveling rubber paste reinforcement in the process of bridge base construction maintenance and reinforcement. the specific steps of the technical analysis scheme are as follows. At present, the main fields of construction technology analysis and technical application of bridge base maintenance and leveling rubber reinforcement in our country include bridge base maintenance and reinforcement construction, bridge bottom glue construction, carbon fiber cloth leveling rubber reinforcement construction in the process of bridge construction and carbon fiber cloth leveling rubber paste reinforcement construction in the process of bridge maintenance and reinforcement. The following article makes some corresponding technical analyses of the bridge from these four technical aspects.

### *3.1. Construction of bridge base*

In the process of bridge concrete construction, a certain degree of depression and inferior layer phenomena such as corrosion, honeycomb and spalling may occur in the treatment and performance of concrete, which should be found and dealt with in time. When the phenomenon of the inferior layer becomes more serious, we can consider using the construction method of polymer cement mortar to repair the concrete effectively after the concrete is eradicated. When there are some cracks in concrete, the construction of concrete sealing members can be carried out in these places. The construction methods of cement grinding wheel and concrete angle grinder can also be used to clean up the oil pollution that may occur in the performance of concrete in the construction of leveling members, as well as concrete such as cement floating paste. after that, the part of the concrete performance on the concrete surface of the flat member is effectively polished, and the part of the concrete slightly protruding from the surface of the flat member is polished in a circular arc focusing on the corner. The key arc concrete of the chamfer can be polished at other locations where the concrete corner is pasted. After the leveling member has completed the above process, a hairdryer is used to effectively clean the performance of the concrete. [5]

### *3.2. In terms of bridge primer coating*
In the process of bridge concrete construction, according to a certain mixing ratio, the curing agent and the main agent of concrete are uniformly mixed under the action of the mixing container, and the curing agent is evenly mixed under the action of the mixer. According to the different air temperature in the working environment of the specific bridge, different amount of curing agent should be used, and the mixing time of the curing agent should be controlled accordingly. The bridge concrete coating and surface components are evenly smeared on the drum under the interaction of the concrete brush and the drum brush, and the coating thickness of the glue is appropriate between 0.4mm. To give full play to its role, the phenomenon of flow and leakage of the mixing glue should be avoided as far as possible, and the mixing work of the bridge construction should be started after the mixing glue is completely solidified [4]. To give full play to its role, we should try to avoid the phenomenon of flow and leakage of the mixing glue, and then start the bridge construction mixing work after the mixing glue is completely solidified.

## 4. Conclusion
The construction of infrastructure and the growth of its national economy are inextricably linked, and the early construction and maintenance of highways and bridges are an important part of the construction of social and national public transport infrastructure. As society and the country attach great importance to the cause of public transport, the cause of public transport in various countries and regions has been significantly promoted and developed. the research and development and application of technologies related to the pre-detection and maintenance of highways and bridges have also been promoted and developed rapidly. In the process of construction and inspection of highway bridges, we should design and carry out strictly by the basic principles of its correctness, scientificness, and practicability. To effectively prolong the construction and service life of highway bridges, we should actively develop and design reasonable plans for the construction and maintenance of highway bridges. Scientific formulation and strict compliance with the construction technology of highway bridges can finally achieve the balance and unity of scientific and technical benefits and social and economic benefits achieved in the construction and construction of highway bridges.

## References
[1] Zhang Jianguo. Application and analysis of bridge inspection and maintenance scheme [J]. Engineering Technology: abstract Edition, 2016 (4): 0021500215.

[2] Wang Jian. Application and analysis of bridge inspection and maintenance scheme [J]. Science and Technology and Enterprises, 2016 (1): 134,135.

[3] Guo Xia. Application and analysis of bridge inspection and maintenance scheme [J]. Engineering and Technology: citation Edition, 2016 (5): 00128-00128.

[4] Zhang Zemin. Detection and Evaluation method and Application Analysis of large Urban Bridges-taking Yanshan overpass in Jinan as an example [D]. Shandong University, 2016

[5] He Shuanhai, Zhao Xiangmo, Ma Jian, Zhao Yu, Song Huansheng, Song Hongxun, Cheng Lei, Yuan Zhuoya, Huang Fuwei, Zhang Jian, Tian Bin, Wang Luyang, Qi Xiuzhen. Summary of Highway Bridge Inspection and Evaluation Technology [J]. 2017, 30 (11): 63-80.

IOP Conf. Series: Materials Science and Engineering **914** (2020) 012041 doi:10.1088/1757-899X/914/1/012041

# Influence research of weak layer on slope sliding mode and stability

**Ang Li[a], Wenli Li**

School of Engineering, Huanghe Science and Technology College, Zhengzhou, Henan 450000, China

[a]293513627@qq.com

**Abstract**. Weak layer is an important factor leading to slope instability, which cannot be ignored in slope stability analysis. In order to explore the influence of weak layers on the slope slip mode and stability of the slope, this paper selects typical open-pit mine slopes containing weak layers as the engineering background, and comprehensively uses geological survey, laboratory tests, theoretical analysis and other methods to verify the characteristics of slope geological structure and physical mechanical properties of soil, at the same time, uses simplified Bishop method to analyze the change law of slope stability under the change of weak layer dip angle, and finally clarifies the influence of weak layer on slope slip mode and stability. The results show that: the initial dip angle of the weak layer is 6°, the corresponding stability factor is 1.553 and the initial landslide mode is the combined failure of shear arc and weak layer. As the dip angle of the weak layer increases, the slope stability coefficient decreases. When the dip angle of the weak layer increases to 56°, the slope slip mode is transformed into the translational failure formed by the weak layer, and the stability factor is reduced to 1.055 at this time.

## 1. Introduction

Weak layer is the weak structural surface in the slope body, due to its weak mechanical properties, it is an important disaster-causing factor that cannot be ignored in slope stability analysis [1]. In the current slope engineering practice, the landslide disasters caused by weak layers are numerous. Therefore, it is great significance to ascertain the influence of weak layers on slope slip patterns and stability for preventing landslide disasters and reducing landslide disaster losses [2].

At this stage, a large number of scholars have carried out research work on the influence of weak layers on slope stability. Huang Runqiu [3] analyzed large-scale landslide accidents in the field of slope engineering in China since the 21st century, and proposed that weak layers are an important factor inducing slope landslides. Ma Jingjia [4] clarified the influence law of weak layer on slope stability by means of mechanical theory analysis, and realized the optimized design of open-pit mine slope shape. Xia Dong [5] used the typical open-pit mine slope as the engineering background, clarified the creep characteristics of the weak layer by means of laboratory tests, and proposed a long-term strength and stability analysis method for the weak layer. Wang Dong [6] explored the influence of the weak slope along the slope on the slope slip pattern through numerical simulation technology on the basis of analyzing the characteristics of weak layers, and optimized the shape of the slope. Xiao Xian [7] analyzed the factors influencing the stability of slopes with weak layers, and explored the influence of weak layer morphology on slope slip patterns and stability through improved genetic algorithms. Obviously, a large number of scholars have conducted extensive research in this field and achieved many scientific research results, but there are still certain deficiencies in this research field. The influence of the weak layer on the slope stability is extremely complex and is controlled by many factors. At the present stage, the research results have relatively little research on the influence of the weak layer dip angle on the slope stability, the instability and damage of the slope is precisely affected by the weak layer.

In response to this problem, this paper takes the typical open-pit mine slope as the engineering background, explores the slope slip mode and stability change law by changing the weak layer inclination angle, and provides important scientific means for the open-pit mine slope stability analysis and landslide disaster management Theoretical basis. The research results in this paper are of guiding significance for the study of slope stability with weak layers, and have theoretical analysis significance and engineering reference value for slope stability analysis of open-pit mines.

## 2. Overview of engineering geology

This article selects the south slope of Shengli East 2nd Open-pit Mine as the engineering background. The slope engineering overview is as follows.

### *2.1. Project overview*

Shengli East 2nd Open-pit Mine is located in the central and eastern part of Shengli Coalfield. Its main structural form is NE-SW, and the two wings are asymmetrical and wide syncline. The mining area is 49.63km2, with an average length of 8.0km from east to west and an average width of 6.6km from north to south.

The designed production capacity of the first phase of the open pit mine is 10Mt/a, and the designed production capacity of the second phase is 30Mt/a. The southern mining boundary is bounded by the coal seam outcrops and the F8 and F8-1 faults, and is checked level by level based on the boundary stripping ratio of $6m^3/t$. The north, east and west sides all use the mineral rights boundary approved by the Ministry of Land and Resources as the surface boundary, and the deep boundary is the No. 6 coal seam floor (partially 6 under 2 coal seam floor). The recoverable reserves in the mining boundary are 3821.98Mt, the recoverable raw coal is 4024.80Mt, and the average stripping ratio is $2.93m^3/t$.

### *2.2. Topography*

Shengli Coalfield is located in the middle and eastern part of the Inner Mongolia Plateau, on the northern slope of the west of the Greater Xing'an Mountains. It is a plateau hilly terrain with an elevation of 970.0m ~ 1326.6m and a relative height difference of 356.6m. The geomorphology in the area is composed of four geomorphic units: structural denuded topography, eroded accumulation topography, eroded accumulation topography, and lava platform. Shengli East No. 2 Opencast Mining Area is located in the transition zone between the eroded accumulation terrain and low gentle hilly terrain in the middle and east of Shengli Coalfield. It is mainly characterized by hilly and rounded terrain and distributed in a large area in the open area, followed by the valley plain terrain, the terrain is relatively the altitude is 990m ~ 1213m, and the relative height difference is 223m.

### *2.3. Stratigraphic lithology*

The south formation of Shengli East 2 Open-pit Mine is composed of the Shengli Formation of the Quaternary, the Cenozoic Upper Pliocene, and the Mesozoic Cretaceous Lower Peyan Flower Group from top to bottom. The Quaternary is distributed throughout the region, and the lithology is mainly composed of light yellow to gray-yellow gravel and coarse, medium and fine silt and sub-clay, sub-sand, and overburden. The thickness of the whole layer is 1.00m ~ 128.92m, and the average thickness is 37.01m. The thickness changes gradually thinning from southwest to east and northeast. The Tertiary system is distributed throughout the Pliocene, and the lithology is mainly composed of alluvial alluvial gray-green gravel, sandstone and brown-red, off-white to gray-green clay. The thickness is generally from 8.40 to 103.00m, and the average thickness is 45.26m. The Bayanhua Group Shengli Formation is the main coal-bearing stratum in this area. The bottom boundary of this group is the bottom plate of the 11th coal group. The main lithology is coal seam, gray to off-white siltstone, mudstone, fine sandstone, coarse sandstone with conglomerate; a total of 13 coal seams. The thickness is 30.02 ~ 720.50m, and the average thickness is 354.83m.

Based on the previous research results of slope stability and the geological results of surrounding open-pit coal mines, the physical and mechanical indexes of the rock mass are finally determined. The physical and mechanical indexes of the selected rock and soil are shown in Table 1.

**Table 1.** Rock mass physical and mechanical indexes

| stratum | cohesion $c$/KPa | internal friction angle $\varphi/°$ | rock weight $\gamma$/KN/m$^{-3}$ |
|---|---|---|---|
| disposal | 25 | 19 | 19 |
| fourth | 20 | 33 | 20.5 |
| third series | 150 | 26 | 18.5 |
| mudstone | 120 | 28 | 22.2 |
| weak layer | 10 | 10 | 10 |

## 3. Slope stability analysis

In this paper, the simplified Bishop method is used to analyze the slope stability. This method assumes that the sliding surface is in the shape of an arc or an approximate arc, and solves the slope stability coefficient based on the static balance condition and the moment balance condition.

In the initial state, the dip angle of the weak layer is 6°. At this time, the corresponding slope stability coefficient is 1.553. At this time, the slope slip mode is the combined failure of the shear arc weak layer. At the same time, in order to find out the influence of the weak layer on the slope stability, the slope stability coefficient is solved by changing the dip angle of the weak layer.

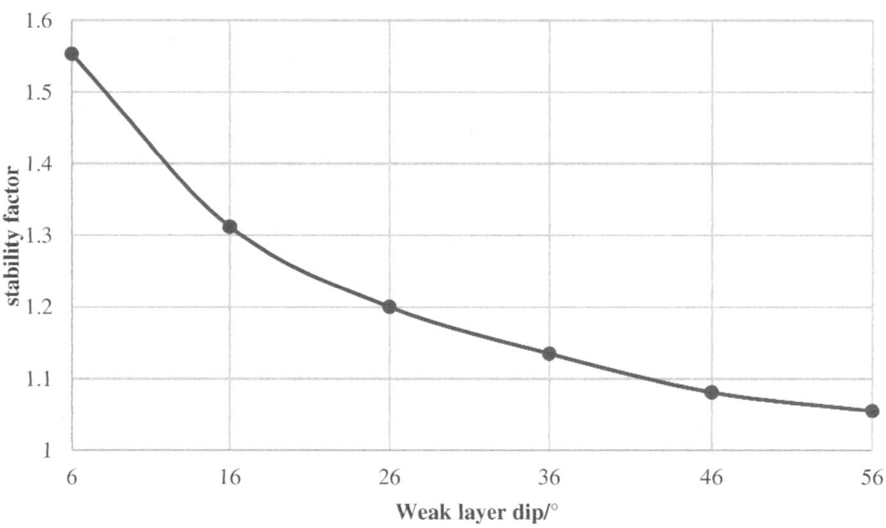


**Fig. 1.** Slope stability coefficient change curve

It can be seen from Fig. 1 that as the dip angle of the weak layer increases, the slope stability coefficient decreases. When the dip angle of the weak layer is between 6°and 56°, the slope slip mode is combined failure of the shear arc weak layer. When the dip angle of the weak layer reaches 56 °, the slope slip mode is converted to the formation of the weak layer Translational damage. The weak layer serves as the bottom sliding surface of the sliding body, and improving or maintaining the strength of the weak layer is a key factor for improving and managing the slope.

## 4. Conclusion

In this paper, based on the engineering geological characteristics of the south slope of Shengli East No. 2 Open-pit Mine, we deeply study the influence of weak layer morphology on the sliding pattern and stability of the slope. This paper mainly formed the following conclusions

(1) The factors influencing the stability of the south slope of Shengli East 2nd Open-pit Mine are analyzed, and the potential slippage pattern of the slope under the action of weak layers is clarified.

(2) The influence of weak layer morphology on slope slip mode and stability is revealed. When the weak layer dip angle is less than 56 °, the slope slip mode is combined failure of shear arc weak layer, on the other hand, When the dip angle of the weak layer is greater than 56 °, the slope slip mode is the translational failure caused by the weak layer.

(3) The weak layer as the bottom sliding surface of the sliding body will greatly reduce the slope stability. The south slope of Shengli East 2nd Open-pit Mine is significantly affected by the weak layer, and maintaining the strength of the weak layer is a key factor in managing the slope.

## Acknowledgments

This paper is supported by the key research project for higher education in Henan province (20B560009), the scientific and technological project in Henan Province (202102310567) and 2019 provincial college student innovation and entrepreneurship training program of education department in Henan Province (201911834036).

## Reference

[1] CAO Lanzhu, WANG Zhen, WANG Dong, et al. Optimization design of slope space shape for multiple coal seams surface mine with weak layer[J]. Journal of Liaoning Technology University, 2018, 37(3): 469-475.

[2] WANG Dong, LI Guanghe, CAO Lanzhu, et al. Mechanical effect of supporting and retaining coal pillar on internal waste dump with weak basement[J]. Journal of China Coal Society, 2019, 44(3): 934-941.

[3] HUANG Runqiu. Large-scale landslides and their sliding mechanisms in China since the 20th century[J]. Chinese Journal of Rock Mechanics and Engineering, 2007, 26(3): 433-454.

[4] MA Jingjia, XU Yongchao, LIU Gan, et al. Stability and morphological optimization of slopes with weak interlayers in Zaha Naoer open-pit mine[J]. Opencast Mining Technology, 2020, 35(2): 88-91.

[5] XIA Dong, WU Chaosong, JIA Yufei, et al. Long-term strength calculation and slope stability analysis of saturated layer of saturated sandy mudstone in Bailitan Coal Mine[J]. Safety in Coal Mines, 2019, 50(10): 198-201.

[6] WANG Dong, GUO Haijian, LV Wenwei, et al. Stability and failure mechanism of bedding layered slope with Y-shaped weak layer[J]. Journal of Safety and Environment, 2019, 19(6): 1940-1946.

[7] XIAO Xian. Application of artificial genetic algorithm in slope stability data processing[J]. Journal of China and Foreign Highway, 2014, 34(3): 41-44.

IOP Conf. Series: Materials Science and Engineering **914** (2020) 012042 doi:10.1088/1757-899X/914/1/012042

# Research on rainfall infiltration law and slope stability analysis

**Ang Li[a], Lingbo Meng**

School of Engineering, Huanghe Science and Technology College, Zhengzhou, Henan 450000, China

[a]293513627@qq.com

**Abstract**. Rainfall is an important factor leading to slope landslide. Aiming at the problem of slope stability under rainfall, this paper uses the unsaturated seepage theory to dynamically describe the rainfall infiltration effect, and analyzes the distribution characteristics of pore water pressure of slope soil and the change law of slope water level under the action of rainfall, at the same time, deeply analyzes the law of slope stability changes based on strength reduction algorithm. The results show that: with the passage of rainfall time and the increase of rainfall intensity, the slope infiltration depth gradually increased. At the same time, rainfall increased the groundwater level in the slope, and leaded to increased pore water pressure in the slope soil, and the transient saturation zone gradually formed and expanded, and finally leaded to a significant reduction in slope stability coefficient.

## 1. Introduction

Rainfall is an important factor leading to landslides. The occurrence of a large number of landslides in engineering practice is closely related to rainfall. Therefore, the problem of slope stability under the action of rainfall has become an important issue to be solved urgently in the field of geotechnical engineering [1-2]. The problem of slope stability under the action of rainfall is a complicated process that dynamically evolves with factors such as the duration and spatial changes of rainfall. The rainfall infiltration process is related to many factors such as rainfall intensity, duration, and soil permeability. When the soil permeability of the slope is greater than the rainfall intensity, the actual rainfall infiltration depends on the degree of precipitation. The increase in the water content of the slope has a positive correlation with the degree of precipitation, and eventually gradually stabilizes [3-5]. On the other hand, when the rainfall intensity is greater than the permeability of the slope soil slope, the actual infiltration depends on the infiltration performance of the slope soil itself. At the beginning of rainfall, the slope soil will first form a transient saturation zone near the shoulder of the slope. With the increase of precipitation time and precipitation intensity, the transient saturation zone gradually expanded to the inside and the lower part of the slope, and gradually showed a saturated shell on the entire slope [6]. As the precipitation increases, it expands to the inside of the slope, the affected area gradually increases, and the saturated crust gradually thickens [7]. With the increase of precipitation infiltration, the water content of the soil quality is increased, and the pore water pressure is increased, so that the matrix suction of the soil slope is reduced, and the shear resistance of the soil quality is reduced, which in turn causes the slope stability to decrease [8].

## 2. Stability analysis of rainfall infiltration slope

The influence of rainfall infiltration on slope stability is mainly reflected in the attenuation of soil strength. In 19th century, the scholar Darcy described the law of water penetration in porous media through a large number of experiments, and pioneered the quantitative study of the law of seepage movement of soil. Darcy's law is still the basic theory for studying the law of soil seepage movement nowadays. In the 1920s, scholars such as Richards and Terzaghi described the basic partial differential equations of water movement in saturated soil based on Darcy's law and continuous equations, forming a classic saturated soil mechanics theory. In the middle of the 20th century, with the gradual maturity of experimental methods, the majority of scholars focused on the theory of unsaturated soil mechanics. Bishop, Jennings and other scholars conducted in-depth research on the mechanical properties and constitutive relationship of unsaturated soil under seepage, and established the effective stress of unsaturated soil based on Mohr-Coulomb strength theory expression.

In view of the characteristics of slope soil strength changes under rainfall, the strength reduction index concept was initially considered in the numerical study of geo-elastoplastic finite element. The shear strength reduction coefficient is interpreted as: The ratio of the highest shear strength that the soil in the slope can express to the true shear stress caused by the external load in the slope is called the strength reserve safety index.

## 3. Engineering case analysis

The engineering example selected in this paper is located in the tributary area of the Jinsha River in Sichuan Province. The slope has a length of 320 meters, a lateral width of 340m, and an overall thickness of the slope aggregate of 30 meters. The slope is a gentle slope structure along the slope. Based on the analysis of the distribution of avalanche deposits and aggregates in the area, the slope is a landslide of avalanche deposits and terrace deposits. The constituent elements of the slope body are generally sand Clay plus crushed stone and some stones.

According to the slope geological situation and soil mechanical properties, this paper uses ANSYS finite element software to establish the slope model, and analyzes the slope stability changes through FLAC$^{3D}$. The application of the two softwares can describe the yield flow of materials under external load by adjusting the deformation and movement of the grid lines in the model, and can truly simulate the stress-strain relationship of slope rock and soil. At the same time, the strength reduction algorithm embedded in the human software is used to cyclically reduce the strength parameters of the rock and soil until the slope reaches the limit equilibrium state, and finally obtain the slope stability coefficient and the most dangerous slip surface shape. Its advantage is that it can obtain the slope shear strain incremental cloud map and the maximum displacement vector cloud map. It can accurately analyze the mechanical mechanism and slip mode of slope failure, and has good applicability to complex ground conditions. The number of units in this model is 47550, and the number of grid nodes is 43410.

Based on the statistics of rainfall in the area in recent years, rainfall in the area has a clear pattern, with the largest rainfall concentrated in summer and the lowest rainfall concentrated in winter. Among them, summer has the highest rainfall in July, about 305mm, but no rainfall in December and January in winter. The overall data results show that regional precipitation is concentrated from April to September each year, accounting for about 85 percent of the annual precipitation. The largest precipitation generally occurs in July and August each year. According to the characteristics of rainfall distribution, the maximum daily rainfall (95.5mm/h) was used as the research condition to analyze the stability of the slope under the action of rainfall.

After research, the initial stability coefficient of the slope is 1.351. Under the rain intensity of 95.5mm / h, after the slope rains, due to the increase in pore water pressure, the safety factor of the slope gradually decreases. It should be noted that After the rain stops, the safety factor of the slope will not reach the minimum, because after the rain stops, the water flow will also seep. After 48 hours of rainfall, the slope stability coefficient was 1.119, and when the rainfall was 72 hours, the slope stability coefficient reached 1.091. At this time, the slope stability coefficient was the smallest. Later, as the rain gradually drained, the slope stability coefficient gradually increased 120 hours after the rain, the slope stability coefficient returned to 1.358, and 168 hours after the rain, the slope stability coefficient returned to the initial stability coefficient approximately, indicating that the rainfall was basically discharged. The slope stability coefficient change curve is shown in Figure 1.

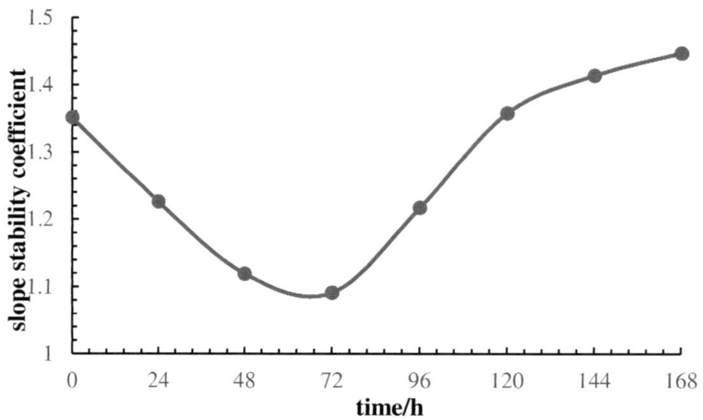


**Fig.1.** Slope stability coefficient change law

## 4. Conclusion

Through the above analysis, the following conclusions can be obtained:

(1) During the rainfall infiltration of the slope, the surface of the slope gradually becomes wet, and as the matrix suction decreases, the pore water pressure on the surface of the slope is increasing and a transient saturation zone is formed, which gradually rises from the initial negative pore water pressure Positive pore water pressure. After the rain stops, although it has the function of a drainage hole, because the water flow continues to infiltrate, the pore water pressure on the surface of the slope does not immediately drop, and will continue to rise for a period of time. Conditions are relevant; as drainage continues, the transient saturation zone on the surface of the slope gradually dissipates, and the pore water pressure on the surface of the slope gradually decreases, resulting in a gradual increase in matrix suction.

(2) During the rainfall infiltration process, the displacement of the slope is increasing. After the rain stops, although there are drainage measures, the displacement of the slope continues to rise.For the upper part of the slope, the displacement rise is more obvious. After infiltration, the drainage effect of the drainage hole began to be obvious, and the displacement of the slope began to decline. In the process of rainfall infiltration, the impact of rainfall infiltration on the upper part of the slope is greater than the lower part of the slope, and the slope displacement caused by rainfall infiltration increases with the increase of elevation.

(3) After the slope rains, due to the increase of pore water pressure, the safety factor of the slope gradually decreases.It should be noted that after the rain stops, the safety factor of the slope may not reach the minimum, because after the rain stops, the water flow may continue Infiltration. As the drainage holes continue to drain, the transient saturation area on the surface of the slope gradually dissipates, resulting in a decrease in pore water pressure or an increase in matrix suction, and the safety factor of the slope is also increasing.

**Acknowledgments**

This paper is supported by the key research project for higher education in Henan province (20B560009), the scientific and technological project in Henan Province (202102310567) and 2019 provincial college student innovation and entrepreneurship training program of education department in Henan Province (201911834034).

IOP Conf. Series: Materials Science and Engineering **914** (2020) 012042 doi:10.1088/1757-899X/914/1/012042

**Reference**


[1] CAO Lanzhu, WANG Zhen, WANG Dong, et al. Optimization design of slope space shape for multiple coal seams surface mine with weak layer[J]. Journal of Liaoning Technology University, 2018, 37(3): 469-475.

[2] WANG Dong, GUO Funing, CAO Lanzhu, et al. Study on stage-by-stage optimization method for slope parameters of soft rock open-pit mine in composite coal seam[J]. Coal Science and Technology, 2019, 47(10): 75-80.

[3] LI Shigui, HUANG Da, SHI Lin, et al. Numerical modeling of the evolution of slope failure using the limit STR criterion and dynamic strength reduction method[J]. Journal of Engineering Geology, 2018, 26(5): 1227-1236.

[4] XUE Tao, GUO Wei, GUAN Weiya, et al. Three dimensional stability analysis of slopes with rotation angles based on strength reduction method[J]. Journal of Water Resources and Architectural Engineering, 2019, 17(1): 154-158.

[5] ZHANG Xiaoyan, ZHANG Lixiang, LI Ze. Reliability analysis of soil slope based on upper bound method of limit analysis[J]. Rock and Soil Mechanics, 2018, 39(5): 1840-1850.

[6] CHEN Zuyu, ZHAN Chengming, YAO Hailin, et al. Safety criteria and standards for stability analysis of gravity retaining walls[J]. Rock and Soil Mechanics, 2016, 37(8): 2129-2137.

[7] HU Weidong,CAO Wengui.Upper limit limit analysis method for ultimate bearing capacity of ground foundation adjacent to slope based on asymmetry failure mode[J].China Journal of Highways And Transport,2014,27 (6):1-9.

[8] WANG Genlong,WU Faquan,ZHANG Junhui.Upper bound approach of rigid elements for inhomogeneous soil slope stability analysis[J].Chinese Journal of Rock Mechanics and Engineering,2008(S2):3425-3430.

IOP Conf. Series: Materials Science and Engineering **914** (2020) 012043 doi:10.1088/1757-899X/914/1/012043

# Paper counting display device based on STM32F103 single chip computer

**Lizhen Gao, Fanmao Jiang, Shihao Wang**

School of Electronics and Electrical Engineering, Xiamen Institute of Technology, Xiamen, Fujian, 362101

**Abstract**. This system uses the STM32F103 single-chip microcomputer as the control center, and the capacitance-to-digital converter FDC2214 collects the current paper value, performs data collection and transmits it to the main control single-chip microcomputer. Methods: Calculate accurate data through Kalman filtering, and accurately calculate and display the number of measured paper sheets on this basis. The device also sets a buzzer to accurately short circuit alarm. The circuit structure of the entire system is simple, and the reliability is high. Results: The system has high precision and fast speed, and can complete the basic requirements and expansion requirements of the problem. Although it is limited by time and venue, it can also have a high reference value for the design of practical applications such as detecting the number of banknotes. Conclusions: Experimental test results show that this design can complete the basic indicators required by the topic and play part of the content. This paper focuses on the hardware design, software design and test result analysis of the system.

## 1. Background

### *1.1. System design requirements*

(1) Basic requirements

The metal electrode parts on the electrode plate A and the electrode plate B are squares with a side length of 50 mm ± 1 mm, and the lengths of the wires a and b are 500 mm ± 5 mm. The measurement display circuit should have the function of "self-calibration", that is, before the formal test, the paper of different sheets placed between the two polar plates is measured to obtain measurement calibration information.

The measurement and display circuit can self-check and report whether the electrodes of plate A and plate B are short-circuited.

Measure the given number of sheets ranging from 1 to 10 sheets placed between the two plates [1]. Each time the paper to be tested is placed between the plates and fixed, one-key measurement is started, the number of papers to be tested is displayed and a buzzer sounds. The time from pressing the same measurement start key to beeping for each measurement shall not exceed 5 seconds, during this period, there shall be no manual intervention on the measurement device [2].

(2) Play part

1) The polar plates and wires are unchanged, and the number of given paper ranging from 15 to 30 sheets placed between the polar plates is measured [3,4]. The relevant requirements for the measurement start key, display beep, measurement time, and no manual intervention are the same as "Basic Requirements (3)".

2) The polar plates and wires are unchanged, and the number of papers given above 30 sheets placed between the polar plates is measured [5]. The relevant requirements for the measurement start key, display beep, measurement time, and no manual intervention are the same as "Basic Requirements (3)".

### *1.2. Overall design*

Due to the problem, the two copper sheets on the top and bottom of the paper can only be connected to the wire, and there can be no other additional components. The following three schemes can be used to detect the number of paper systems:

Option 1: Mechanical pressure, using a screw stepper motor to apply a fixed downward pressure on the plane of data collection, and can accurately obtain the increased paper thickness between the two plates, use a single-chip microcomputer to record this data and divide by a single sheet 70gA4 The thickness of the paper, and data analysis of the collected multiple sets of data to accurately obtain the number of paper.

Solution 2: Hall sensor, through which the thickness of the paper can be accurately measured.

Solution 3: Collect the capacitance value between the two plates through the capacitance-to-digital converter FDC2214, and analyze the collected data. After collecting a large amount of data and fitting the data, we can approximate an exponential function, but this The range covered by the function is limited. If we want to further improve the measurement range, we must process the collected data in sections. The initial test can be obtained within the range of 30 sheets. The data obtained is approximately linear, but after exceeding the range of 30 sheets The trend of increasing data is small or even decreasing. Through experiments, it is found that adding a weight above the pole plate can shift this peak backwards. It can be obtained by analyzing the formula of capacitance ($C=\varepsilon S/4\pi kd$). Under other circumstances where the other factors remain unchanged, the smaller the value of d, The larger the value of C, the more convenient it is for us to collect data. The final peak value is roughly set at about 40 sheets. After this range, we use Kalman filtering to process the data collected by each channel to accurately measure large amounts of paper.

In summary, the data collected and analyzed by the capacitance-to-digital converter can have very accurate results. And the circuit is simple and easy, and can also display the data of each acquisition channel in real time, so we choose the software timing in scheme three.

## 2. Theoretical analysis and calculation

### *2.1. Analysis and calculation of measurement principle*

First, self-calibration reads the data of each piece of paper, reads 80 pieces of data per piece of paper, and applies Kalman filtering to these data to obtain accurate values, which are stored in an array. When testing the number of papers, the data of the current paper is obtained in the same way, and compared with the data in the array to obtain the accurate number of papers.

### *2.2. Anti-interference analysis*

Shielding, grounding, filtering, isolation and absorption. Under normal circumstances, the spatial interference is much smaller than the interference from other channels, and can be effectively solved with good shielding, correct grounding, and filter filtering.

Therefore, the data collection system should focus on preventing interference between the power supply system and the process channel. The interference sources of each part of the system will have some emphasis and different impacts, so the anti-interference measures taken will be different.

We use data filtering to reduce or weaken the impact of the environment on metal electrodes through certain calculation or judgment procedures.

IOP Conf. Series: Materials Science and Engineering **914** (2020) 012043 doi:10.1088/1757-899X/914/1/012043

### 2.3. Error Analysis

When the contact surface of the electrode plate is different, the capacitance is different, so errors will occur in the measurement, so make sure that the contact surface of the electrode plate is the same for each measurement;

When the paper is loaded, the data generated by the different gravity will also be different, so the gravity should be fixed to ensure that the external gravity generated each time the paper is measured is the same.

Then carry on Kalman filtering to the data to reduce the error of each measurement.

## 3. Circuit and programming

### 3.1. Circuit design

*3.1.1. Data collection section.* Due to the design problem, the metal electrode parts on the polar plate A and the polar plate B are squares with a side length of 50mm±1mm, and it is stipulated that if a double-sided board is used, a square with a side length of 50mm±1mm is processed on one side of the double-sided board On the other side of the copper-clad electrode plate, only via pads and leads for soldering wires a and b are allowed, no copper-clad mesh is allowed, and it is forbidden to use multi-layer plates to make polar plates. Therefore, we use a brass plate as the polar plate, which has the advantage of being flat and not easily deformed.

FDC2214 accurately collects multiple channels of data and performs Kalman filtering. The FDC2214 series has many advantages over other capacitive sensing integrated circuits. FDC2x1x-Q1 is an anti-noise and EMI, high-resolution, high-speed, multi-channel capacitor-to-digital converter series for capacitive sensing solutions. This series of devices supports a wide excitation frequency range, which can bring flexibility to system design.

The FDC221x has been optimized with a resolution of up to 28 bits and a maximum measurement of 250nF. The FDC211x has a sampling rate of up to 13.3ksps, making it easy to implement applications that use fast moving targets.

*3.1.2. Display circuit.* Use LCD to display two-way FDC data and display learning.

*3.1.3. Alarm circuit.* The alarm signal is output from the P3.0 port of the single-chip microcomputer, and a short-circuit alarm is issued through the buzzer. The circuit diagram is shown in Figure 1 below.

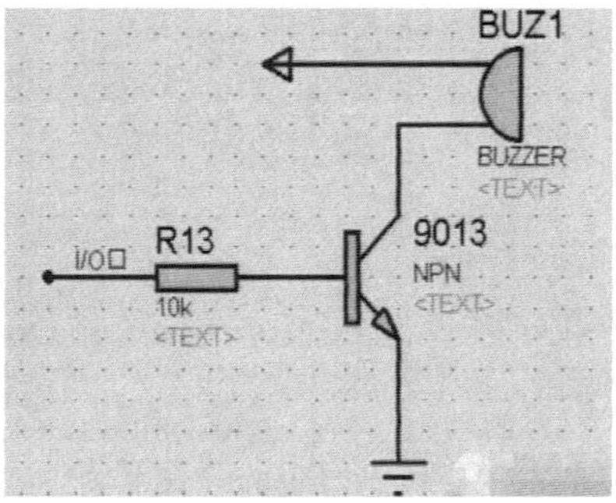


**Figure 1.** Alarm circuit diagram

### 3.2. Software design

The device is divided into two working states:

State 1: Self-calibration of the whole machine, reading the capacitance value of each paper superimposed multiple times, and adopting the Kalman filter algorithm to obtain the most reliable value in the middle.

State 2: Read the paper number multiple times, use the Kalman filter algorithm to obtain the most reliable value in the middle, determine the range of the value, determine the number of papers placed at this time and display on the LCD screen.

## 4. Test methods and test results

### *4.1. Test methods and test instruments*

Learn by sampling one by one, and test after completing self-calibration.

(1) First determine the paper in the range of 1-15 sheets, and repeatedly test to determine its accuracy.

(2) Re-test the paper in the range of 16-30 sheets, and repeatedly test to determine its accuracy.

(3) Wait for 30 sheets of paper that can be accurately measured within a test range of 60 sheets, and repeatedly test to determine their accuracy.

Because the difference between the two values measured above 60 sheets is not obvious, the value was not continued.

**Table 1.** Paper detection record

| frequency | Number of sheets | Accuracy% |
|---|---|---|
| 10 | 1 | 100 |
| 10 | 2 | 100 |
| 10 | 3 | 100 |
| 10 | 4 | 100 |
| 10 | 5 | 100 |
| 10 | 6 | 90 |
| 10 | 7 | 100 |
| 10 | 8 | 100 |
| 10 | 9 | 100 |
| 10 | 10 | 100 |
| 10 | 11 | 100 |
| 10 | 12 | 100 |
| 10 | 13 | 100 |
| 10 | 14 | 100 |
| 10 | 15 | 100 |
| 10 | 16 | 100 |
| 10 | 17 | 100 |
| 10 | 18 | 90 |
| 10 | 19 | 100 |
| 10 | 20 | 100 |
| 10 | 44 | 100 |
| 10 | 45 | 90 |
| 10 | 46 | 100 |
| 10 | 47 | 100 |
| 10 | 48 | 100 |
| 10 | 49 | 100 |
| 10 | 50 | 100 |
| 10 | 51 | 100 |
| 10 | 52 | 100 |
| 10 | 53 | 100 |
| 10 | 54 | 100 |
| 10 | 55 | 100 |
| 10 | 56 | 100 |
| 10 | 57 | 100 |
| 10 | 58 | 100 |
| 10 | 59 | 100 |
| 10 | 60 | 100 |

Error analysis: The upper and lower plates do not cause a great error in the device measurement.

IOP Conf. Series: Materials Science and Engineering **914** (2020) 012043 doi:10.1088/1757-899X/914/1/012043

## 5. Summary

This system has completed the design of a paper count display device. The characteristic of this design is that through the design of a stable electrode plate design, the capacitance value of different papers is obtained through a high-precision capacitance value sensor, and the accurate value is obtained through Kalman filtering, thereby calculating the number of papers to be measured. The system has high precision and fast speed, and can complete the basic requirements and expansion requirements of the problem. Although it is limited by time and venue, it can also have a high reference value for the design of practical applications such as detecting the number of banknotes.

## Reference:


[1] Huang Haichao, Huang tangsen, Yin Xiangdong. Design and application of paper counting device [J]. Information technology and informatization, 2020 (05): 105-110

[2] Lou Yingying, Huang Xiansheng, Zheng Sikai, Huang Yiming, Wang Zhenyu, Wang Gang, Jin Baogen, Lu Xinxiang. Design of paper counter based on capacitance sensor [J]. Electronic test, 2020 (02): 8-10

[3] Chen Rui. Design of paper counting display device based on STM32F103 [J]. Journal of Pu'er University, 2019,35 (06): 42-44

[4] Ma Qicai, Liu Yibo, Wu tangsheng. Design of paper counting display device [J]. China Science and technology information, 2019 (21): 62-63

[5] Shi Wei, long Yonghong, Zhong Yunfei, Yang danjun, Zhou Shuqing. Research and application of counting method of hardboard paper based on gray projection [J]. Journal of packaging, 2017,9 (01): 66-71

# AUTHOR INDEX


Chen, Depin ... 169
Chen, Guang ... 50
Chen, Tingyao ... 186
Chen, Wenjing ... 111
Chen, Xiujuan ... 246
Chen, Yao ... 83
Cheng, Jie ... 151
Cheng, Wenxiang ... 151
Cui, Chun li ... 205
Cui, Chunli ... 92
Deng, Bowen ... 176
Deng, Weixiang ... 151
Fang, Rongze ... 55
Feng, Junchuang ... 35
Gao, Lizhen ... 264
Gao, Yuqiu ... 169
Gu, Xiaohui ... 209
Guo, Bokai ... 1
Guo, Jie ... 199
Guo, Meng ... 111
Guo, Shumei ... 199
Guo, Xiaohui ... 6
Guo, Yanhui ... 227
Han, Zhixuan ... 169
He, Anping ... 151
He, Jin ... 227
Hu, Chang ... 121
Hu, Shaodong ... 180
Huang, Changjiang ... 191
Huang, Kun ... 35
Jia, Jianwei ... 221
Jiang, Fanmao ... 264
Jiang, Xingnan ... 176
Jianga, Shouli ... 240
Li, Ang ... 256, 260
Li, Cheng ... 96
Li, HuiJuan ... 72
Li, Lei ... 35
Li, Ling ... 55
Li, Lin ... 92
Li, Wenli ... 256
Li, Xinna ... 199
Li, Yong ... 50
Li, Zhu ... 1
Liang, Hui ... 140
Lin, Qiao ... 83, 96
Liu, Hong bo ... 205
Liu, Hongbo ... 92
Liu, Junhua ... 96
Liu, Kewen ... 140
Liu, Shi jun ... 205
Liu, Shijun ... 22, 92
Liu, Xiaoyan ... 96
Liu, Xiao ... 28
Liu, Yalei ... 209
Liu, Yao ... 6
Liu, Yile ... 35
Longa, Le Xuan ... 233
Luo, Bangzhou ... 83
Luo, Jun ... 55
Lv, Yanping ... 22
Ma, Lei ... 6
Ma, Yuan ... 140
Mei, Ying ... 62
Mei, Zaimei ... 115
Meng, Lingbo ... 260
Ni, Leibin ... 151
Pu, Xiang ... 62
Qiu, Ting ... 96
Qu, Peng ... 78
Qu, Xiangling ... 62
Ren, Shuyu ... 169
Ren, Yanling ... 83
Rong, Jiachao ... 78
Ru, Han ... 111
Shi, Xiongfei ... 41
Song, Zhong xing ... 205
Song, Zhongxing ... 92
Sui, Ming ... 78
Sun, Huanhuan ... 22
Sun, Jing ... 92, 205
Sun, Ran ... 1
Tan, Lingye ... 252
Tand, Hoang Anh ... 233
Tang, Mingguo ... 35
Tang, Xiusheng ... 55
Tang, Zhi shu ... 205
Tang, Zhishu ... 92
Tao, Min ... 128, 134
Van Cuong, Bui ... 233
Van Liemb, Nguyen ... 233
Van Quynh, Le ... 233
Wan, Ke ... 1
Wang, Chen ... 159
Wang, Dong ... 246
Wang, Huan ... 115

Wang, Shaocheng ........ 22
Wang, Shihao ........ 264
Wang, Xin ........ 252
Wang, Yimeng ........ 221
Wang, Yinzhe ........ 159
Wang, Yueying ........ 176
Wang, Yueyuan ........ 140
Wang, Yu ........ 128, 134
Wei, Honghe ........ 246
Wu, Wanli ........ 6
Wu, Xuechen ........ 180
Wu, Yanan ........ 121
Xiao, Rui ........ 240
Xiao, Yang ........ 1
Xu, Hong bo ........ 205
Xu, Hongbo ........ 92
Xue, Bo ........ 180, 221
Yang, Jiandong ........ 199
Yang, Lili ........ 111
Yang, Liuwei ........ 83
Yang, Long ........ 1
Yang, Yuxin ........ 83
Yu, Xue ........ 106
Zhang, Hanping ........ 50
Zhang, Hao ........ 128, 134
Zhang, Jinhui ........ 147
Zhang, Junjie ........ 17
Zhang, Lili ........ 199
Zhang, Pan ........ 205
Zhang, Xiao ........ 121
Zhang, Yang ........ 78
Zhang, Yu ........ 92, 111, 205
Zhang, Zhuoqun ........ 159
Zhang, Ziyang ........ 252
Zhao, Haiyu ........ 215
Zhao, Jinyi ........ 1
Zhao, Shudong ........ 83
Zheng, Bixuan ........ 35
Zheng, Jiashuo ........ 215
Zheng, Ying ........ 121
Zhongb, Jianfeng ........ 240
Zhu, Yusong ........ 78

**Institute of Physics**
Dirac House, Temple Back
Bristol BS1 6BE UK

ISSN: 1757-8981
ISBN 978-1-7138-3188-4
9 781713 831884